Advanced Permanent Magnet Machines and Drives

Advanced Permanent Magnet Machines and Drives

Editors

Quntao An
Bing Tian
Xinghe Fu

MDPI • Basel • Beijing • Wuhan • Barcelona • Belgrade • Manchester • Tokyo • Cluj • Tianjin

Editors
Quntao An
Harbin Institute of
Technology
Harbin
China

Bing Tian
Nanjing University of
Aeronautics and Astronautics
Nanjing
China

Xinghe Fu
Southeast University
Nanjing
China

Editorial Office
MDPI
St. Alban-Anlage 66
4052 Basel, Switzerland

This is a reprint of articles from the Special Issue published online in the open access journal *Energies* (ISSN 1996-1073) (available at: https://www.mdpi.com/journal/energies/special_issues/PMMD).

For citation purposes, cite each article independently as indicated on the article page online and as indicated below:

LastName, A.A.; LastName, B.B.; LastName, C.C. Article Title. *Journal Name* **Year**, *Volume Number*, Page Range.

ISBN 978-3-0365-6303-9 (Hbk)
ISBN 978-3-0365-6304-6 (PDF)

Contents

Pawel Prajzendanc and Piotr Paplicki
Performance Evaluation of an Axial Flux Machine with a Hybrid Excitation Design
Reprinted from: *Energies* **2022**, *15*, 2733, doi:10.3390/en15082733 . . . . **1**

Bing Tian, Runze Lu and Jiasongyu Hu
Single Line/Phase Open Fault-Tolerant Decoupling Control of a Five-Phase Permanent Magnet
Synchronous Motor under Different Stator Connections
Reprinted from: *Energies* **2022**, *15*, 3366, doi:10.3390/en15093366 . . . . **13**

Dong-Youn Shin, Min-Jae Jung, Kang-Been Lee, Ki-Doek Lee and Won-Ho Kim
A Study on the Improvement of Torque Density of an Axial Slot-Less Flux Permanent Magnet
Synchronous Motor for Collaborative Robot
Reprinted from: *Energies* **2022**, *15*, 3464, doi:10.3390/en15093464 . . . . **31**

Shuangshuang Guo, Bo Zhao, Cunshan Zhang, Binglin Lu, Yukang Chu and Peng Yang
Research on a Limit Analytical Method for a Low-Speed Micro Permanent Magnet Torque
Motor with Back Winding
Reprinted from: *Energies* **2022**, *15*, 4662, doi:10.3390/en15134662 . . . . **45**

Mengji Zhao, Quntao An, Changqing Chen, Fuqiang Cao and Siwen Li
Observer Based Improved Position Estimation in Field-Oriented Controlled PMSM with
Misplaced Hall-Effect Sensors
Reprinted from: *Energies* **2022**, *15*, 5985, doi:10.3390/en15165985 . . . . **65**

Fuqiang Cao, Quntao An, Jianqiu Zhang, Mengji Zhao and Siwen Li
Variable Weighting Coefficient of EMF-Based Enhanced Sliding Mode Observer for Sensorless
PMSM Drives
Reprinted from: *Energies* **2022**, *15*, 6001, doi:10.3390/en15166001 . . . . **79**

Chengjun Liu, Jianfei Sun, Yifei Zhang and Jing Shang
Analysis and Error Separation of Capacitive Potential in the Inductosyn
Reprinted from: *Energies* **2022**, *15*, 6910, doi:10.3390/en15196910 . . . . **93**

Jian-Ya Zhang, Qiang Zhou and Kai Wang
Dual Three-Phase Permanent Magnet Synchronous Machines Vector Control Based on Triple
Rotating Reference Frame
Reprinted from: *Energies* **2022**, *15*, 7286, doi:10.3390/en15197286 . . . . **105**

Sebastian Berhausen, Tomasz Jarek and Petr Orság
Influence of the Shielding Winding on the Bearing Voltage in a Permanent Magnet Synchronous
Machine
Reprinted from: *Energies* **2022**, *15*, 8001, doi:10.3390/en15218001 . . . . **119**

**Jonathan Muñoz Tabora, Bendict Katukula Tshoombe, Wellington da Silva Fonseca, Maria
Emília de Lima Tostes, Edson Ortiz de Matos, Ubiratan Holanda Bezerra and Marcelo de
Oliveira e Silva**
Virtual Modeling and Experimental Validation of the Line-Start Permanent Magnet Motor in
the Presence of Harmonics
Reprinted from: *Energies* **2022**, *15*, 8603, doi:10.3390/en15228603 . . . . **139**

Pierpaolo Dini and Sergio Saponara
Review on Model Based Design of Advanced Control Algorithms for Cogging Torque
Reduction in Power Drive Systems
Reprinted from: *Energies* **2022**, *15*, 8990, doi:10.3390/en15238990 . . . . **157**

Article

Performance Evaluation of an Axial Flux Machine with a Hybrid Excitation Design

Pawel Prajzendanc * and Piotr Paplicki

Faculty of Electrical Engineering, West Pomeranian University of Technology, 70-313 Szczecin, Poland; piotr.paplicki@zut.edu.pl
* Correspondence: pawel.prajzendanc@zut.edu.pl

Abstract: Variable speed, permanent magnet synchronous machines with hybrid excitation have attracted much attention due to their flux-control potential. In this paper, a design of permanent magnet axial flux machines with iron poles in the rotor and an additional electrically controlled source of excitation fixed on the stator is presented. This paper shows results pertaining to air-gap flux control, electromagnetic losses, electromagnetic torque, back emf and efficiency maps obtained through field-strengthening and weakening operations and investigated by 3D finite element analysis. Moreover, the temperature distribution of the machine was analyzed according to the fluid–thermal coupling method. The presented machine was prototyped and experimentally tested to validate the effectiveness of numerical models and achieved results.

Keywords: permanent magnet machines; axial flux machine; hybrid excitation; variable speed machines

Citation: Prajzendanc, P.; Paplicki, P. Performance Evaluation of an Axial Flux Machine with a Hybrid Excitation Design. *Energies* **2022**, *15*, 2733. https://doi.org/10.3390/en15082733

Academic Editors: Quntao An, Bing Tian and Xinghe Fu

Received: 25 February 2022
Accepted: 4 April 2022
Published: 8 April 2022

Publisher's Note: MDPI stays neutral with regard to jurisdictional claims in published maps and institutional affiliations.

1. Introduction

Nowadays, electrical machines with high efficiency and good reliability are required. Although permanent magnet machines are suitable and have excellent efficiency, they have some flux-control limitations, especially in the field-weakening region at high rotor speed [1]. Permanent magnet (PM) machines with flux-weakening features are desirable in drives with a wide range of rotational speeds (e.g., in electrical vehicles or in wind turbine generators operated under different weather conditions).

Alternative design solutions for permanent magnet machines with hybrid excitation (HE) have been proposed, in which there is an additional source of field excitation, usually in the form of an additional coil electrically controlled by DC current and fixed on the machine stator [2,3]. The air-gap magnetic flux density is created by PMs and an additional magnetic field excited by the DC coil. In contrast to a parallel HE system, a serial one has an additional magnetic flux that directly affects the PM and changes the operating point of the magnet. Hence, the parallel system is more suitable for magnets from a demagnetization point of view.

Significant progress in hybrid excitation machines can also be observed in reluctance or synRM machines [4–7] and permanent magnet synchronous machines [8–12]. An additional excitation coil can be fixed on the stator [13,14] or in the rotor [15,16], which depends on the concept. In the literature, axial flux machines with hybrid excitation concepts [15–23] can be found where an additional coil is most often fixed on the stator.

In this paper, a concept of axial flux machines with permanent magnets, iron poles and an additional DC coil for magnetic flux control fixed in the middle on the stator is presented. The advantages of the proposed machine design solution are: a good range of magnetic flux regulation, brushless supplying stator windings and an additional coil, no demagnetization risk for the magnets, low cost and volume of the magnets compared to conventional machines and a high efficiency of up to 95%. The main drawbacks are: extra space needed to place the additional coil, additional losses from the supply DC coil and increased cooling requirements.

2. Dual-Disc Axial Flux Machine Design with Hybrid Excitation Concept

Figure 1 presents a design of a 3-phase, 12-pole dual-disc axial flux machine with a hybrid excitation parallel system using an additional DC coil. The presented machine concept has been partially analyzed in [24,25]. An additional DC coil with the number of turns N_{DC} = 500 allows control of the air-gap magnetic flux density. The stator is built from the toroidal core with armature windings. Every core's side has 36 half-opened slots, in which three-phase windings are placed. The DC coil creates a magnetomotive force (MMF) of 2500 AT by a DC coil current of 5 A. The rotor is built from two steel discs with permanent magnets of type N38H with a remanence of 1.23 T and coercivity of 947.8 kA/m. The iron poles are made from solid steel. The rotor discs are connected together with ferromagnetic bushing. The stator's core is built from two similar, one-side-slotted toroids based on grain-oriented electrical steel wound cores.

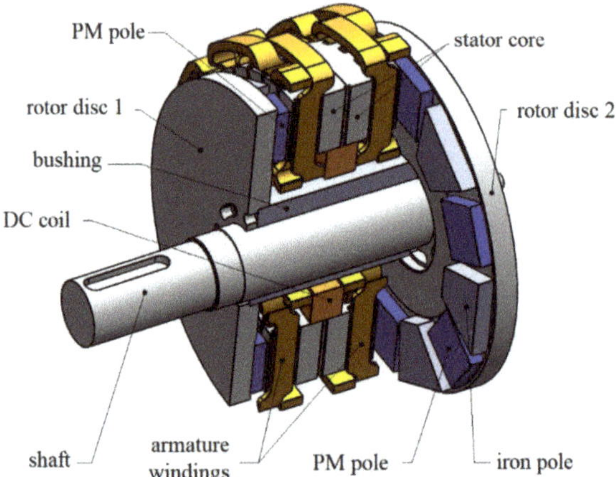

Figure 1. Field-Controlled Axial Flux Permanent Magnet Machine (FCAFPMM) design.

Figure 2a shows that the main parts of the rotor were made of two steel discs connected with ferromagnetic bushing, which form the yoke of the machine. On each rotor's disc, six neodymium magnets with uniform polarization are placed with iron poles (IP), which do not have polarity on their own. The solid magnet's shape is square and the IP's shape is trapezoidal; the IP's size is given in Figure 2b.

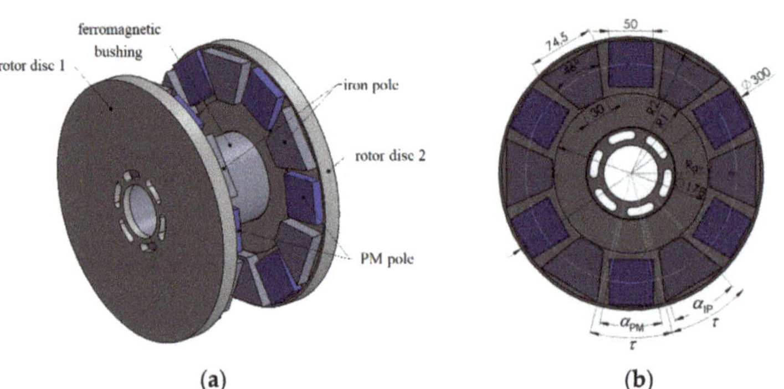

Figure 2. FCAFPMM's rotor (**a**), dimensions of rotor's disc with magnets and iron poles (**b**).

3. Principle of Flux Control

Since the rotor poles of the machine are different shapes, as can be seen in Figure 2, and they have dissimilar magnetic properties, the air-gap flux ϕ_g of the machine will be strongly dependent on the difference between the air-gap flux of the PM pole $\phi_{g\text{PM}}$ and the air-gap flux of the iron pole $\phi_{g\text{IP}}$. Moreover, the influence of an additional field DC coil excited by DC current I_{DC} on the air-gap flux density distribution is different for the PM pole and the iron pole. The magnetic pole fluxes, $\phi_{g\text{PM}}$ and $\phi_{g\text{IP}}$, can be calculated:

$$\phi_{g\text{PM}} = \frac{\alpha_{\text{PM}}}{\tau} B_{mg\text{PM}} \frac{\pi}{p} \left(R_2^2 - R_1^2 \right) \tag{1}$$

$$\phi_{g\text{IP}} = \frac{\alpha_{\text{IP}}}{\tau} B_{mg\text{IP}} \frac{\pi}{p} \left(R_2^2 - R_1^2 \right) \tag{2}$$

where α_{PM}, α_{IP} are the PM pole and iron pole arc length, R_1, R_2 are the rotor inner and outer radius, τ is the pitch of the pole, p is the pole-pair number and $B_{mg\text{PM}}$, $B_{mg\text{IP}}$ are the maximum magnetic flux density in the air gap of the PM pole and the iron pole, respectively.

Neglecting flux leakage, a no-load magnetic flux component generated by PMs ($\phi_{g\text{PM}/\text{PM}}$) crosses an air gap over the PM pole in the axial direction and flows through the stator core (Figure 3a). A significant part of the flux $\phi_{g\text{PM}/\text{PM}}$ crosses the second air gap of the PM, but some part of the flux is returned through an air gap over the iron pole in the form of a flux $\phi_{g\text{IP}/\text{PM}}$. The effect of the no-load air-gap magnetic flux density of the iron pole excited by the PM can clearly be observed in Figure 4a in the iron pole section (pitch of the iron pole).

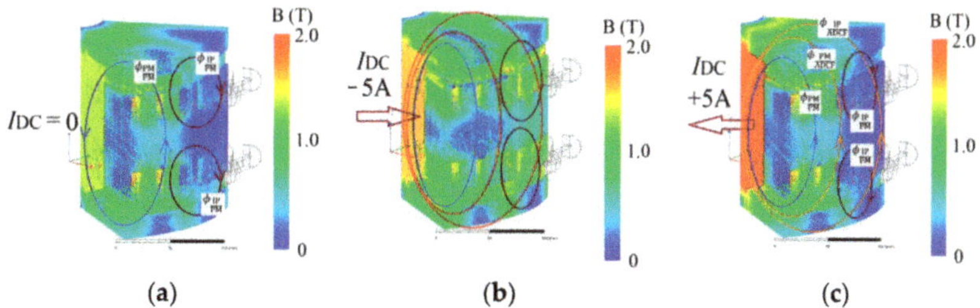

Figure 3. Idealized magnetic flux distribution in FCAFPMM at DC coil current $I_{\text{DC}} = 0$ (**a**), $I_{\text{DC}} -5$ A (**b**) and $I_{\text{DC}} +5$ A (**c**).

When an additional magnetomotive force (mmf) is excited by the current I_{DC}, an additional DC flux (ADCF) of the control coil ϕ_{ADCF} is increased (Figure 3b,c). Depending on the direction of the DC coil current, air gap fluxes will strengthen or weaken. In the case of coil current $I_{\text{DC}} < 0$, there will be air-gap flux strengthening for the iron pole $\phi_{g\text{IP}}$ and air-gap flux weakening for the pole with a magnet $\phi_{g\text{PM}}$. In the case of $I_{\text{DC}} > 0$, there will be gap flux extenuation for the iron pole $\phi_{g\text{IP}}$ and gap flux amplification for the pole with a magnet $\phi_{g\text{PM}}$. The flux ϕ_{ADCF} is conducted by a ferromagnetic bushing, the rotor's yokes and the stator core of the machine; it passes through the air gap where it is divided into two components: flux $\phi_{g\text{PM}/\text{ADCF}}$ crosses the air gap of the magnet pole and flux $\phi_{g\text{IP}/\text{ADCF}}$ crosses the air gap of the iron pole. The flux ϕ_{ADCF} can be expressed as:

$$\phi_{\text{ADCF}} = \phi_{g\text{PM}/\text{ADCF}} + \phi_{g\text{IP}/\text{ADCF}}$$

Neglecting saturation, flux leakage and armature effects, the magnetic air-gap flux of the PM pole $\phi_{g\text{PM}}$ in the linear model can be calculated as:

$$\phi_{g\text{PM}} = \phi_{g\text{PM}/\text{PM}} + \phi_{g\text{PM}/\text{ADCF}}$$

The magnetic air-gap flux of the iron pole ϕ_{gIP} can be expressed as:

$$\phi_{gIP} = \phi_{gIP/ADCF} - \phi_{gIP/PM}$$

The magnetic flux passing through the DC coil ϕ_{DCF} can be expressed as:

$$\phi_{DCF} = p(\phi_{PM/PM} - \phi_{IP/PM} + \phi_{ADCF})$$

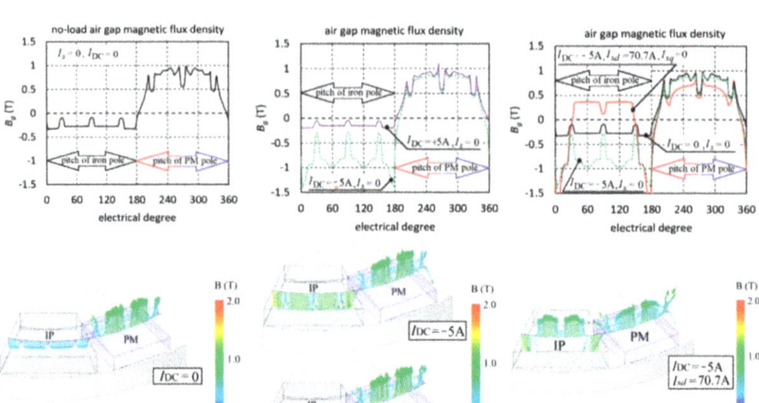

Figure 4. Air-gap magnetic flux density at the different magnetic field distributions $I_{DC} = 0$, $I_s = 0$ (**a**), $I_{DC} = \pm 5$ A, $I_s = 0$ (**b**) and $I_{DC} = -5$ A, $I_{sd} = 70.7$ A (**c**).

Based on the 3D FEA results presented in Figure 4b, it can be seen that the ADCF excited by $I_{DC} = -5$ A changes the amplitude of the magnetic flux density more effectively than that excited by $I_{DC} = +5$ A, and the influence of the ADCF on the magnetic air-gap flux distribution is greater in the air gap of the iron pole than in the air gap of the PM pole. In order to show the effect more clearly, Figure 5 shows the air-gap magnetic flux density distribution achieved at different DC load conditions: at $I_{DC} = +5$ A (Figure 5a) and at $I_{DC} = -5$ A (Figure 5b) as a special case of study where PMs of the machine were replaced with air boxes.

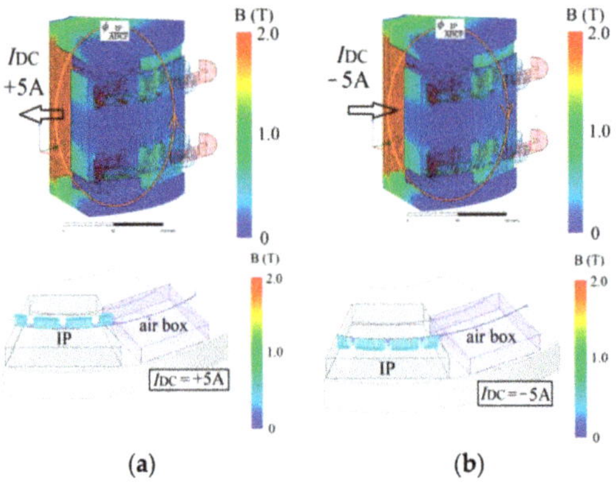

Figure 5. Air-gap magnetic flux density at the different magnetic field distributions in the model without PMs at DC coil current $I_{DC}+$ (**a**) and $I_{DC}-$ (**b**).

Moreover, in order to evaluate the demagnetization risk of magnets, the air-gap magnetic flux density distribution is performed at the most unfavorable load conditions (at $I_{sd} = 70.7\,\text{A}$ and $I_{sq} = 0$, where I_{sd} is the d-axis stator current and I_{sq} is the q-axis stator current). From the results shown in Figure 4c, it can be concluded that at operation conditions, there is no demagnetization risk of magnets in high current overload.

4. Magnetic Flux Density Distribution and Back EMF Analysis

In order to perform an analysis of the magnetic field distribution of the machine under different additional DC coil current excitations, a 3D model of the FCAFPMM in Ansys Electronics Desktop was developed considering symmetry conditions with respect to 1/6 part of the machine. For three cases of DC coil current excitations, magnetic field distribution results are shown in Figure 6a. Additionally, for these cases, the air-gap magnetic flux density distributions (in the middle of the gap for one pole-pair) are also shown in Figure 6b.

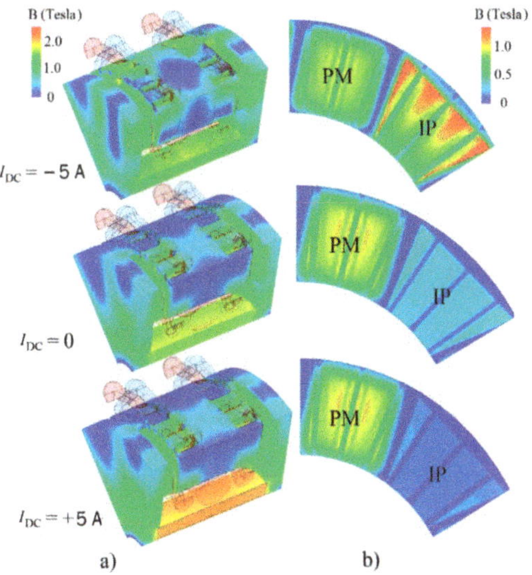

Figure 6. Magnetic flux density distribution on the 3-D FEA model (**a**) and in the middle of the air gap of the machine (**b**).

In the presented figures, it is clearly possible to see the strengthening effect of a current load of $I_{DC} < 0$ and significant magnetic flux of the iron pole section. Moreover, in this case, there is a slight but not significant decrease in flux in the air gap for the PM pole section. In the second case, where the DC coil current changed direction ($I_{DC} > 0$), there is a slight decrease in flux in the iron pole section due to the saturation effect. The effect can also clearly be seen on the rotor bushing, even in the no-load case ($I_{DC} = 0$), where the maximum magnetic flux density is approximately 1.5 T. This means that the dimensions of the rotor bushing should also be optimized. The presented results also show that the additional DC field excited by the DC coil current is mostly passing through the iron pole section, and it does not significantly change the value of the air-gap flux of the PM pole, which is a desirable effect.

Finally, in order to investigate the effect of the additional DC field on the magnetic flux linked with the stator windings, initial experiments were conducted. Figure 7 shows characteristics of the no-load-induced phase voltage in the stator windings, which were obtained for three cases of the additional DC coil current excitation.

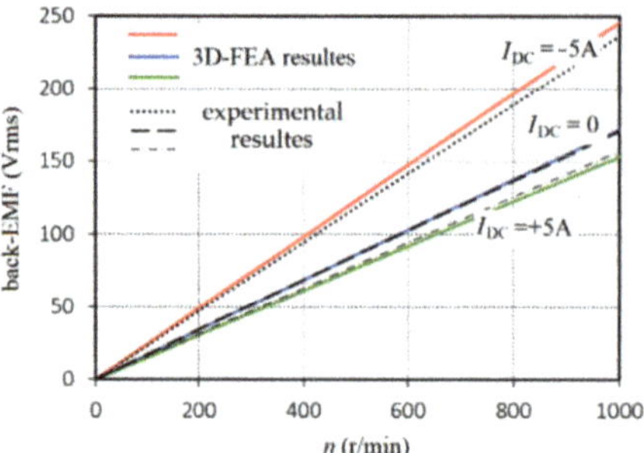

Figure 7. No-load phase voltage characteristics for three cases of I_{DC} current load.

In the presented Figure 7, one can clearly observe the effect of increasing phase terminal voltages with a DC coil current of $I_{DC} = -5$ A, which causes an increased rms voltage of approximately 50% from 172 V to 253 V. When the current is flowing in the opposite direction, it means at $I_{DC} = +5$ A, the induced voltage is decreased by approximately 10%. It should be noted that the experimental results were similar to 3D FEA predictions.

5. Electromagnetic Torque

The flux created by the DC coil affects the electromagnetic torque and torque-ripple characteristics of the machine. The influence of an additional DC field on the static torque characteristics is presented in Figure 8. The results show that, although maximum electromagnetic torque is increased two-fold with a load current of $I_{DC} = -5$ A (Figure 8a), the pulsation of the torque, which is directly related to air-gap flux density, is also increased. In other cases, the torque pulsations are comparable and relatively small (Figure 8b,c). It is worth observing that the no-load cogging torque with field-strengthening is rapidly increased. Publication [21] discussed a method of cogging torque reduction by optimizing the geometry of the rotor pole pieces.

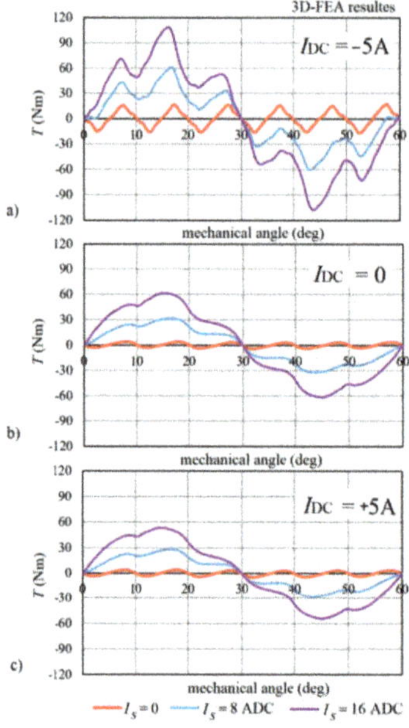

Figure 8. Electromagnetic torque waveforms versus rotor position under different stator currents for three I_{DC} current excitations I_{DC} −5 A (**a**), I_{DC} = 0 A (**b**) and I_{DC} +5 A (**c**).

Figure 9 shows the characteristics of maximum electromagnetic static torque obtained at three different currents of the DC coil versus stator current I_s.

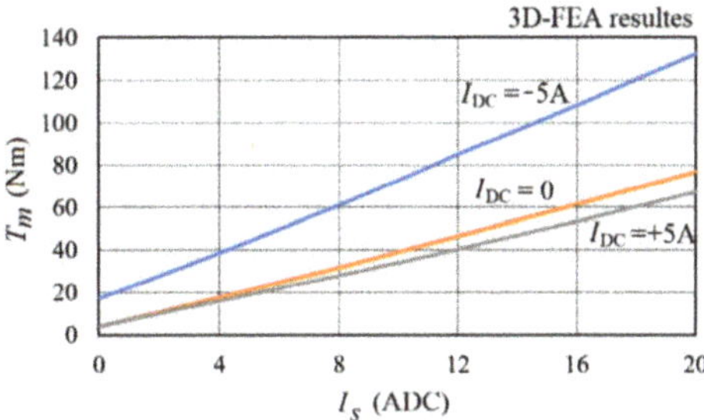

Figure 9. Maximum electromagnetic stator torque characteristics at the different control coil currents.

6. Performance of FCAFPMM

Based on the 3D FEA results, efficiency maps (Figure 10a–c) and total losses of the machine (Figure 10d–f) are calculated based on the mapping of core and windings losses, flux-linkage and torque as a function of the d- and q-axis currents and speed. The effect of additional field losses on the shape of the efficiency map is explored in specified conditions and limits: voltage power supply $U = 200$ Vrms; stator current $I_s \leq 50$ Arms; MTPA (maximum torque per ampere) control strategy; windings star connection; phase resistance $R_f = 0.37\ \Omega$.

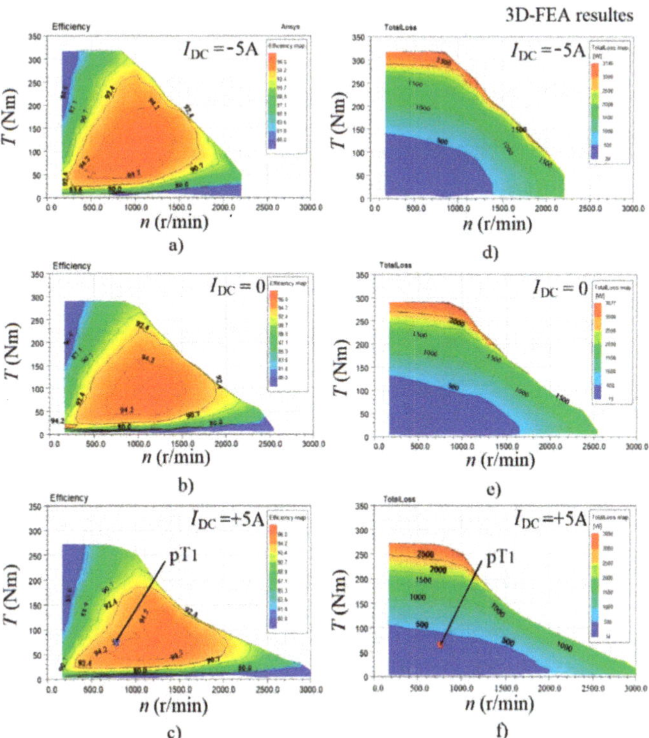

Figure 10. Maps of machine efficiency (**a–c**) and total losses (**d–f**) for three I_{DC} current excitations.

It should be noted that, although the field-strengthening and weakening effect can be seen in Figure 10, additional sources of excitation increase additional losses. The consequent power loss density is high, and heat transfer is considered a challenge. This is particularly the case for high-speed operations in field-weakening regions when an additional source of excitation is activated.

Attention was also paid to the influence of additional DC field excitation on a startup torque and speed limit. Presented results show that magnetic field-strengthening with a DC coil current of $I_{DC} = -5$ A increased startup torque by approximately 10%, and a value of electromagnetic torque of 320 Nm was reached. In field-weakening operations with a current of $I_{DC} = +5$ A, the limit of speed was increased from 2500 rpm to 3500 rpm. In Figure 11, torque–speed curves of the machine calculated under DC coil currents of $I_{DC} = \pm 5$ A are shown.

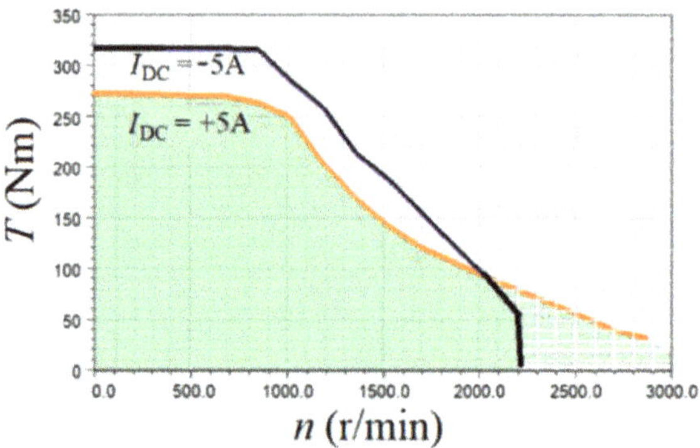

Figure 11. Torque–speed envelopes compared under different DC coil currents, at DC coil current $I_{DC} = -5$ A (black line—field-strengthening operation region) and $I_{DC} = +5$ A (orange line—field-weakening operation region).

7. Thermal Analysis

Developing PM motors within a particular temperature limit is a key factor affecting the efficiency of the overall design. Based on the 3D thermal model of the machine without the cooling system presented in Figure 12a, steady-state and transient thermal analyses were carried out to evaluate the distribution of temperature in various parts of the stator. The model was mainly applied to predict the temperature rise in the stator (Figure 12b) of the machine. The temperature distribution was determined by considering convection from the back of stator core surfaces, air gaps, stator windings and epoxy resin surfaces. The heat sources, such as copper loss and stator core loss, were calculated using the 3D FEA transient method. In the first case study, the copper losses of 218 W in armature windings at an initial temperature of 40 °C were obtained through FE analysis. The losses were sent to CFD fluid–thermal analysis as heat sources, which correspond to copper losses of armature windings generated with a current density of $J = 5$ A/mm^2. The thermal conductivity of copper windings and insulation, both with epoxy resin, were taken together in the slot for simplification of calculation.

Figure 12b presents the CFD calculation results of the stator at the rated load condition. As can be seen, the hottest spot at the highest temperature of 66.25 °C occurs in the middle of the stator core near the air gap of the machine. In the same current loading conditions, the experimental tests were performed mainly to verify simulation results, where a maximum temperature of 65.6 °C (Figure 12c) was clearly observed. During the test, the rotor was fixed and a thermal steady state of armature windings at the rated DC load conditions was reached.

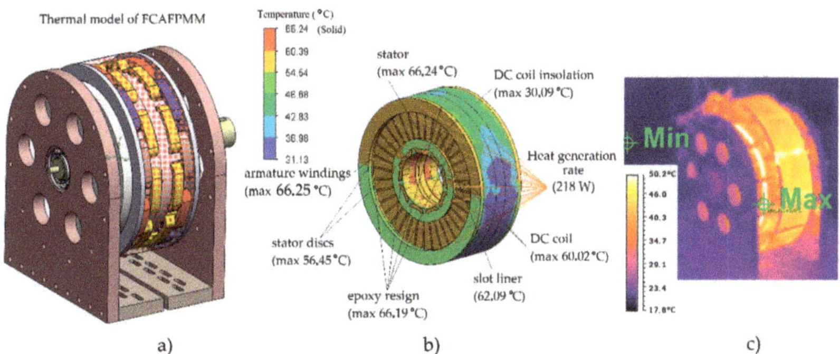

Figure 12. Three-dimensional thermal model of FCAFPMM (**a**); 3D FEA results of thermal distribution on the stator (**b**); thermal imaging result (**c**).

It should be noted that the stator of the machine under transient conditions was designed to maintain all temperatures below class B insulation limits of a 120 °C hot spot. Hence, thermal transient curves for all parts of the stator were calculated and shown in Figure 13. In this case, total power losses of approximately 430 W were obtained at the rated operating point of the machine; this means that n = 750 r/min, I_s = 12 A and I_{DC} = 5 A (without a cooling system) were adopted for the calculation of the curves. The rated operating point as the starting point for the thermal analysis (pT1) is depicted in Figure 10c,f. The total losses consist of the copper losses in the armature AC windings 218.1 W, core losses in the stator 90.5 W and copper losses in the DC coil 121.2 W.

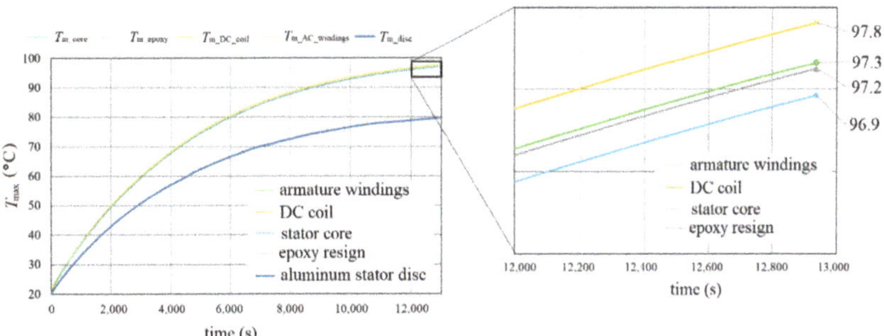

Figure 13. Heat characteristics.

The results show that under the rated load conditions, all stator components will not be overheated. The maximal temperature of 97.8 °C was observed in the DC coil, and it was much below the limit.

8. Conclusions

In this paper, a design of a permanent magnet axial flux machine with iron poles and an additional DC coil was analyzed with detailed electromagnetic and thermal performance. Three-dimensional FEA results showed that flux control of the proposed machine in the range of approximately 50% was successfully obtained.

The paper evaluated torque and speed loops at a current of 50 A and a voltage of 200 V limits, where the maximum efficiency of 95% and the increased rotational speed control range of 0–3000 RPM were reached.

The temperature distribution of the presented machine model under the rated conditions was calculated based on fluid–thermal coupling analysis by CFD. The air cooling effectiveness of the CFD models was verified by experimental measurements on a prototype machine. Full thermal analyses and tests will be performed and presented in further publications.

Further scientific studies on the FCAFPMM machine will also be focused on optimizing the rotor pole geometry to reduce the cogging torque, which is particularly disadvantageous in the low-speed region while the magnetic flux is increased.

Author Contributions: Conceptualization, P.P. (Pawel Prajzendanc); Methodology, P.P. (Pawel Prajzendanc) and P.P. (Piotr Paplicki); Supervision, P.P. (Piotr Paplicki); Writing—original draft, P.P. (Pawel Prajzendanc); Writing—review & editing, P.P. (Piotr Paplicki). All authors have read and agreed to the published version of the manuscript.

Funding: This research received no external funding.

Institutional Review Board Statement: Not applicable.

Informed Consent Statement: Not applicable.

Data Availability Statement: Study did not report any data.

Conflicts of Interest: The authors declare no conflict of interest.

References

1. Nedjar, B.; Hlioui, S.; Lecrivain, M.; Amara, Y.; Vido, L.; Gabsi, M. Study of a new hybrid excitation synchronous machine. In Proceedings of the XXth International Conference on Electrical Machines 2012, Marseille, France, 2–5 September 2012; pp. 2927–2932. [CrossRef]
2. Gieras, J.F. PM synchronous generators with hybrid excitation systems and voltage control Capabilities: A review. In Proceedings of the XXth International Conference on Electrical Machines 2012, Marseille, France, 2–5 September 2012; pp. 2573–2579. [CrossRef]
3. Wardach, M.; Palka, R.; Paplicki, P.; Prajzendanc, P.; Zarebski, T. Modern Hybrid Excited Electric Machines. *Energies* **2020**, *13*, 5910. [CrossRef]
4. Geng, W.; Zhang, Z.; Jiang, K.; Yan, Y. A New Parallel Hybrid Excitation Machine: Permanent-Magnet/Variable-Reluctance Machine with Bidirectional Field-Regulating Capability. *IEEE Trans. Ind. Electron.* **2015**, *62*, 1372–1381. [CrossRef]
5. Andrada, P.; Blanqué, B.; Martinez, E.; Torrent, M. A Novel Type of Hybrid Reluctance Motor Drive. *IEEE Trans. Ind. Electron.* **2014**, *61*, 4337–4345. [CrossRef]
6. Zhao, W.; Shen, H.; Lipo, T.A.; Wang, X. A New Hybrid Permanent Magnet Synchronous Reluctance Machine with Axially Sandwiched Magnets for Performance Improvement. *IEEE Trans. Energy Convers.* **2018**, *33*, 2018–2029. [CrossRef]
7. Zhao, X.; Niu, S.; Zhang, X.; Fu, W. Design of a New Relieving-DC-Saturation Hybrid Reluctance Machine for Fault-Tolerant In-Wheel Direct Drive. *IEEE Trans. Ind. Electron.* **2020**, *67*, 9571–9581. [CrossRef]
8. Luo, X.; Lipo, T.A. A synchronous/permanent magnet hybrid AC machine. *IEEE Trans. Energy Convers.* **2000**, *15*, 203–210. [CrossRef]
9. Xia, Y.; Yi, X.; Wen, Z.; Chen, Y.; Zhang, J. Direct Torque Control of Hybrid Excitation Permanent Magnet Synchronous Motor. In Proceedings of the 2019 22nd International Conference on Electrical Machines and Systems (ICEMS), Harbin, China, 11–14 August 2019; pp. 1–5. [CrossRef]
10. Fodorean, D.; Djerdir, A.; Viorel, I.-A.; Miraoui, A. A Double Excited Synchronous Machine for Direct Drive Application—Design and Prototype Tests. *IEEE Trans. Energy Convers.* **2007**, *22*, 656–665. [CrossRef]
11. Paplicki, P. Influence of Magnet and Flux-Barrier Arrange-ment on Flux Control Characteristics of Hybrid Excited ECPMS-machine. *Elektron. Ir Elektrotechnika* **2017**, *23*, 15–20. [CrossRef]
12. Paplicki, P.; Palka, R.; Wardach, M.; Prajzendanc, P.; Mognaschi, M.E. Hybrid excited electric machine with axial flux bridges. *Int. J. Appl. Electromagn. Mech.* **2019**, *59*, 703–711. [CrossRef]
13. May, H.; Meins, J.; Canders, W.R.; Pałka, R. New permanent magnet excited synchronous machine with extended, stator fixed auxilary excitation coil. In Proceedings of the XIV International Symposium on Electromagnetic Fields in Mechatronics, Electrical and Electronic Engineering, Arras, France, 10–12 September 2009.
14. Liwen, L.; Hongmei, L.; Zhiwei, C.; Tian, Y.; Rundong, L.; Jingkui, M. Design of a Hybrid Excited Permanent Magnet Machine with AC Field Winding Excitation. In Proceedings of the 22nd International Conference on Electrical Machines and Systems (ICEMS), Harbin, China, 11–14 August 2019; pp. 1–4. [CrossRef]
15. Gao, Y.; Li, D.; Qu, R.; Fan, X.; Li, J.; Ding, H. A Novel Hybrid Excitation Flux Reversal Machine for Electric Vehicle Propulsion. *IEEE Trans. Veh. Technol.* **2018**, *67*, 171–182. [CrossRef]
16. Wardach, M.; Bonislawski, M.; Palka, R.; Paplicki, P.; Prajzendanc, P. Hybrid Excited Synchronous Machine with Wireless Supply Control System. *Energies* **2019**, *12*, 3153. [CrossRef]
17. Aydin, M.; Huang, S.; Lipo, T.A. Design, Analysis, and Control of a Hybrid Field-Controlled Axial-Flux Permanent-Magnet Motor. *IEEE Trans. Ind. Electron.* **2010**, *57*, 78–87. [CrossRef]

18. Capponi, F.G.; De Donato, G.; Borocci, G.; Caricchi, F. Axial-Flux Hybrid-Excitation Synchronous Machine: Analysis, Design, and Experimental Evaluation. *IEEE Trans. Ind. Appl.* **2014**, *50*, 3173–3184. [CrossRef]
19. Ostroverkhov, M.; Chumack, V.; Monakhov, E. Synchronous Axial-Flux Generator with Hybrid Excitation in Stand Alone Mode. In Proceedings of the 2019 IEEE 2nd Ukraine Conference on Electrical and Computer Engineering (UKRCON), Lviv, Ukraine, 2–6 July 2019; pp. 455–459.
20. Chen, Z.; Zhou, N. Flux regulation ability of a hybrid excitation doubly salient machine. *IET Electr. Power Appl.* **2011**, *5*, 224–229. [CrossRef]
21. Hou, J.; Geng, W.; Zhu, T.; Li, Q. Topological Principle and Electromagnetic Performance of a Novel Axial-Flux Hybrid-Excitation In-wheel Motor. In Proceedings of the 23rd European Conference on Power Electronics and Applications (EPE'21 ECCE Europe) 2021, Ghent, Belgium, 6–10 September 2021; pp. 1–8.
22. Geng, W.; Zhang, Z.; Li, Q. Concept and Electromagnetic Design of a New Axial Flux Hybrid Excitation Motor for In-wheel Motor Driven Electric Vehicle. In Proceedings of the 2019 22nd International Conference on Electrical Machines and Systems (ICEMS), Harbin, China, 11–14 August 2019; pp. 1–6. [CrossRef]
23. Hsu, J. Direct control of air-gap flux in permanent-magnet machines. *IEEE Trans. Energy Convers.* **2000**, *15*, 361–365. [CrossRef]
24. Paplicki, P.; Prajzendanc, P. The influence of permanent magnet length and magnet type on flux-control of axial flux hybrid excited electrical machine. In Proceedings of the 14th Selected Issues of Electrical Engineering and Electronics (WZEE), Szczecin, Poland, 19–21 November 2018; pp. 1–4. [CrossRef]
25. Paplicki, P.; Prajzendanc, P.; Wardach, M.; Palka, R.; Cierzniewski, K.; Pstrokonski, R. Influence of Geometry of Iron Poles on the Cogging Torque of a Field Control Axial Flux Permanent Magnet Machine. *Int. J. Appl. Electromagn. Mech.* **2022**, 1–5, *in press*.

Article

Single Line/Phase Open Fault-Tolerant Decoupling Control of a Five-Phase Permanent Magnet Synchronous Motor under Different Stator Connections

Bing Tian *, Runze Lu and Jiasongyu Hu

Department of Electrical Engineering, College of Automation Engineering,
Nanjing University of Aeronautics and Astronautics, Nanjing 210016, China; nhlrz@nuaa.edu.cn (R.L.);
hu_jiasongyu@nuaa.edu.cn (J.H.)
* Correspondence: tian.bing@nuaa.edu.cn

Abstract: Fault-tolerant control (FTC) of a star-connected Five-phase Permanent Magnet Synchronous Motor (5Φ-PMSM) under open-circuit faults has been extensively studied, among which the decoupled control is attractive and finds a broad application in many fields. Pentacle- and pentagon-connected (generally known as "Penta-connected") 5Φ-PMSMs are popular due to their low voltage and high-power density, and especially, the demanded DC voltage level for the pentacle-connection mode accounting for merely 1/1.9021 of the star-connection mode, which is very appealing today. On the other hand, as one of the recent advances, the fault-tolerant decoupling control for penta-connections is still yet to be reviewed, and so this study investigates this issue and attempts to find the similarities and dissimilarities between star- and penta-connections under either single-line or single-phase open faults. Torque behavior analysis under, respectively, the fault-tolerant MPPT and $i_\mathrm{d} = 0$ is conducted to confirm the validity of the presented FTC, and it is expected to provide a reference for selecting a 5Φ-PMSM for practical use.

Keywords: field-oriented control; five-phase permanent magnet synchronous motor; fault-tolerant control; open-circuited fault; pentagon and pentacle connections

Citation: Tian, B.; Lu, R.; Hu, J. Single Line/Phase Open Fault-Tolerant Decoupling Control of a Five-Phase Permanent Magnet Synchronous Motor under Different Stator Connections. *Energies* **2022**, *15*, 3366. https://doi.org/10.3390/en15093366

Academic Editor: Mario Marchesoni

Received: 6 April 2022
Accepted: 3 May 2022
Published: 5 May 2022

Publisher's Note: MDPI stays neutral with regard to jurisdictional claims in published maps and institutional affiliations.

1. Introduction

Reliability is a constant topic in electrified transportation [1]. To improve the reliability of electric drives, a dual-motor system, where the two motors are coupled on the same shaft, was once popular in some safe-crucial applications [2]. Due to it being costly and bulky, this dual-motor system has been gradually substituted with the latest advances. The rapid progress in semiconductor techniques has stimulated the development of multi-phase windings within one motor to gain enhanced reliability [3]. As such, the multi-phase motor drive has been at the heart of engineering practice and successfully finds its use in more electric aircrafts, electric vehicles, wind energy conversion systems, and ship propulsion [4]. A five-phase permanent magnet synchronous motor (5Φ-PMSM) can be an outstanding candidate for using decent numbers of power switches [5], in addition to its high torque density and small torque ripples [6]. It is well known that a 3Φ-PMSM can be wired into star-connection, delta-connection, and star–delta connection modes, as well as open-end winding, regardless of whether the windings are series- or parallel-connected [7,8]. However, regarding a 5Φ-PMSM, except for star-connection and open-end winding modes, it can also be wired into the star-, pentagon-, and pentacle-connection (known as penta-connection) modes, as well as the combination of two of them [9]. Overall, the demanded DC link voltage for a pentacle-connected PMSM is merely 1/1.9021 of its star-connected counterpart [10]. Given this, the penta-connection mode is very attractive and may find a broad future in many high-power and low-voltage applications.

An open circuit fault is one of the most common types of fault in an electric drive [9]. To cope with this fault, the remaining winding phases of a 5Φ-PMSM have to be re-energized properly in the post-fault operation [10]. Regarding an open-winding five-phase machine, Pengye Wang, in [11], introduces a new inverter topology reconfiguration method along with a fault-tolerant control (FTC) strategy, which utilizes the bidirectional switching devices to respond to the open-circuit event on different phases. Zuosheng Yin [12] proposes a short-circuited FTC also for an open winding machine, and it features an equal current amplitude and sinusoidal waveform current, as well as maximum average torque. Concerning the torque ripple reduction, Qian Chen in [13] proposes a third harmonic injection method to reduce the torque ripple caused by the shorted phase for the star-connection mode. Li Zhang in [14] comes up with a generalized FTC scheme for both field-oriented control (FOC) and direct torque control (DTC) under open-line faults, avoiding the use of different transforming matrices under different fault conditions. Recently, model predictive fault-tolerant control has attracted much attention, however, most of them are dedicated to a star-connected motor. For instance, reference [15] proposes a multi-vector-based model predictive control with voltage error-tracking short-circuit faults, including a single-phase short-circuit fault and inter-phase short-circuit fault. Reference [16] proposes a virtual voltage vector-based model predictive control for an open line fault, which claims to reduce the computational burden.

On the other hand, reference [17] provides a comprehensive comparison of the post-fault torque capability for 5Φ-PMSMs under all connection conditions with open-circuit phase and line faults. The mirror symmetry theory in combination with the balance of power theory is used, however, the reluctance torque is not considered. Reference [18] states that some stator winding configurations other than star-connection yield a small copper loss and increased torque capacity under the same torque command for a dual three-phase drive. Among star-, pentagon-, and combined Star/Pentagon-connection modes, reference [19] suggests that the star-connection mode produces the lowest average torque and highest ripple torque, whereas the pentagon connection mode gives the highest average torque and minimum ripple torque, and the star/pentagon-connection mode mediates between both. Reference [20] concludes the maximum available torque that a five-phase induction motor (5Φ-IM) can deliver under open-circuit phase faults, where all connection conditions are tested using a unified balanced power source. Reference [21] also investigates the torque ripple-free operation of all connection modes under an open-circuit phase and line fault, and it confirms that a penta-connection mode outputs a higher torque than a star-connection mode. Reference [22] put forward an open line FTC to maximize the average torque and minimize the ripple torque for a combined star/pentagon-connected SynRM fed from a matrix converter. Reference [10] concludes that the performance of a pentagon-connection mode with a single-line open fault under open-loop control is advantageous over the star-connection mode. For optimal current control, although the star-connection mode results in a better current waveform, the copper loss can be less using a pentagon-connection mode. Regarding a pentacle-connected induction motor under one phase supply failure, Reference [23] proves that a 5Φ-PMSM can switch to a three-phase operation state without an electromagnetic torque ripple. Reference [24] investigates the phase transposition on the five-phase induction machine with different stator connection modes, namely, star, pentagon, and pentacle. Owing to the lower derating factor of the pentacle-connection mode under the open phase, Reference [25] proposes a fault-tolerant motor using a combined star/pentacle-connection.

Given the previous achievements, specifically, those under penta-connection modes, the optimal current references are usually solved under the natural coordinate frame, and the hysteresis current controller or resonant current controller is therefore adopted. However, the line/phase voltage modulation does not take care of the zero-sequence component, and this constitutes a huge disturbance to the closed-loop current regulation. On the other hand, the decoupled control, which relies on decoupled modeling, plays an important role in today's variable-speed drive and has become a standard technique in the industry sector [13,26,27]. To enable the continuous use of a decoupled FTC under penta-connection modes, this work firstly explores the similarity and dissimilarity between star-connection and penta-connection, and then raises a unified FTC under, respectively, the single open-line fault and the single open-phase fault, and also elaborated are the key technique details to unlock this decoupled FTC. To be specific, for a single open-line fault, the transformation of resistance, inductance, and back-EMF from the penta-connection to the star-connection is conducted, along with the angular of vectorial control properly fixed. Regarding the single open-phase fault, voltage and current transformation from a line-to-line frame to a winding-oriented frame is proposed by incorporating the zero-sequence voltage and current. Finally, experimental results on $i_d = 0$ and Maximum Torque Per Ampere (MTPA) are carried out, which confirms the validity of the presented decoupled FTC for all connection modes.

2. Single Line Open FTC

The 5Φ-PMSM's stator windings can be wired into either star-, pentagon-, or pentacle-connection modes, as revealed by Figure 1, and among them, the pentagon-connection features a series connection of two adjacent sets of windings, while the pentacle-connection features a series connection of two non-adjacent sets of windings. Since the pentagon can be viewed as a convex polygon and the pentacle is a kind of concave polygon, the latter two connection modes in Figure 1 can be unified, which is termed a penta-connection in this paper. Since one cannot differentiate a motor's connection mode from the outside, from the inverter's perspective, a generalized control scheme for all connection modes is therefore possible.

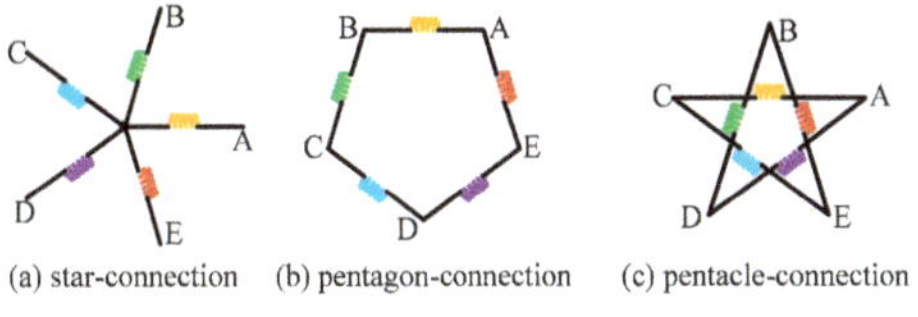

Figure 1. Stator winding connection modes.

On the other hand, the withstanding voltage of one winding phase is different from the star- to penta-connection modes, even when powered by the same DC power source. Overall, the winding phase voltage for a pentagon-connection mode is $1/1.1172$ of that under a star-connection mode, and this ratio reduces to $1/1.902$ for a pentacle-connection mode. Therefore, a penta-connection mode is more suitable for low-voltage and high-power applications, such as electrified transportation, and this property resembles a conventional delta-connected three-phase motor.

2.1. Star-Connection

The open-circuit fault accounts for one of the most common faults in an electric drive. The open-circuit fault comprises an open-line fault and an open-phase fault, and to avoid confusion, they are illustrated in Figures 2 and 3, respectively.

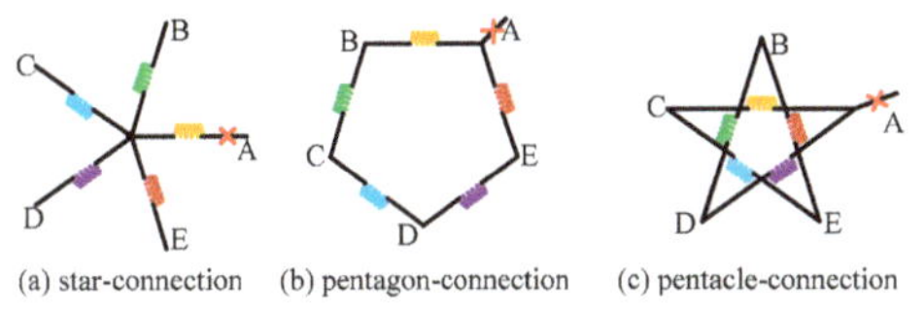

Figure 2. Graphical illustration for a single open-line fault.

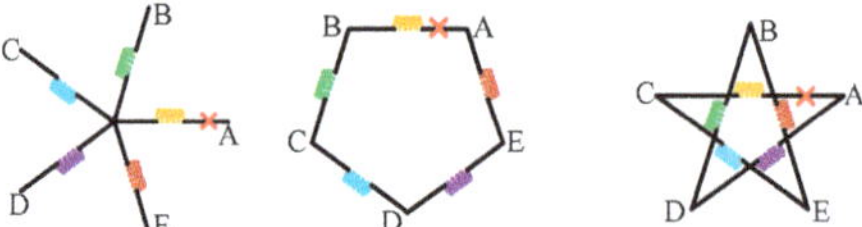

Figure 3. Graphical illustration for a single-phase open fault.

For a star-connection mode, the open-line fault and open-phase fault create the same effect since the phase and line are the same. Thus, their fault-tolerant decoupling controls are the same, and hereby termed "star-control". Readers may refer to Figure 4 for an overview of the star-control under a single-line open fault, where the adopted Clarke transformation is no longer a unitary matrix to offer unbalanced voltages even with a balanced u_d and u_q. Notice that superscript * indicates the reference value.

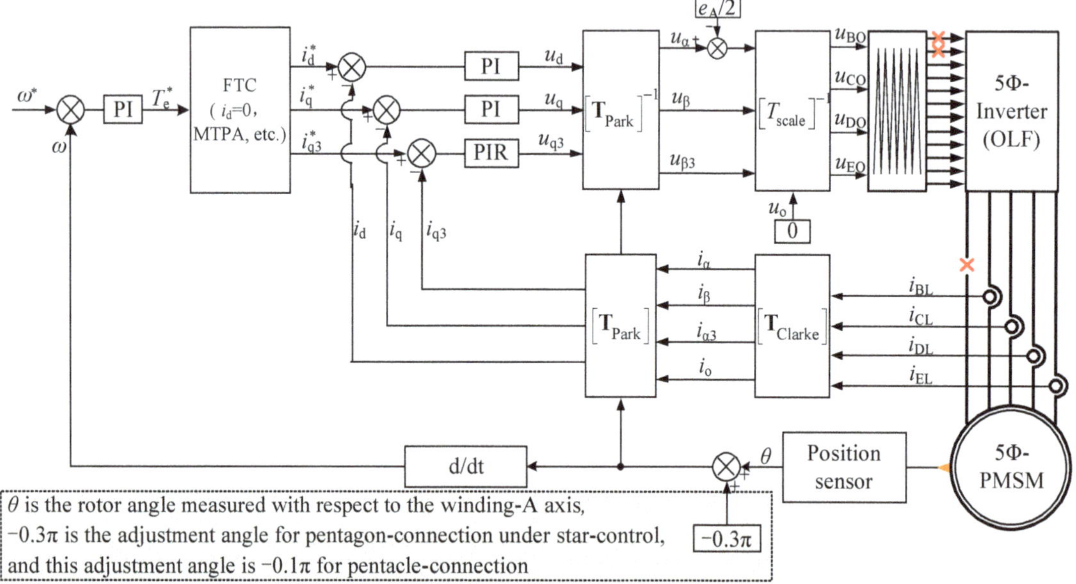

Figure 4. Block diagram of unified open-line fault-tolerant control for all connections.

In this way, the post-fault torque performance can be maintained as stable enough by merely using open-loop voltage control. The involved matrices for this decoupled FTC are referred to in (1)–(3) [28]. In addition, in the figure, $e_A/2$ is introduced to compensate for the oscillating neutral voltage post fault, which must be used in combination with T_{scalar} to ensure the line voltage modulation work as anticipated [29]. It has to be emphasized that e_A corresponds to the line-to-neutral back-EMF, and it is easy to acquire/estimate with merely rotor flux linkage parameters. The acquisition of e_A under penta-connection modes is detailed in the section below.

$$T_{\text{scalar}} = \frac{2}{5} \begin{bmatrix} \cos\delta + \frac{1}{4} & \cos 2\delta + \frac{1}{4} & \cos 3\delta + \frac{1}{4} & \cos 4\delta + \frac{1}{4} \\ \sin\delta & \sin 2\delta & \sin 3\delta & \sin 4\delta \\ \sin 3\delta & \sin 6\delta & \sin 9\delta & \sin 12\delta \\ 1 & 1 & 1 & 1 \end{bmatrix} \tag{1}$$

$$\mathbf{T}_{\text{Clarke}} = \frac{2}{5} \begin{bmatrix} \cos\delta - 1 & \cos 2\delta - 1 & \cos 3\delta - 1 & \cos 4\delta - 1 \\ \sin\delta & \sin 2\delta & \sin 3\delta & \sin 4\delta \\ \sin 3\delta & \sin 6\delta & \sin 9\delta & \sin 12\delta \\ 1 & 1 & 1 & 1 \end{bmatrix} \tag{2}$$

$$\mathbf{T}_{\text{Park}} = \begin{bmatrix} \cos\theta & \sin\theta & 0 & 0 \\ -\sin\theta & \cos\theta & 0 & 0 \\ 0 & 0 & 1 & 0 \\ 0 & 0 & 0 & 1 \end{bmatrix} \tag{3}$$

2.2. Pentagon-Connection

The mathematical model of each connection mode can be equivalent to each other as long as the motor is not damaged from the inside. From the inverter's perspective, the fault-tolerant current for all connection modes has the potential to be unified. This means the star-control can be also available to the penta-connection mode under the open line fault. It must be mentioned that the use of star-control is conditional, and several important control parameters have to be transformed from the penta-connection mode into the star-connection mode. A total of four paramount parameters need rectification.

Take the pentagon-connection mode as an example. According to Figure 1a,b, the line-to-neutral resistance of a pentacle-connected motor is calculated by

$$2R = \frac{4R_\phi R_\phi}{4R_\phi + R_\phi} \tag{4}$$

where R herein indicates the line-to-neutral resistance of an equivalent star-connection, and R_ϕ is the winding phase resistance.

The simplification of the above equation yields

$$R = 0.4R_\phi \tag{5}$$

Similarly, the relationship of inductance between the line-to-neutral frame and a winding-oriented frame is shown below

$$L = 0.4L_\phi \tag{6}$$

where L herein indicates the self- or mutual-inductance (instantaneous value) measured between one of the motor's five terminals and the neutral point, and L_ϕ is the winding phase inductance (also an instantaneous value).

The line-to-neutral back-EMF is manifested as the line-to-neutral voltage under no-load conditions, and thanks to the matrix that transforms the line voltages to line-to-neutral voltages of the health condition, one can obtain the detailed representation of the back-EMFs in the sense of a star-connection, as below

$$\begin{cases} e_A = -\dfrac{\omega\lambda_{r1-\phi}\sin(\theta-0.3\pi)}{1.1172} + \dfrac{3\omega\lambda_{r3-\phi}\sin(3(\theta-0.3\pi))}{1.9021} \\[2mm] e_B = -\dfrac{\omega\lambda_{r1-\phi}\sin(\theta-0.3\pi+\delta)}{1.1172} + \dfrac{3\omega\lambda_{r3-\phi}\sin(3(\theta-0.3\pi+\delta))}{1.9021} \\[2mm] e_C = -\dfrac{\omega\lambda_{r1-\phi}\sin(\theta-0.3\pi+2\delta)}{1.1172} + \dfrac{3\omega\lambda_{r3-\phi}\sin(3(\theta-0.3\pi+2\delta))}{1.9021} \\[2mm] e_D = -\dfrac{\omega\lambda_{r1-\phi}\sin(\theta-0.3\pi+3\delta)}{1.1172} + \dfrac{3\omega\lambda_{r3-\phi}\sin(3(\theta-0.3\pi+3\delta))}{1.9021} \\[2mm] e_E = -\dfrac{\omega\lambda_{r1-\phi}\sin(\theta-0.3\pi+4\delta)}{1.1172} + \dfrac{3\omega\lambda_{r3-\phi}\sin(3(\theta-0.3\pi+4\delta))}{1.9021} \end{cases} \tag{7}$$

where $\lambda_{r1-\phi}$ and $\lambda_{r3-\phi}$ are the magnitudes of, respectively, the 1st and 3rd rotor flux linkages that are measured under the winding-oriented frame. From (7), the ratio of the 3rd to 1st harmonic back-EMFs shrinks to 0.587, i.e., 1.1172/1.9021, of that under the real star-connection mode, which means the 3rd rotor flux is dampened with a pentagon-connection, while comparing the star-connection mode with the base value per unit.

Finally, to correctly use FTC, the position angle for FOC also needs modification, which can be adjusted by biasing the existing position with a constant value of -0.3π (this angular can be inferred from the line-to-neutral back-EMF's representation). The above manipulation enables us to continually use a star-control even under the pentagon-connection.

The post-fault torque can be given by the following equation on the premise that all parameters have been transformed into the star-connection.

$$T_e = \frac{5p}{2}\left[i_d i_q(L_d - L_q) + i_q\lambda_{r1}\right] + \frac{5p\lambda_{r3}}{2}\left[3i_{q3} + \frac{3(i_d\sin 2\theta + i_d\sin 4\theta - i_q\cos 2\theta + i_q\cos 4\theta)}{2}\right] \quad (8)$$

Here, L_d, L_q, L_{q3} are the inductances under a fault-tolerant synchronously rotating frame, and they shrink to 0.4 of those under a star-connection mode as per (6). λ_{r1} and λ_{r3} are the magnitudes of, respectively, the 1st and 3rd rotor flux linkages under the line-to-neutral frame which shrinks to, respectively, $1/1.1172$, and $1/1.9021$ of those under a star-connection mode, as per (7).

Notice also that the impact of the magnetic saturation and cross saturation on the d-q-q_3 frame inductances are not considered in FTC currently, and this may affect the torque ripple to some extent while evaluating the decoupled FTC experimentally. However, this effect is within a controllable range as FTC is usually not targeting a precise torque control.

2.3. Pentacle-Connection

As a summary, parameters from the pentacle-connection sense to the star-connection sense are also derived as below

$$R = 0.6R_\phi \quad (9)$$

$$L = 0.6L_\phi \quad (10)$$

Likewise, under the pentagon-connection mode, the line-to-neutral back-EMF is derived as below

$$\begin{cases} e_A = -\dfrac{\omega\lambda_{r1-\phi}\sin(\theta-0.1\pi)}{1.9021} - \dfrac{3\omega\lambda_{r3-\phi}\sin(3(\theta-0.1\pi))}{1.1172} \\[2mm] e_B = -\dfrac{\omega\lambda_{r1-\phi}\sin(\theta-0.1\pi+\delta)}{1.9021} - \dfrac{3\omega\lambda_{r3-\phi}\sin(3(\theta-0.1\pi+\delta))}{1.1172} \\[2mm] e_C = -\dfrac{\omega\lambda_{r1-\phi}\sin(\theta-0.1\pi+2\delta)}{1.9021} - \dfrac{3\omega\lambda_{r3-\phi}\sin(3(\theta-0.1\pi+2\delta))}{1.1172} \\[2mm] e_D = -\dfrac{\omega\lambda_{r1-\phi}\sin(\theta-0.1\pi+3\delta)}{1.9021} - \dfrac{3\omega\lambda_{r3-\phi}\sin(3(\theta-0.1\pi+3\delta))}{1.1172} \\[2mm] e_E = -\dfrac{\omega\lambda_{r1-\phi}\sin(\theta-0.1\pi+4\delta)}{1.9021} - \dfrac{3\omega\lambda_{r3-\phi}\sin(3(\theta-0.1\pi+4\delta))}{1.1172} \end{cases} \quad (11)$$

From (11), the ratio of the 3rd to 1st harmonic back-EMFs increases to 1.703, i.e., $1.9021/1.1172$, compared with the base value 1 set by the star-connection mode.

Moreover, the adjustment angle for using a star-control is supposed -0.1π under the pentacle-connection mode (this bias angle can also be inferred from the line-to-neutral back-EMF). Furthermore, the L_d, L_q, L_{q3}, and λ_{r1}, as well as λ_{r3} also need calibration for the torque control, and the scaling factor can be deduced from (10) and (11).

An experiment will be carried out to demonstrate the validity of the FTC using the same star-control for all connection modes.

3. Single Phase Open FTC

3.1. Pentagon-Connection

The FTC of a star-connection mode under an open-phase fault maintains the same as the previous open-line fault. However, regarding the penta-connection mode, an open-phase fault and open-line fault make a big difference; this is because the equivalent conversion between motor models does not hold under this condition. To achieve a unified open-phase FTC for all connection modes, the FTC should be built onto a winding-oriented frame.

To this end, the difference between a star-connected motor and a penta-connected motor should be firstly figured out. For all connection modes, should we observe the motor from the winding-oriented frame, the models under all connections, even with an open-phase fault, are no different from each other, and this lays a theoretical foundation for achieving a unified FTC. To this end, one has to change the control feedbacks, i.e., the sampled currents, from a line-to-line frame to a winding-oriented frame, as well as the control outputs, i.e., voltage, from the line-to-line frame to a pole-voltage frame. In this way, the fault-tolerant decoupling control, which is originally derived from a star-connected motor, still has a chance to be applied under the penta-connection mode.

Without a loss of generality, assume the stator is pentagon-connected, and let i_{AL}, i_{BL}, i_{CL}, i_{DL}, and i_{EL} denote the line-to-line currents, and i_A, i_B, i_C, i_D, and i_E the winding phase currents.

By applying Kirchhoff's Current Law to Figure 3b, one can obtain

$$
\begin{bmatrix} i_{AL} \\ i_{BL} \\ i_{CL} \\ i_{DL} \\ i_{EL} \end{bmatrix} = \begin{bmatrix} 0 & 0 & 0 & 0 & -1 \\ 0 & 1 & 0 & 0 & 0 \\ 0 & -1 & 1 & 0 & 0 \\ 0 & 0 & -1 & 1 & 0 \\ 0 & 0 & 0 & -1 & 1 \end{bmatrix} \begin{bmatrix} i_A \\ i_B \\ i_C \\ i_D \\ i_E \end{bmatrix}, \ i_A = 0 \tag{12}
$$

The above formula involves a singular matrix, and this means that the solutions for i_B, i_C, i_D, and i_E can be multiple under an already known i_{AL}, i_{BL}, i_{CL}, i_{DL}, and i_{EL}. To obtain a unique solution, some extra constraints have to be imposed.

A zero-sequence circulating current (ZSCC) can be the next dimension to decide the winding phase currents. A ZSCC is common in a delta-wired, as well as penta-wired drive. Usually, a ZSCC is harmful as it generates heat and makes the torque fluctuate. Therefore, a ZSCC is not desired in most situations.

This faulty drive also expects a null ZSCC, which means

$$
i_B + i_C + i_D + i_E = 0, \ i_A = 0 \tag{13}
$$

It has to be emphasized that the winding phase currents estimated from the measured line currents contain no ZSCC now. However, in reality, a ZSCC still appears in the actual winding phase currents even with winding-A open-circuited. Rather than circulating among five balanced winding phases under the health condition, the circulating path of a ZSCC includes the inverter's active legs in fault mode. A ZSCC is difficult to suppress without further knowledge of the winding phase current in a penta-wired system.

Combining (13) with (12), one can estimate the winding phase currents without a ZSCC, and the analytical solutions are shown below.

$$
\begin{bmatrix} i_B & i_C & i_D & i_E \end{bmatrix}^T = C_{L2\Phi} \begin{bmatrix} i_{AL} & i_{BL} & i_{CL} & i_{DL} & i_{EL} \end{bmatrix}^T \tag{14}
$$

where

$$
C_{L2\Phi} = \begin{bmatrix} 0 & \frac{-1}{3} & -1 & \frac{-2}{3} & \frac{-1}{3} \\ 0 & \frac{-1}{3} & 0 & \frac{-2}{3} & \frac{-1}{3} \\ 0 & \frac{-1}{3} & 0 & \frac{1}{3} & \frac{-1}{3} \\ 0 & \frac{-1}{3} & 0 & \frac{1}{3} & \frac{-2}{3} \end{bmatrix} \tag{15}
$$

In this way, the winding phase currents, which are not easy to measure in a pentagon-wired drive, can be acquired now, and they will be sent to the current controller for monitoring the torque behavior.

Simultaneously, the modulation of the pole voltage (relative to the midpoint of the DC bus) also needs rectification to correctly generate the wanted line-to-line voltages. The relationship between the line-to-line voltage and the pole voltage can be always represented, as below, under either healthy or fault conditions,

$$
\begin{bmatrix} u_\mathrm{A} \\ u_\mathrm{B} \\ u_\mathrm{C} \\ u_\mathrm{D} \\ u_\mathrm{E} \end{bmatrix} = \begin{bmatrix} 1 & -1 & 0 & 0 & 0 \\ 0 & 1 & -1 & 0 & 0 \\ 0 & 0 & 1 & -1 & 0 \\ 0 & 0 & 0 & 1 & -1 \\ -1 & 0 & 0 & 0 & 1 \end{bmatrix} \begin{bmatrix} u_\mathrm{AO} \\ u_\mathrm{BO} \\ u_\mathrm{CO} \\ u_\mathrm{DO} \\ u_\mathrm{EO} \end{bmatrix}
\tag{16}
$$

where u_x, with x = A, B, C, D, and E represents the line-to-line voltage. The above matrix is also singular, and to obtain a unique solution for the demanded pole voltage, one needs to raise some additional constraints.

Likewise, the common-mode voltage (averaged value) can be used as another dimension to decide the pole voltage solutions. The common-mode voltage often takes a switching pattern in an inverter-driven system, however, it is often averaged to zero concerning the fundamental components of the pole voltages, and hereby the 'pole' point refers to the midpoint of the DC-bus. It is also expected that this property could still hold, and thus the following constraint is imposed

$$
u_\mathrm{AO} + u_\mathrm{BO} + u_\mathrm{CO} + u_\mathrm{DO} + u_\mathrm{EO} = 0
\tag{17}
$$

By incorporating (17) into (16), and then applying the inverse operation, one can have

$$
\begin{bmatrix} u_\mathrm{AO} & u_\mathrm{BO} & u_\mathrm{CO} & u_\mathrm{DO} & u_\mathrm{EO} \end{bmatrix}^\mathrm{T} = \mathrm{V_{L2P}} \begin{bmatrix} u_\mathrm{A} & u_\mathrm{B} & u_\mathrm{C} & u_\mathrm{D} & u_\mathrm{E} \end{bmatrix}^\mathrm{T}
\tag{18}
$$

$$
\mathrm{V_{L2P}} = \begin{bmatrix} \frac{2}{5} & \frac{1}{5} & 0 & \frac{-1}{5} & \frac{-2}{5} \\ \frac{-2}{5} & \frac{2}{5} & \frac{1}{5} & 0 & \frac{-1}{5} \\ \frac{-1}{5} & \frac{-2}{5} & \frac{2}{5} & \frac{1}{5} & 0 \\ 0 & \frac{-1}{5} & \frac{-2}{5} & \frac{2}{5} & \frac{1}{5} \\ \frac{1}{5} & 0 & \frac{-1}{5} & \frac{-2}{5} & \frac{2}{5} \end{bmatrix}
\tag{19}
$$

where u_B, u_C, u_D, and u_E can be obtained by

$$
\begin{bmatrix} u_\mathrm{B} \\ u_\mathrm{C} \\ u_\mathrm{D} \\ u_\mathrm{E} \end{bmatrix} = [\mathrm{T_{Clarke}}]^{-1} [\mathrm{T_{Park}}]^{-1} \begin{bmatrix} u_\mathrm{d} \\ u_\mathrm{q} \\ u_\mathrm{q3} \\ 0 \end{bmatrix}
\tag{20}
$$

where u_d, u_q, and u_q3 are the output actions of the current PI controller and denote the line-to-line voltage references under a fault-tolerant synchronously rotating frame. Evidently, with the presented line-to-phase/pole matrices, i.e., $\mathrm{C_{L2\Phi}}$ and $\mathrm{V_{L2P}}$, the current feedback and pole voltage modulation can be rectified to perfectly match the decoupled model that was originally developed for a star-connected motor.

From (18), u_A is pivotal while solving the pole voltage solutions. As aforementioned in the star-connection case, u_A can be replaced by the back-EMF of winding-A because of there being no current flowing through the open-circuited phase. Up to this point, it may remain questionable whether the same approximation applies to a penta-wired drive. As another contribution, this paper proves that the above hypothesis can still find its use in this context.

In response to the missing representation of u_A, its definition has to be clarified first. As u_A represents the line-to-line voltage between motor terminals-A and -B, it is no longer identical to winding-A's phase voltage when phase-A is open-circuited.

Figure 5 depicts the details of a pentagon-wired drive with phase-A being open-circuited. From the figure, the winding-A's phase voltage, i.e., u_Aa, is not equal to the corresponding line-to-line voltage, i.e., u_A, any longer, and the difference lies in Δu_A which is supposed to vary along with the ZSCC. As per Figure 5, the relationship between u_{Aa} and u_{A} can be represented by

$$u_{\text{A}} = u_{\text{Aa}} + \Delta u_{\text{A}} \tag{21}$$

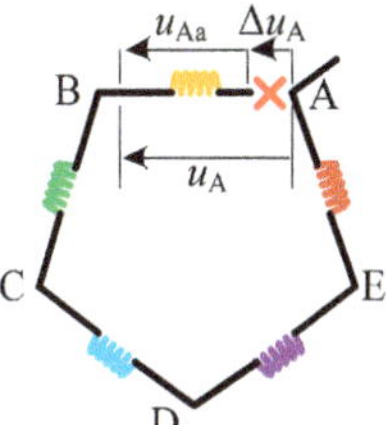

Figure 5. Single open-phase fault in pentagon-connection mode.

The zero-sequence voltage is defined as

$$u_{\text{o}} = u_{\text{Aa}} + u_{\text{B}} + u_{\text{C}} + u_{\text{D}} + u_{\text{E}} \tag{22}$$

where u_{o} is the phase zero-sequence voltage in the context of winding-A open-circuited, and it is calculated by summing all five winding phase voltages.

By substituting (21) into (22), it can be proved that

$$u_{\text{o}} = u_{\text{Aa}} + u_{\text{B}} + u_{\text{C}} + u_{\text{D}} + u_{\text{E}} = u_{\text{A}} - \Delta u_{\text{A}} + u_{\text{B}} + u_{\text{C}} + u_{\text{D}} + u_{\text{E}} = -\Delta u_{\text{A}} \tag{23}$$

As per the Appendix A, u_{o}, or rather Δu_{A}, usually comes in a pair with the ZSCC, and the actual ZSCC depends on the zero-sequence impendence in this unbalanced system.

In engineering practice, u_{o} is unmeasurable, and thus it is overlooked in this work. On the other hand, by substituting u_{A} with its back-EMF because of there being no current circulating through the open-circuited winding, one can have

$$u_{\text{A}} \approx u_{\text{Aa}} \approx e_{\text{A}} \tag{24}$$

The proposed fault-tolerant decoupling control of a pentagon-connected drive is illustrated in the next section, which is based on a star-connection drive by including two matrices in connection with the feedback and output rectification.

3.2. Pentacle-Connection

Similarly, in the context of pentacle-connection, both line-to-line currents and the line-to-pole voltage need rectification.

Regarding the line-to-line currents, one can have

$$\begin{bmatrix} i_{\text{AL}} \\ i_{\text{BL}} \\ i_{\text{CL}} \\ i_{\text{DL}} \\ i_{\text{EL}} \end{bmatrix} = \begin{bmatrix} 0 & 0 & 0 & -1 & 0 \\ 0 & 1 & 0 & 0 & -1 \\ 0 & 0 & 1 & 0 & 0 \\ 0 & -1 & 0 & 1 & 0 \\ 0 & 0 & -1 & 0 & 1 \end{bmatrix} \begin{bmatrix} i_{\text{A}} \\ i_{\text{B}} \\ i_{\text{C}} \\ i_{\text{D}} \\ i_{\text{E}} \end{bmatrix}, \; i_{\text{A}} = 0 \tag{25}$$

By imposing zero ZSCC constraints, as below

$$i_{\text{B}} + i_{\text{C}} + i_{\text{D}} + i_{\text{E}} = 0, \; i_{\text{A}} = 0 \tag{26}$$

one can obtain the phase currents

$$\begin{bmatrix} i_{\text{B}} & i_{\text{C}} & i_{\text{D}} & i_{\text{E}} \end{bmatrix}^{\text{T}} = C_{\text{L2}\Phi} \begin{bmatrix} i_{\text{AL}} & i_{\text{BL}} & i_{\text{CL}} & i_{\text{DL}} & i_{\text{EL}} \end{bmatrix}^{\text{T}} \tag{27}$$

where $C_{L2\Phi}$ is represented by

$$C_{L2\Phi} = \begin{bmatrix} 0 & \frac{1}{3} & \frac{-1}{3} & \frac{-1}{3} & 0 \\ 0 & \frac{-2}{3} & \frac{-1}{3} & \frac{-1}{3} & -1 \\ 0 & \frac{1}{3} & \frac{-1}{3} & \frac{2}{3} & 0 \\ 0 & \frac{-2}{3} & \frac{-1}{3} & \frac{-1}{3} & 0 \end{bmatrix} \tag{28}$$

With regard to the line-to-pole voltage transformation, one can always have

$$\begin{bmatrix} u_A \\ u_B \\ u_C \\ u_D \\ u_E \end{bmatrix} = \begin{bmatrix} 1 & 0 & -1 & 0 & 0 \\ 0 & 1 & 0 & -1 & 0 \\ 0 & 0 & 1 & 0 & -1 \\ -1 & 0 & 0 & 1 & 0 \\ 0 & -1 & 0 & 0 & 1 \end{bmatrix} \begin{bmatrix} u_{AO} \\ u_{BO} \\ u_{CO} \\ u_{DO} \\ u_{EO} \end{bmatrix} \tag{29}$$

Additionally, imposing a zero common-mode voltage constraint as below

$$u_{AO} + u_{BO} + u_{CO} + u_{DO} + u_{EO} = 0 \tag{30}$$

can yield

$$\begin{bmatrix} u_{AO} & u_{BO} & u_{CO} & u_{DO} & u_{EO} \end{bmatrix}^T = V_{L2P} \begin{bmatrix} u_A & u_B & u_C & u_D & u_E \end{bmatrix}^T \tag{31}$$

where V_{L2P} can be calculated by

$$V_{L2P} = \begin{bmatrix} \frac{2}{5} & \frac{-1}{5} & \frac{1}{5} & \frac{-2}{5} & 0 \\ 0 & \frac{2}{5} & \frac{-1}{5} & \frac{1}{5} & \frac{-2}{5} \\ \frac{-2}{5} & 0 & \frac{2}{5} & \frac{-1}{5} & \frac{1}{5} \\ \frac{1}{5} & \frac{-2}{5} & 0 & \frac{2}{5} & \frac{-1}{5} \\ \frac{-1}{5} & \frac{1}{5} & \frac{-2}{5} & 0 & \frac{2}{5} \end{bmatrix} \tag{32}$$

Figure 6 shows the graphical relationship between u_A and u_{Aa} under a pentacle-connection, where u_{Aa} is the winding phase voltage and can be approximated as a back-EMF component when the corresponding phase winding is open-circuited. By neglecting Δu_A, which is usually not measured in practice, u_A is finally substituted by e_A.

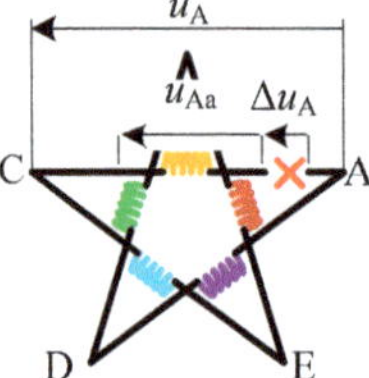

Figure 6. Single open-phase fault in a pentacle-connection mode.

Figure 7 refers to a unified open phase fault-tolerant decoupling control for all connections, where one only has to change $C_{L2\varphi}$ and V_{L2P} to enable the FTC under the preferred connection. Notice that all five IGBT bridges are still available in this context, whereas only four bridges are controllable in an open-phase fault, as revealed in Figure 4. Therefore, the torque derating will be less under single-phase open FTC, and it is not possible to merge the two control schemes into one.

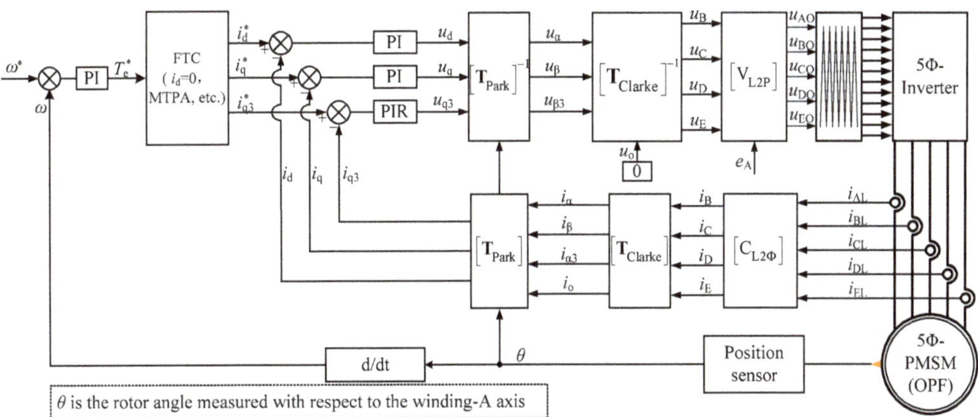

Figure 7. Block diagram of unified open-phase fault-tolerant control for all connection modes.

4. Experimental Results

A laboratory-scale 5Φ-PMSM, having its shaft coupled to a three-phase PM synchronous generator, has been developed to validate the presented fault-tolerant decoupling control for all connection modes under a single-line/phase open fault. The investigated 5Φ-PMSM is fed by an IGBT-based five-phase half-bridge inverter, with its parameters referred to Table 1. The 1st and 3rd harmonics of the rotor flux are measured by letting the 5Φ-PMSM work in generation mode, and deduced from its winding phase back-EMFs.

Table 1. Prototype motor parameters.

Symbol	Description	Value
$\lambda_{r1-\phi}$	1st order rotor flux	0.17 Wb
$\lambda_{r3-\phi}$	3rd order rotor flux	0.0062 Wb
R	Resistance per phase	0.4 Ω
p	Pole Pairs	2
L_d	d-inductance of 1st subspace	20.66 mH
L_q	q-inductance of 1st subspace	25.18 mH
I_r	Rated phase current (rms)	10 A
J	Rotor inertia	0.1 kg·m^2
n_r	Rated speed	1500 rpm

Two control schemes, i.e., $i_d = 0$ to minimize the torque ripples and MTPA to maximize the torque capacity, are evaluated in the experiment, where the speed is regulated to a constant 750 rpm with a single PI controller. The output action of the speed PI, i.e., a torque command, is fed to a q-axis current regulator, and i_{q3} is subject to open-loop voltage control, which helps to reduce the torque ripples. The FTC for all connection modes under an open-line fault is conducted firstly, and then the open-phase FTC comes after. As phase-A and line-A are the same for the star-connection, the torque and current behaviors of the FTC under an open-phase fault are referred to in the case of an open-line fault. In addition, the torque and current performance post-fault under a star-connection also serves as a benchmark for comparing with a penta-connection mode under the same fault conditions.

4.1. Single-Line Open Fault

Figure 8 refers to the behavior of d-q-q3 frame currents with line-A open-circuited. From Figure 8a–c, which are evaluated under the control of $i_d = 0$, the optimal value of i_q to achieve a constant speed of 750 rpm differentiates under different connection modes, and it decreases from a pentacle-connection to pentagon-connection, and is then followed by a star-connection mode. This is because their respective line-to-neutral back-EMFs increase and this phenomenon can be explained by (7) and (11). On the other hand, the feedbacks, i_d and i_q, are DCs, and evidently, the star-control philosophy works well for all connection modes under a single open-line fault.

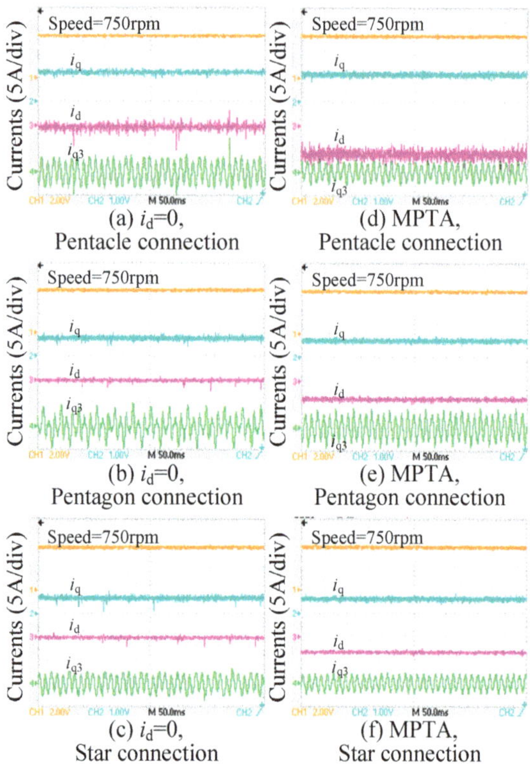

(a) i_d=0,
Pentacle connection

(d) MPTA,
Pentacle connection

(b) i_d=0,
Pentagon connection

(e) MPTA,
Pentagon connection

(c) i_d=0,
Star connection

(f) MPTA,
Star connection

Figure 8. Waveforms of d-q-q₃ frame current as well as the rotor speed under a single open-line fault.

Figure 8d–f show the same currents under an MTPA control, and the value of i_q decreases to preserve the motor at a constant 750 rpm, compared with their counterparts under the control of $i_d = 0$. This is because reluctance torque is utilized in this case which boosts the torque capacity to some extent. Under either $i_d = 0$ or MTPA control, the speed profile is smooth, and it proves the capability of the star-control for all connection modes under a single open-line fault.

Figure 9 shows the waveforms of the torque, Δu_A, and ZSCC, where i_o denotes the sum of all five winding phase currents, and herein is calculated by sampling four line-currents together with one winding phase current. Figure 9a–c corresponds to the control of $i_d = 0$, and from the sub-figures, amounts of the ZSCC are visible under a penta-connection, whereas the ZSCC under a star-connection remains null. The ZSCC primarily fluctuates at the 5th frequency, and this is due to the imbalanced properties of this faulty drive. In a summary, ZSCC takes responsibility for part of the torque ripples, whereas 3rd harmonics of rotor fluxes account for the most. This explains why the torque ripple is smallest under a pentagon-connection. The torque ripple worsens under a pentacle-connection as the ratio between the 3rd and 1st harmonic rotor flux is greatest among all connection modes.

Figure 9d–f illustrate the same quantities under MTPA control. Although the use of the reluctance torque contributes to the average torque, it also increases the torque ripples slightly when compared with their counterparts of $i_d = 0$ control.

From the figure, Δu_A can be considerable, and it is compensated for by using $e_A/2$ together with a scalar matrix, which is a completely different way from the penta-connection mode with an open-phase fault.

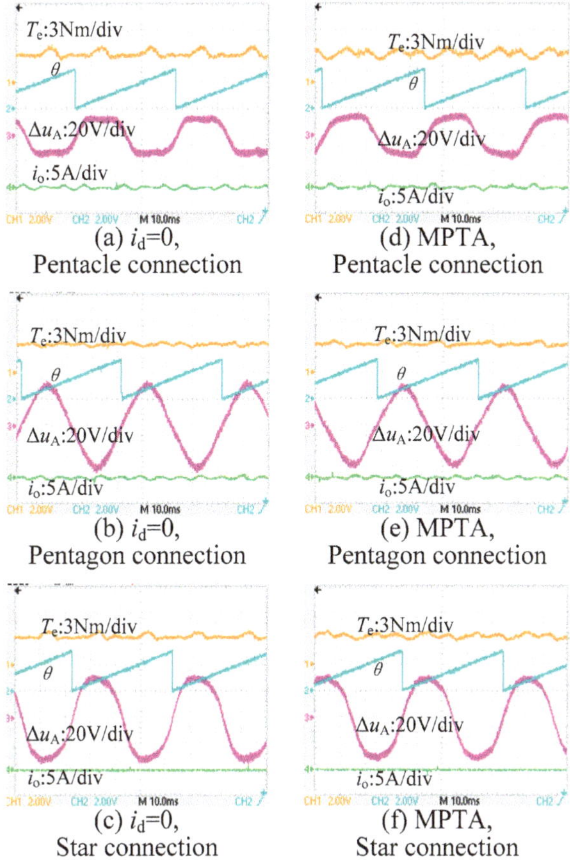

Figure 9. Waveforms of Δu_A, i_o, T_e, and θ under a single open-line fault.

Figure 10 shows the waveforms of the actual winding phase currents. It should be noted that the line currents and winding phase currents are two different quantities for pentacle- and pentagon-connection modes. From Figure 10a–c, which correspond to the results under the control of $i_d = 0$, the peak current to achieve the identical average torque is largest for the pentagon-connection mode, however, they decrease under the pentagon- and star-connection modes and are very close to each other.

Figure 10d–f reveal the peak currents under the control of MTPA, however, the peak current under the pentagon-connection mode tends to be smallest under the same torque command, which indicates that the pentagon-connection mode prevails at a low torque derating. On the other hand, the phase current magnitude is more balanced under a star-connection, thus a pentagon-connection mode can be promising in terms of fault-tolerant capability.

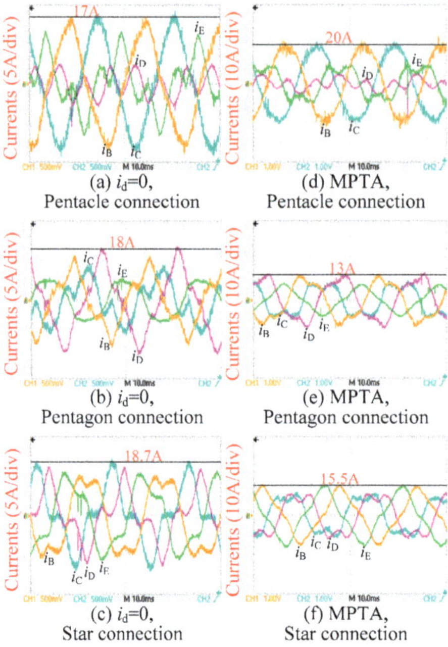

(a) $i_d=0$,
Pentacle connection

(d) MPTA,
Pentacle connection

(b) $i_d=0$,
Pentagon connection

(e) MPTA,
Pentagon connection

(c) $i_d=0$,
Star connection

(f) MPTA,
Star connection

Figure 10. Actual winding phase currents under a single open-line fault.

4.2. Open-Phase Fault

Figure 11 shows the waveforms of the winding phase currents under a single open-phase fault. For a penta-connection mode, a different control philosophy is used, rather than the previous star-control. From Figure 11a,b, both i_d and i_q are well regulated, which confirms the effectiveness of the presented control scheme. On the other hand, i_{q3} under a pentagon-connection is rather small when the voltage of the q_3-axis is subject to an open-loop control, and this implies i_{q3} can be well regulated without consuming too much harmonic power.

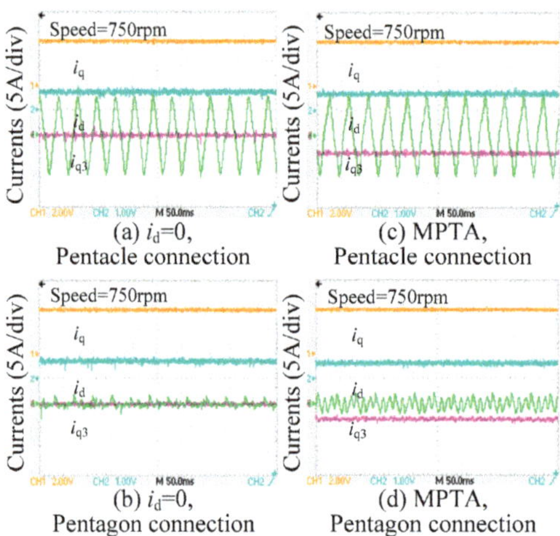

(a) $i_d=0$,
Pentacle connection

(c) MPTA,
Pentacle connection

(b) $i_d=0$,
Pentagon connection

(d) MPTA,
Pentagon connection

Figure 11. Waveforms of d-q-q3 frame currents and rotor speed under a single open-phase fault.

The control of MTPA contributes to the average torque, as evident in Figure 11c,d; however, i_{q3} increases a bit because of MTPA. When comparing with the d-q-q$_3$ frame currents under a star-connection, as shown in Figure 8c, i_{q3} under a pentagon-connection is the smallest, which means a pentagon-connection mode can be a good choice for FTC.

Figure 12 refers to the profiles of the torque, Δu_A, and ZSCC under a single open-phase fault. From Figure 12a,b, the ZSCC can be remarkable for both pentacle- and pentagon-connection modes, and it interacts with the zero-sequence voltage, introducing additional torque ripples, which is one of the drawbacks of a penta-connection mode. From Figure 12c,d, the use of MTPA increases the torque ripple at the expense of maximizing the torque capacity. The ZSCC also increases in this procedure under pentacle-connection. On the contrary, the power spectrum of the ZSCC disperses to a high frequency under a pentagon-connection because of MTPA, and this implies the zero sequence components can be indirectly controlled with the use of line currents. On the other hand, Δu_A can be small relative to the case under a single-line open fault, and it illustrates that the omission of Δu_A, while modulating the line voltages, is in line with the expectation.

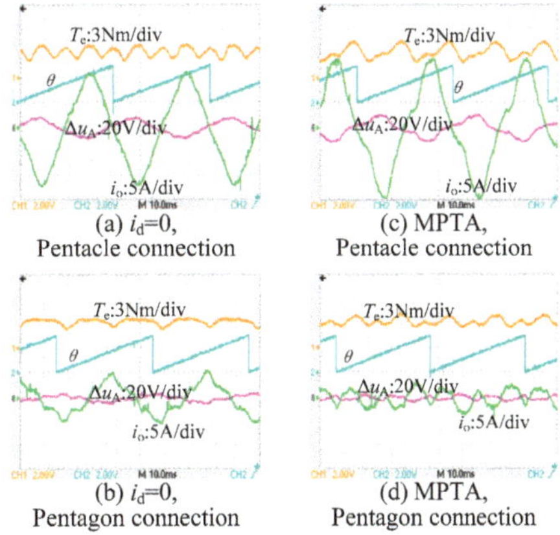

(a) i_d=0,
Pentacle connection

(c) MPTA,
Pentacle connection

(b) i_d=0,
Pentagon connection

(d) MPTA,
Pentagon connection

Figure 12. Waveforms of Δu_A, i_o, T_e, and θ under a single open-phase fault.

Figure 13 shows the waveforms of the actual winding phase currents under a single open-phase fault. Attributed to a considerable ZSCC, the remaining healthy phase currents are seriously imbalanced under a pentacle-connection, and a pentagon-connection mode necessitates a larger phase current to produce an identical average torque, which means the torque derating of the pentagon-connection mode is inferior to the pentacle-connection mode.

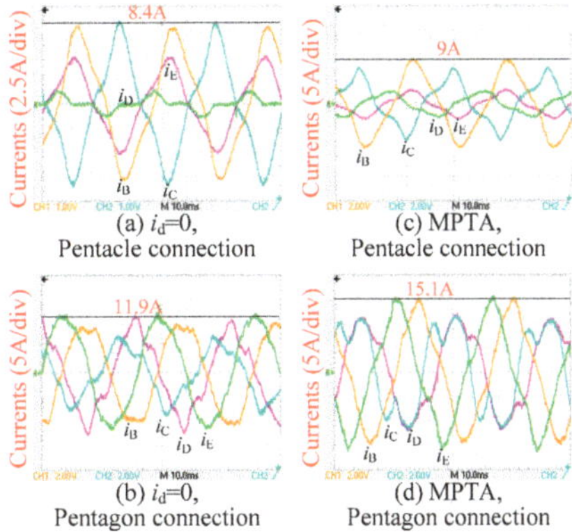

(a) i_d=0,
Pentacle connection

(c) MPTA,
Pentacle connection

(b) i_d=0,
Pentagon connection

(d) MPTA,
Pentagon connection

Figure 13. Actual winding phase currents under a single open-phase fault.

From Figure 13, even though the actual winding phase currents are seriously distorted with the ZSCC, they still manifest themselves as DCs under the presented transformation, and this enables the continuous use of the linear controller and paves its way to engineering practices. This also constitutes one of the main contributions of this work.

5. Conclusions

This work presents a unified fault-tolerant decoupling control for the star-, pentagon-, and pentacle-connected 5Φ-PMSMs. Single-line and -phase open faults are investigated, and key techniques to continuously use a previous decoupled model are elaborated. For a single open-line fault, the transformation of resistance, inductance, and the back-EMF from a penta-connection mode to a star-connection mode is desired, with the angular of vectorial control properly fixed. Concerning a single-phase open fault, voltage and current transformation from a line-to-neutral frame to a winding-oriented frame is suggested by taking care of the zero-sequence voltage and current.

Two FTC schemes, i.e., $i_d = 0$ to minimize the torque ripples and MTPA to maximize the torque capacity, are evaluated under all connection modes. The experimental results confirm the validity of the proposed unified control schemes, and simultaneously suggest that a pentagon-connection mode prevails under a single open-line fault for its small torque ripple and lower torque derating, and the star-connection mode takes the second place. However, with respect to the torque derating under an open-phase fault, the pentacle-connection mode performs best. Overall, under either an open-line fault or open-phase fault, the performance of a pentagon-connection mode is more balanced in terms of the harmonic current's controllability; the ZSCC, torque derating, and torque ripple, and, therefore, this type of connection can find a bright future in engineering practices.

The saturation effect was not considered while performing the FTC, which may have affected the torque performance to some extent. For the practice of some more advanced controls, e.g., model predictive control and position sensorless control, this phenomenon should be incorporated. However, concerning fault-tolerant control, the saturation effect can be secondary to the zero-sequence circulating current. In the future, how to attenuate the zero-sequence current by regulating the remaining healthy phase currents deserves more investigation and will be addressed in our next phase.

Author Contributions: Conceptualization, B.T.; methodology, B.T. and R.L.; software, R.L. and J.H.; validation, R.L. and J.H.; formal analysis, R.L.; investigation, B.T.; resources, B.T.; data curation, B.T.; writing—original draft preparation, B.T. and R.L.; writing—review and editing, B.T. and R.L.; visualization, R.L. and J.H.; supervision, B.T.; project administration, B.T.; funding acquisition, B.T. All authors have read and agreed to the published version of the manuscript.

Funding: This research was funded by [Jiangsu Provincial Innovation and Entrepreneurship Doctor Program] grant number [JSSCBS20210175].

Institutional Review Board Statement: Not applicable.

Informed Consent Statement: Not applicable.

Data Availability Statement: Data available on request from the authors.

Conflicts of Interest: The authors declare no conflict of interest.

Appendix A

In this paper, suppose the stator is undamaged under an open-phase/line fault, and simultaneously, the iron core is unsaturated. Given this, the motor model in Figure 1 can still be presented by

$$
\begin{bmatrix} u_{\text{Aa}} \\ u_{\text{B}} \\ u_{\text{C}} \\ u_{\text{D}} \\ u_{\text{E}} \end{bmatrix} = R \begin{bmatrix} i_{\text{A}} \\ i_{\text{B}} \\ i_{\text{C}} \\ i_{\text{D}} \\ i_{\text{E}} \end{bmatrix} + \frac{\mathrm{d}}{\mathrm{d}t} \left(\begin{bmatrix} L_{\text{AA}} & L_{\text{AB}} & L_{\text{AC}} & L_{\text{AD}} & L_{\text{AE}} \\ L_{\text{BA}} & L_{\text{BB}} & L_{\text{BC}} & L_{\text{BD}} & L_{\text{BE}} \\ L_{\text{CA}} & L_{\text{CB}} & L_{\text{CC}} & L_{\text{CD}} & L_{\text{CE}} \\ L_{\text{DA}} & L_{\text{DB}} & L_{\text{DC}} & L_{\text{DD}} & L_{\text{DE}} \\ L_{\text{EA}} & L_{\text{EB}} & L_{\text{EC}} & L_{\text{ED}} & L_{\text{EE}} \end{bmatrix} \begin{bmatrix} i_{\text{A}} \\ i_{\text{B}} \\ i_{\text{C}} \\ i_{\text{D}} \\ i_{\text{E}} \end{bmatrix} \right) + \begin{bmatrix} e_{\text{A}} \\ e_{\text{B}} \\ e_{\text{C}} \\ e_{\text{D}} \\ e_{\text{E}} \end{bmatrix} \tag{A1}
$$

where L_{xy}, $x = y = A, B, C, D, E$, denotes the self- or mutual-inductance which possesses the following properties

$$
\begin{aligned}
L_{AA} + L_{BA} + L_{CA} + L_{DA} + L_{EA} &= L_{sum} \\
L_{AB} + L_{BB} + L_{CB} + L_{DB} + L_{EB} &= L_{sum} \\
L_{AC} + L_{BC} + L_{CC} + L_{DC} + L_{EC} &= L_{sum} \\
L_{AD} + L_{BD} + L_{CD} + L_{DD} + L_{ED} &= L_{sum} \\
L_{AE} + L_{BE} + L_{CE} + L_{DE} + L_{EE} &= L_{sum}
\end{aligned}
\tag{A2}
$$

with L_{sum} as an instantaneous quantity which is the function of the rotor position. Regarding the back-EMFs, one can have

$$
e_A + e_B + e_C + e_D + e_E = 0 \tag{A3}
$$

The actual phase currents obey following the rule

$$
i_A = 0,\ i_B + i_C + i_D + i_E = i_o \tag{A4}
$$

Note that herein i_A, i_B, i_C, i_D, and i_E stand for the actual winding phase currents, rather than the observed ones, and i_o indicates the actual ZSCC.

Combining (A2)–(A4) with (A1) yields

$$
u_o = u_{Aa} + u_B + u_C + u_D + u_E = Ri_o + d(L_{sum}i_o) \tag{A5}
$$

From (A5), the zero-sequence voltage is associated with the ZSCC and is non-zero in a penta-wired system. The zero-sequence voltage reduces to the minimum while i_o is minimal.

References

1. Sayed, E.; Abdalmagid, M.; Pietrini, G.; Sa'adeh, N.M.; Callegaro, A.D.; Goldstein, C.; Emadi, A. Review of Electric Machines in More-/Hybrid-/Turbo-Electric Aircraft. *IEEE Trans. Transp. Electrif.* **2021**, *7*, 2976–3005. [CrossRef]
2. Zhao, W.; Cheng, M.; Chau, K.T.; Cao, R.; Ji, J. Remedial Injected-Harmonic-Current Operation of Redundant Flux-Switching Permanent-Magnet Motor Drives. *IEEE Trans. Ind. Electron.* **2013**, *60*, 151–159. [CrossRef]
3. Sun, J.; Li, C.; Zheng, Z.; Wang, K.; Li, Y. A Generalized, Fast and Robust Open-Circuit Fault Diagnosis Technique for Star-connected Symmetrical Multiphase Drives. *IEEE Trans. Energy Convers.* **2022**, 1. [CrossRef]
4. Barrero, F.; Duran, M.J. Recent Advances in the Design, Modeling, and Control of Multiphase Machines—Part I. *IEEE Trans. Ind. Electron.* **2016**, *63*, 449–458. [CrossRef]
5. Ghosh, B.C.; Habibullah; Ali, E. Performance Comparison of Five and Six Phase Induction Motors Operating under Normal and Faulty Conditions. In Proceedings of the 2019 4th International Conference on Electrical Information and Communication Technology (EICT), Khulna, Bangladesh, 20–22 December 2019; pp. 1–6. [CrossRef]
6. Bensalem, Y.; Kouzou, A.; Abbassi, R.; Jerbi, H.; Kennel, R.; Abdelrahem, M. Sliding-Mode-Based Current and Speed Sensors Fault Diagnosis for Five-Phase PMSM. *Energies* **2022**, *15*, 71. [CrossRef]
7. Ferreira, F.J.T.E.; Baoming, G.; Almeida, A.T.D. Stator Winding Connection-Mode Management in Line-Start Permanent Magnet Motors to Improve Their Efficiency and Power Factor. *IEEE Trans. Energy Convers.* **2013**, *28*, 523–534. [CrossRef]
8. Cistelecan, M.V.; Ferreira, F.J.T.E.; Popescu, M. Adjustable Flux Three-Phase AC Machines with Combined Multiple-Step Star-Delta Winding Connections. *IEEE Trans. Energy Convers.* **2010**, *25*, 348–355. [CrossRef]
9. Mavila, P.C.; Rajeevan, P. A Space Vector Based PWM Scheme for Realization of Virtual Pentagon, Pentacle Connections in open-end Winding Five Phase Machine Drives with Single DC Source. In Proceedings of the 2018 IEEE International Conference on Power Electronics, Drives and Energy Systems (PEDES), Chennai, India, 18–21 December 2018; pp. 1–6. [CrossRef]
10. Abdel-Khalik, A.S.; Ahmed, S.; Elserougi, A.A.; Massoud, A. Effect of Stator Winding Connection of Five-Phase Induction Machines on Torque Ripples Under Open Line Condition. *IEEE/ASME Trans. Mechatron.* **2015**, *20*, 580–593. [CrossRef]
11. Wang, P.; Gong, S.; Sun, X.; Liu, Z.; Jiang, D.; Qu, R. Fault-Tolerant Reconfiguration Topology and Control Strategy for Symmetric Open-Winding Multiphase Machines. *IEEE Trans. Ind. Electron.* **2022**, *69*, 8656–8666. [CrossRef]
12. Yin, Z.; Sui, Y.; Zheng, P.; Yang, S.; Zheng, Z.; Huang, J. Short-Circuit Fault-Tolerant Control without Constraint on the D-Axis Armature Magnetomotive Force for Five-Phase PMSM. *IEEE Trans. Ind. Electron.* **2022**, *69*, 4472–4483. [CrossRef]
13. Chen, Q.; Gu, L.; Wang, J.; Zhao, W.; Liu, G. Remedy Strategy for Five-Phase FTPMMs Under Single-Phase Short-Circuit Fault by Injecting Harmonic Currents from Third Space. *IEEE Trans. Power Electron.* **2022**, 1. [CrossRef]
14. Zhang, L.; Zhu, X.; Cui, R.; Han, S. A Generalized Open-Circuit Fault-Tolerant Control Strategy for FOC and DTC of Five-Phase Fault-Tolerant Permanent-Magnet Motor. *IEEE Trans. Ind. Electron.* **2022**, *69*, 7825–7836. [CrossRef]

15. Wang, X.; Liu, G.; Chen, Q.; Farahat, A.; Song, X. Multivectors Model Predictive Control with Voltage Error Tracking for Five-Phase PMSM Short-Circuit Fault-Tolerant Operation. *IEEE Trans. Transp. Electrif.* **2022**, *8*, 675–687. [CrossRef]
16. Wang, H.; Zhao, W.; Tang, H.; Tao, T.; Saeed, S. Improved Fault-Tolerant Model Predictive Torque Control of Five-Phase PMSM by Using Deadbeat Solution. *IEEE Trans. Energy Convers.* **2022**, *37*, 210–219. [CrossRef]
17. Mohammadpour, A.; Parsa, L. A Unified Fault-Tolerant Current Control Approach for Five-Phase PM Motors With Trapezoidal Back EMF Under Different Stator Winding Connections. *IEEE Trans. Power Electron.* **2013**, *28*, 3517–3527. [CrossRef]
18. Yepes, A.G.; Doval-Gandoy, J. Study and Active Enhancement by Converter Reconfiguration of the Performance in Terms of Stator Copper Loss, Derating Factor and Converter Rating of Multiphase Drives Under Two Open Legs with Different Stator Winding Connections. *IEEE Access* **2021**, *9*, 63356–63376. [CrossRef]
19. Abdel-Khalik, A.S.; Ahmed, S.; Massoud, A.M. Steady-State Mathematical Modeling of a Five-Phase Induction Machine with a Combined Star/Pentagon Stator Winding Connection. *IEEE Trans. Ind. Electron.* **2016**, *63*, 1331–1343. [CrossRef]
20. Masoud, M.I.; Abdel-Khalik, A.S.; Al-Abri, R.S.; Williams, B.W. Effects of unbalanced voltage on the steady-state performance of a five-phase induction motor with three different stator winding connections. In Proceedings of the 2014 International Conference on Electrical Machines (ICEM), Berlin, Germany, 2–5 September 2014; pp. 1583–1589. [CrossRef]
21. Mohammadpour, A.; Sadeghi, S.; Parsa, L. A Generalized Fault-Tolerant Control Strategy for Five-Phase PM Motor Drives Considering Star, Pentagon, and Pentacle Connections of Stator Windings. *IEEE Trans. Ind. Electron.* **2014**, *61*, 63–75. [CrossRef]
22. Tawfiq, K.B.; Ibrahim, M.N.; Sergeant, P. An Enhanced Fault-Tolerant Control of a Five-Phase Synchronous Reluctance Motor Fed from a Three-to-Five-phase Matrix Converter. *IEEE J. Emerg. Sel. Top. Power Electron.* **2022**, 1. [CrossRef]
23. Zaskalicky, P. Behavior of a Five-Phase Pentacle Connected IM Operated under One-Phase Fault. In Proceedings of the 2019 International Aegean Conference on Electrical Machines and Power Electronics (ACEMP) & 2019 International Conference on Optimization of Electrical and Electronic Equipment (OPTIM), Istanbul, Turkey, 27–29 August 2019; pp. 126–131.
24. Masoud, M.I. Five phase induction motor: Phase transposition effect with different stator winding connections. In Proceedings of the IECON 2016-42nd Annual Conference of the IEEE Industrial Electronics Society, Florence, Italy, 23–26 October 2016; pp. 1648–1655.
25. Abdel-Khalik, A.S.; Elgenedy, M.A.; Ahmed, S.; Massoud, A.M. An Improved Fault-Tolerant Five-Phase Induction Machine Using a Combined Star/Pentagon Single Layer Stator Winding Connection. *IEEE Trans. Ind. Electron.* **2016**, *63*, 618–628. [CrossRef]
26. Zhou, H.; Zhao, W.; Liu, G.; Cheng, R.; Xie, Y. Remedial Field-Oriented Control of Five-Phase Fault-Tolerant Permanent-Magnet Motor by Using Reduced-Order Transformation Matrices. *IEEE Trans. Ind. Electron.* **2017**, *64*, 169–178. [CrossRef]
27. Zhang, G.; Xiang, R.; Wang, G.; Li, C.; Bi, G.; Zhao, N.; Xu, D. Hybrid Pseudorandom Signal Injection for Position Sensorless SynRM Drives with Acoustic Noise Reduction. *IEEE Trans. Transp. Electrif.* **2022**, *8*, 1313–1325. [CrossRef]
28. Tian, B.; An, Q.-T.; Duan, J.-D.; Sun, D.-Y.; Sun, L.; Semenov, D. Decoupled Modeling and Nonlinear Speed Control for Five-Phase PM Motor Under Single-Phase Open Fault. *IEEE Trans. Power Electron.* **2017**, *32*, 5473–5486. [CrossRef]
29. Tian, B.; Sun, L.; Molinas, M.; An, Q.-T. Repetitive Control Based Phase Voltage Modulation Amendment for FOC-Based Five-Phase PMSMs under Single-Phase Open Fault. *IEEE Trans. Ind. Electron.* **2021**, *68*, 1949–1960. [CrossRef]

Article

A Study on the Improvement of Torque Density of an Axial Slot-Less Flux Permanent Magnet Synchronous Motor for Collaborative Robot

Dong-Youn Shin [1], Min-Jae Jung [2], Kang-Been Lee [3], Ki-Doek Lee [1] and Won-Ho Kim [2,*]

[1] Intelligent Mechatronics Research Center, Korea Electronics Technology Institute, Seongnam-si 13509, Gyeonggi-do, Korea; tlsehddbs147@keti.re.kr (D.-Y.S.); kdlee@keti.re.kr (K.-D.L.)
[2] Department of Electrical Engineering, Gachon University, Seongnam-si 13120, Gyeonggi-do, Korea; wjdalswo12@naver.com
[3] Department of Electrical and Computer Engineering, Michigan State University, East Lansing, MI 48824, USA; leekangb@msu.edu
* Correspondence: wh15@gachon.ac.kr; Tel.: +82-31-750-5881

Abstract: In this paper, an axial slot-less permanent magnet synchronous motor (ASFPMSM) was designed to increase the power density. The iron core of the stator was replaced with block coils, and the stator back yoke was removed because 3D printing can provide a wide range of structures of the stator. The proposed model also significantly impacts efficiency because it can reduce iron loss. To meet size and performance requirements, coil thickness and number of winding layers in the block, the total amount of magnet, and pole/slot combinations were considered. The validity of the proposed model was proved via finite elements analysis (FEA).

Keywords: axial flux permanent magnet synchronous motor; SPMSM; slot-less; block-coil; collaborative robot (cobot); robot joint; 3D-print; Somaloy Core; torque density; output power density

Citation: Shin, D.-Y.; Jung, M.-J.; Lee, K.-B.; Lee, K.-D.; Kim, W.-H. A Study on the Improvement of Torque Density of an Axial Slot-Less Flux Permanent Magnet Synchronous Motor for Collaborative Robot. *Energies* **2022**, *15*, 3464. https://doi.org/10.3390/en15093464

Academic Editors: Quntao An, Bing Tian and Xinghe Fu

Received: 6 April 2022
Accepted: 6 May 2022
Published: 9 May 2022

Publisher's Note: MDPI stays neutral with regard to jurisdictional claims in published maps and institutional affiliations.

1. Introduction

Many studies on radial flux permanent magnet synchronous motor (RFPMSM), applied in collaborative robots (cobots) and which can physically interact and share a workspace with humans, have been conducted [1–3]. However, the RFPMSM should be developed with consideration of material developments to meet high-performance requirements in industry because the performance and efficiency of the RFPMSM are saturated due to pioneer studies [4,5]. Furthermore, since there is an unfavorable structural problem of multi-joint motors applied in cobots, bilateral axial slot-less flux permanent magnet synchronous motor (ASFPMSM) with high power density has been addressed as a new motor model [6–8].

Theoretically, there is a limitation that it is hard to manufacture with a slot-less structure because the press die manufacturing method can provide only conventional structure [9]. Therefore, a new structure of the motor can be developed when the motor is manufactured using 3D printing technology with the core powder as Somaloy material. However, when the core powder is used, the B-H characteristics are lower than that of laminated electrical steel sheets, and iron loss, which influences the efficiency of the motor, is higher than that of laminated electrical steel sheets.

To compensate for the drawbacks of the 3D printing technology, a block coil, which can remove the slot of the stator where the coil is wound, and mold the coil with plastic, can be used to reduce iron loss. In addition, it is possible to reduce iron loss with a shape without the back yoke of the stator because of the closed magnetic flux loop made by facing the same permanent magnet poles during operation of the motor [10–13]. In addition, the output power and torque density of ASFPMSM manufactured by 3D printing can be higher compared to RFPMSM [14–16]. In short, the advantages of eliminating the slot and the back yoke of the stator have a significant impact on the efficiency of the motor [17,18].

In this paper, improved output power density and torque density of an ASFPMSM applied in cobots were proposed via optimization design.

2. Target Model Specification

Among the permanent magnet synchronous motor types for collaborative robots, the radial flux type SPMSM (surface permanent magnet synchronous motors) motors are commonly used. Since the magnetic flux density distribution in the air gap is uniform in the SPMSM, the control is accurate. These advantages have a significant impact on minimizing the size of the motor. In addition, it is much easier to obtain more spaces for winding and weight reduction than in the case of the internal permanent magnet synchronous motor (IPMSM).

The general cobots used in a narrower working space should be considered their size requirements during machine design. Therefore, an optimization design process is required to maximize performance with limited requirements.

RFPMSM applied in actual cooperative robots was selected as the target model, as shown in Figure 1, to meet the performance and size requirements when a new model is designed.

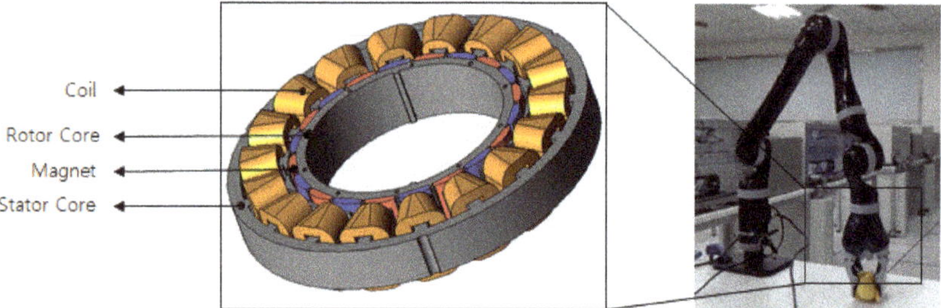

Figure 1. Radial Flux Permanent Magnet Synchronous Motor target model of cobots [1].

To ensure equal performance between the target model and an ASFPMSM, torque, no-load back EMF, and efficiency were considered. Based on the specifications, the ASFPMSM with 52.2 mm of outer rotor diameter was proposed with 0.55 Nm torque at 3500 rpm and 89.6% efficiency, as shown in Tables 1 and 2. The torque curve of the target model is presented in Figure 2. More detailed information related to the dimensions is presented later in the manuscript.

Table 1. The specifications of the target model.

Parameter	Value (mm)
Stator Outer Diameter	82
Stator Inner Diameter	54
Stator Teeth Width	3.3
Rotor Outer Diameter	52.2
Rotor Inner Diameter	44.8
Magnet Thickness	5
Air Gap	1
Stator Stack Length	10
Rotor Stack Length	13
Magnet Thickness	2

Table 2. Performance and materials of the target model.

	Parameter	Value	Unit
	Pole/Slot	20/18	-
	Rated Speed	3500	RPM
	Torque @3500 Rotating Per Minutes	0.55	Nm
	Current	5.1	A_{rms}
Performance	No-load L-L Back Electric Motive Force @3500 Rotating Per Minutes	26.28	V_{rms}
	Copper Loss	11.1	W
	Iron Core Loss	12.5	W
	Efficiency	89.6	%
	Direct Current Link	48	V
	Material	Copper	-
Winding	Diameter	0.85	mm
	Series Turns Per Phase	96	Turns
	Material	N42SH	-
Magnet	B_r	1.33	T
	H_c	1592	kA/m

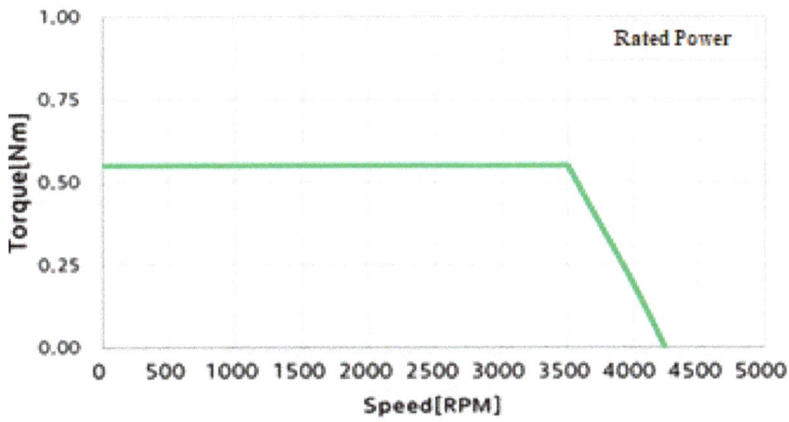

Figure 2. Torque-Nominal Speed curve of the target model.

3. ASFPMSM Proposal Model Concept

3.1. Proposed ASFPMSM Concept

ASFPMSM applied in cobots should provide high output power density because their thin and small structures are needed to meet space limitations in cobots. Therefore, a new model of ASFPMSM is proposed in this paper. As mentioned above, since there are requirements related to size and power density in cobots, it is hard to meet the requirements with the existing winding method. A type of block coil using 3D printing technology can overcome the limitations. The block coil can be manufactured by molding the coil part with plastic using 3D printing technology, as shown in Figure 3. The winding part of the motor can be designed because the 3D printing technology provides the freedom to create the desired model, as shown in Figure 3a,b.

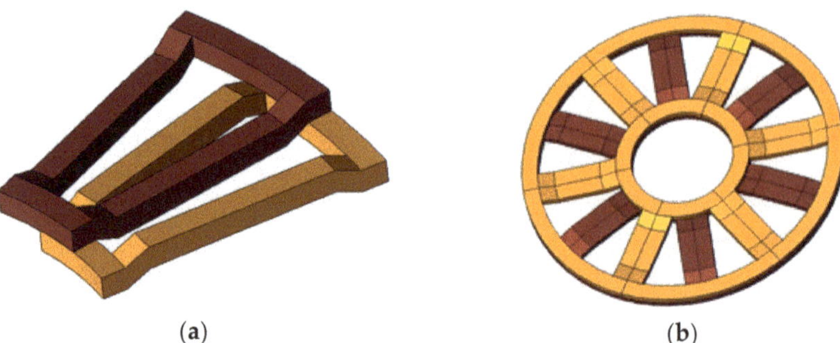

(a)   (b)

Figure 3. (**a**) Part where the coil is wound; (**b**) shape of assembling block coil.

Furthermore, when the rotor N and S poles are designed to face each other in both directions, the facing permanent magnet and the rotor back yoke form a closed loop of magnetic flux. This structure can eliminate the stator back yoke. Finally, multi-poles and multi-slot structure were selected to design a thin motor.

An amount of magnetic flux per pole, which influences the performance of the motor, can be presented as shown in Equation (1). This equation provides some facts that magnetic flux is proportional to the difference between the square of the inner diameter and the square of the outer diameter of the ASFPMSM. Furthermore, the magnetic flux per pole can be reduced as the number of poles increases. This correlation proved that the width of the back yoke of the rotor of ASFPMSM could be decreased as thin as possible when a big number of pole-slot combination is applied in the motor because the magnetic flux per pole can be reduced.

$$\Phi_f = \int_{R_{in}}^{R_{out}} a_i B_{mg} \frac{2\pi}{2p} r \, dr = a_i B_{mg} \frac{\pi}{2p} (R_{out}^2 - R_{in}^2) \tag{1}$$

3.2. Analysis of Proposed ASFPMSM

To make a comparative analysis based on the same criteria as the target model, the total laminated length of ASFPMSM was set to 13 mm. The rotor radius was adjusted according to the stator radius of RFPMSM, and the same winding specifications were designed. In addition, since the ASFPMSM does not have a stator shoe and teeth, the laminated length, excluding the air gap length (block coil height) and core back yoke thickness, is designed to be the same length as the magnet thickness to increase the back electromotive force (back EMF). The ASFPMSM is presented with the same laminated length and winding specifications as the target model in Figure 4a,b.

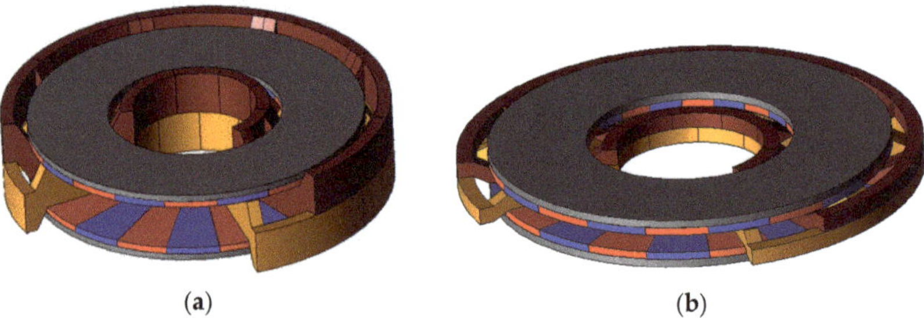

(a)   (b)

Figure 4. (**a**) Same laminated length model as the target model; (**b**) a same number of turns model as the target model.

A comparative analysis of the vector plot of the target model under no load and the proposed ASFPMSM model under no-load was performed at a rated speed of 3500 rpm, as shown in Figure 5.

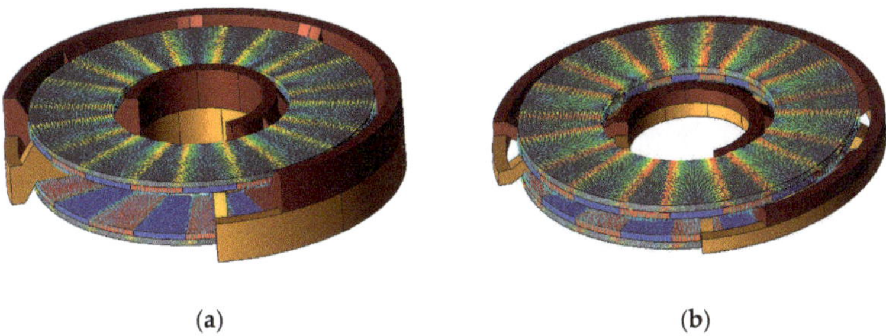

(a)   (b)

Figure 5. (**a**) Flux line in the same laminated length model as the target model, (**b**) flux line of the same turn number model as the target model.

The vector diagram confirmed that the low back EMF is due to the air gap length increasing as the coil height increases. The magnetic flux of the permanent magnet did not reach the back yoke of the stator and leaked to the next magnet. As a result, it is necessary to find the minimum air gap length for the magnetic flux to reach the back yoke of the stator and to make the magnetic flux closed-loop possible, and to design a design with the maximum number of turns within the air gap length while minimizing the use of magnets.

3.3. ASFPMSM Optimal Coil Height Analysis

In ASFPMSM, the airgap length value is increased by the height of the block coil because there are no stator teeth and shoes. Therefore, airgap length has a significant influence on the performance of the motor compared to other types of motors. Thus, flux paths were analyzed at various block coil heights (1~6 mm), respectively, to increase the performance, as shown in Figure 6.

In Figure 6, the coils are located in the empty space between the rotor and the stator, and the coils are molded with plastic material using 3D printing. Therefore, since the coil part becomes an air gap, after removing the coil part, the vector diagram of the magnetic flux was checked to make it easier to check the flow of the magnetic flux.

The results of analyzing the magnetic flux vector diagram for each coil height are as follows. When the block coil height is 1~2 mm, the leakage flux is minimized, and the magnetic flux path forms a closed loop with the back yoke of the stator. Models with a block coil height of 3 mm or more could not produce the desired performance due to large magnetic flux leakage, as shown in Figure 6. Therefore, since the winding height of the target model is 0.85 mm, a value between 1 mm and 2 mm, a height of 1.7 mm block coil, which is a multiple of 0.85 mm, was selected as the optimal height.

Since 0.85 mm is close to 1 mm, even though it is not within the optimal height range (1~2 mm), both 0.85 mm and 1.7 mm of the block coil height were selected for 1-layer winding and 2-layer winding to conduct accurate performance analysis, as shown in Figure 7a–d. Both models have their pros and cons. Although the 1-layer winding model has a short airgap length, when winding four turns per coil side, the number of serial turns is 24 turns, so it is difficult to expect a good performance. The 2-layer model has better performance because it can wind up to 48 turns, despite the longer airgap length compared to the 1-layer model. No-load back EMFs are presented to validate that the 2-layer model, which has 1.7 coil height, has a better performance compared to the 1-layer model, as shown in Figure 8. As a result, it was found that 1.7 mm is the most optimal value for coil height.

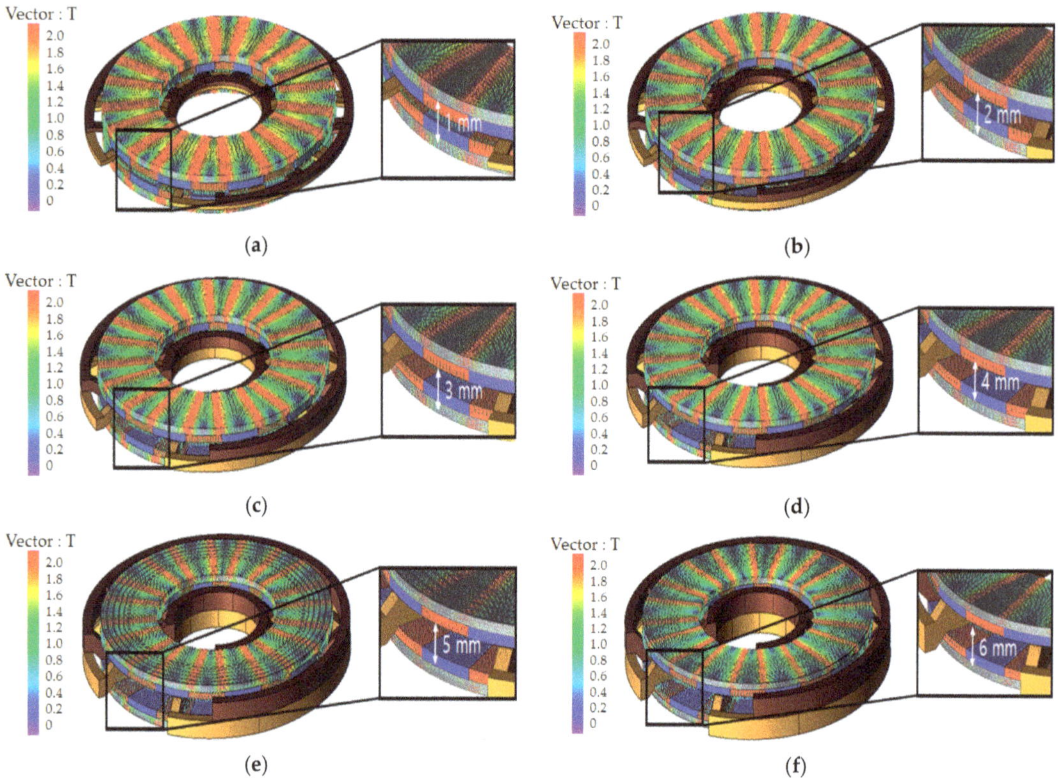

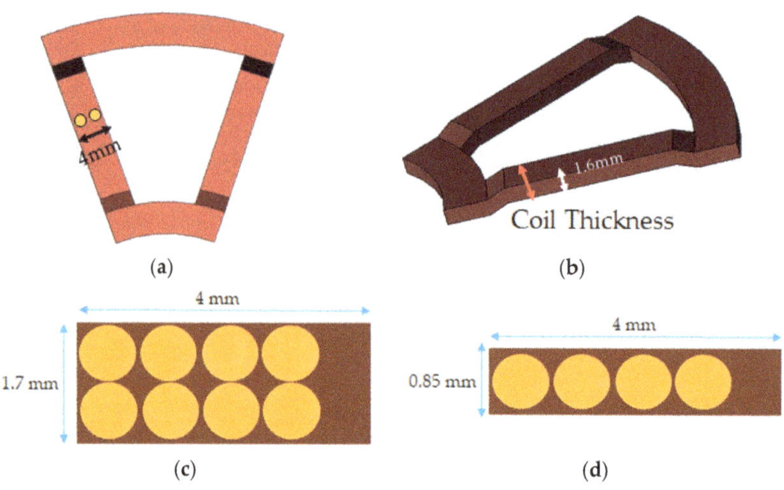

Figure 6. Flux line analysis according to block coil height. (**a**) Block-Coil Thickness 1 mm, (**b**) Block-Coil Thickness 2 mm, (**c**) Block-Coil Thickness 3 mm, (**d**) Block-Coil Thickness 4 mm, (**e**) Block-Coil Thickness 5 mm, (**f**) Block-Coil Thickness 6 mm.

Figure 7. (**a**) Shape of coil width, (**b**) coil height, (**c**) coil area of coil height 1.7 mm, (**d**) coil area of coil height 0.85 mm.

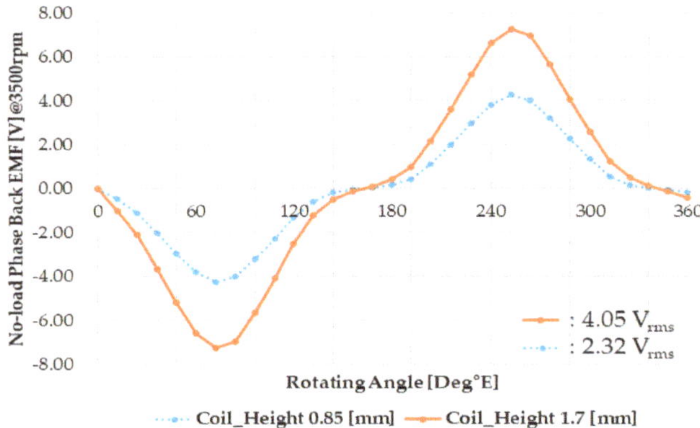

Figure 8. No load phase back-EMF compared to coil height 0.85 mm (Blue) and coil height 1.7 mm (Red).

3.4. ASFPMSM Optimal Magnet Size Analysis

In order to select a reasonable amount of magnet while minimizing leakage flux from the optimal block coil thickness analyzed previously, the performance analysis was conducted by changing the amount of magnet in the axial direction. Back EMFs were compared and analyzed between models with the total magnet thickness from 2 mm to 10 mm, as shown in Figure 9.

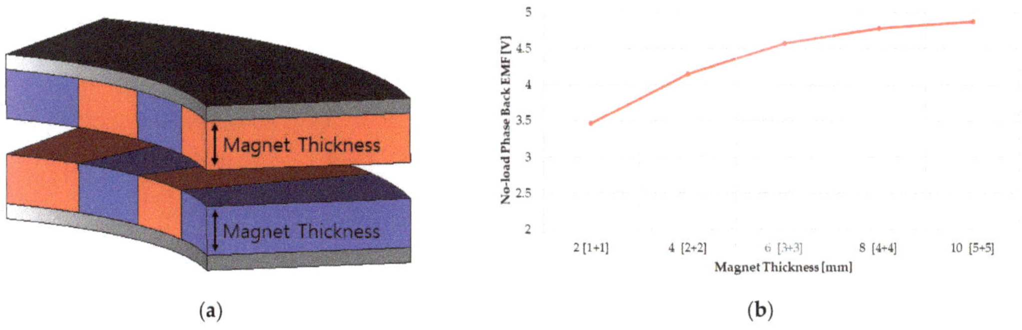

Figure 9. Analysis of no load phase back EMF according to magnet thickness. (**a**) Magnet Thickness, (**b**) No load phase Back EMF according to Magnet Thickness.

As a result of the analysis of no-load back EMF according to various amounts of magnet, it was confirmed that back EMF increases as magnet thickness increases until 6 mm. However, the performance was saturated when the total magnet thickness is larger than 6 mm. It is possible that back EMF does not increase because of saturation of the core. Therefore, whether no-load back EMF saturation is from core saturation or not should be checked from simulation results with various core thicknesses, as shown in Figure 10.

From the result presented in Figure 10, the fact that no-load back EMF does not increase as magnet thickness increases is not from the saturation of the core. There is a correlation between the core thickness and the total amount of magnet. Since the laminated length was fixed to 13 mm, after determining the thickness of the coil, the remaining laminated length is used as the core thickness and the magnet thickness. Therefore, the thinner the core, the greater the amount of magnet can be used. Considering that the stacking length of the target model is 13 mm, the optimal core thickness was selected as 2 mm to minimize the size of the proposed motor.

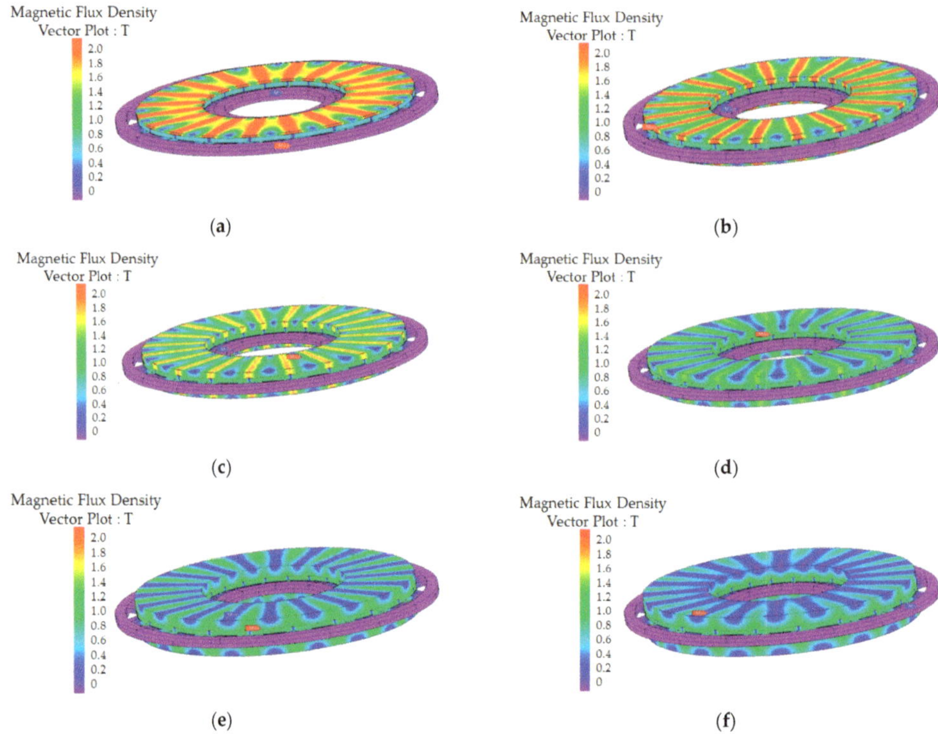

Figure 10. Core saturation analysis according to core thickness: (**a**) core thickness 1 mm, (**b**) core thickness 2 mm, (**c**) core thickness 3 mm, (**d**) core thickness 4 mm, (**e**) core thickness 5 mm, and (**f**) core thickness 6 mm.

3.5. ASFPMSM Optimal Pole Slot Combination Analysis

Based on the above analysis, it was confirmed that the back EMF of the ASFPMSM was significantly lower than that of the target model. As a result of analyzing the reasons, the motor of the ASFPMSM generally adopts a 2:3 structure or a 4:3 structure of pole number and slot number combination to use the concentrated winding to reduce the bulk of end turns. Furthermore, multi-pole and multi-slot combinations are applied to the motor to make the back yoke of the core thinner. Therefore, it is essential to analyze the combination of poles and the number of slots. The analysis of the number of poles and the number of slot combinations were performed based on the proposed model explained above, as shown in Table 3 and Figure 11.

Table 3. Analysis of poles/slots combination (2:3 structure).

Poles/Slots	No-Load Phase Back EMF	Number of Serial Turns	Magnet Volume
24p/36s	14.87 V_{rms}	48 Turn	19,739 mm^3
28p/42s	16.62 V_{rms}	56 Turn	19,739 mm^3
32p/48s	9.11 V_{rms}	32 Turn	19,739 mm^3
36p/54s	7.74 V_{rms}	36 Turn	19,739 mm^3
48p/72s	5.93 V_{rms}	48 Turn	19,739 mm^3

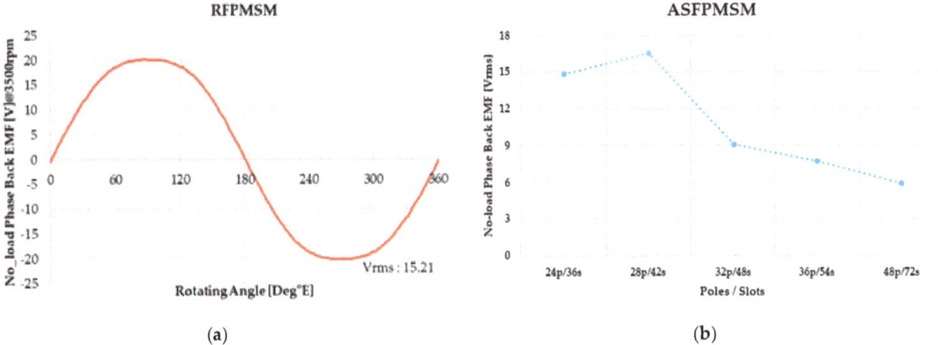

(**a**)　　　　　　　　　　　　　　(**b**)

Figure 11. Comparative analysis of target RFPMSM and ASFPMSM pole/slot combination. (**a**) RFPMSM No load phase Back EMF, (**b**) ASFPMSM No load phase Back EMF according to Poles/Slots Combinations.

In cases of the block coil models, it is difficult to analyze the models while keeping the same number of turns in series per pole because there is no slot, and the area of the block on which the coil is wound varies according to the number of slots. Therefore, the size of the winding used in the target model was used according to the number of slots, and the number of windings according to the number of slots was used, as shown in Figure 12. Therefore, when the 0.85 mm winding is used, it is advantageous to minimize the gap by applying the (c) and (d) pole/slot combination in Figure 12.

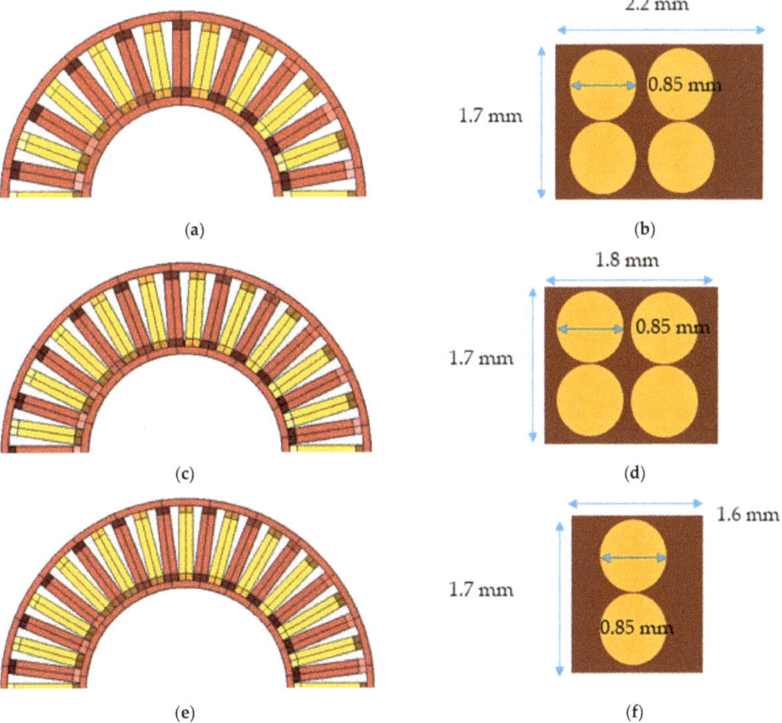

(**a**)　　　　　　　　　　　　　　(**b**)

(**c**)　　　　　　　　　　　　　　(**d**)

(**e**)　　　　　　　　　　　　　　(**f**)

Figure 12. Coil area by pole/slot combination: (**a**) 24p/36s coil shape, (**b**) 24p/36s coil area, (**c**) 28p/42s coil shape, (**d**) 28p/42s coil area, (**e**) 32p/48s coil shape, and (**f**) 32p/48s coil area.

However, in the case of 28 poles/42 slots, there are spaces between the winding and the outline of the block when 0.85 mm thickness winding is used, as shown in Figure 13. Therefore, 0.9 mm thickness winding was checked as an optimistic design. Then, 0.9 mm thickness was shown to deliver 12% more current with the same current density (Table 4). Finally, the final model of the ASFPMSM of 28p/42s was selected based on the previously analyzed optimal design and used to comparatively analyze the performance of the existing RFPMSM load.

(**a**)   (**b**)

Figure 13. Coil empty space comparison: (**a**) 0.85 m, (**b**) 0.9 mm.

Table 4. Current density comparison.

Parameter	Coil 0.85	Coil 0.9	Unit
Current	5.1	5.7	A_{rms}
Coil Diameter	0.85	0.9	mm
Coil Area	0.57	0.64	mm^2
Current Density	8.98	8.98	A/mm^2

4. Optimal Designed Model Comparison Analysis

In this chapter, the final ASFPMSM model was proposed based on the above optimum design and analysis, and this was compared and analyzed with the target model, as shown in Figure 14.

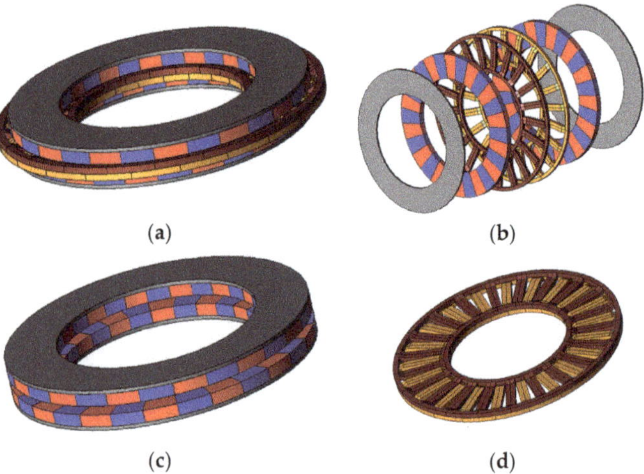

(**a**)   (**b**)

(**c**)   (**d**)

Figure 14. (**a**) Optimal design analysis model; (**b**) optimal design model exploded view; (**c**) optimally designed rotor shape; (**d**) optimally designed block coil shape.

The optimal model was proposed with the same stacking length as the target model RFPMSM with a core thickness of 2 mm, a magnet thickness of 3.3 mm, a coil thickness of 1.8 mm, and an air gap length of 0.3 mm, as shown in Table 5. More details are presented in Table 6.

Table 5. Specification of optimally designed ASFPMSM.

Parameter	Value (mm)
Rotor Outer Diameter	52.2
Rotor Inner Diameter	44.8
Magnet Length	7.4
Air Gap	0.3
Motor Stack Length	13
Magnet Thickness	3.3

Table 6. Materials of optimally designed ASFPMSM.

	Parameter	Value	Unit
	Poles/Slots	28/42	-
Performance	Rated Speed	3500	RPM
	DC Link	48	V
	Material	Copper	-
Winding	Diameter	0.9	mm
	Series Turns Per Phase	56	Turns
	Phase Resistance	0.032	Ω
	Material	N42SH	-
Magnet	B_r	1.33	T
	H_c	1592	KA/m

For comparative analysis, the no-load back EMF of the target model and the no-load back EMF of the ASFPMSM are shown in the Figure 15, and both were analyzed at the rated speed of 3500 Rpm.

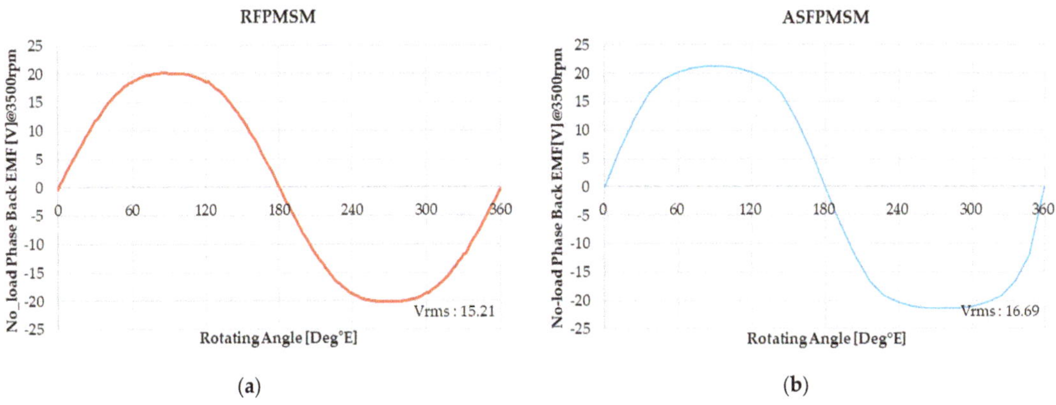

Figure 15. No load phase back EMF comparison: (**a**) no load phase back EMF of RFPMSM; (**b**) no-load phase back EMF of ASFPMSM.

As shown in Figure 15, Table 7 it was confirmed that the back EMF of the proposed model is 10% larger than that of the target model. It is validated that the performance of the proposed model is 10% higher than that of the target model at the same current density.

Table 7. Performance and materials of optimally designed ASFPMSM.

Parameter	RFPMSM	ASFPMSM	Unit
RPM	3500		r/min
DC Link	48		V
Current Density	8.98	8.98	A/mm^2
Torque	0.55	0.63	N·m
Input	226.2	235.4	W
Copper Loss	11.1	4.48	W
Iron Core Loss	12.5	0.5	W
Output	202.0	230.4	W
Efficiency	89.6	97.9	%

From load torque comparison, as shown in Figure 16, the torque of the proposed model is 20% higher than that of the target model. It was confirmed that the output power and the efficiency increased by 15% and 8.3%, respectively. In addition, it can be seen that the iron loss is very low because only iron loss from the rotor remains at no back yoke of the stator structure. Therefore, it is possible to reduce the total iron loss significantly. In the case of an axial flux motor, it is possible to reduce the surface magnetic flux density per pole by using a multi-pole structure to increase the output density in a thin structure, and design a thin rotor back yoke. However, as the back yoke becomes thinner, the rotor back yoke is saturated, and torque ripple may occur.

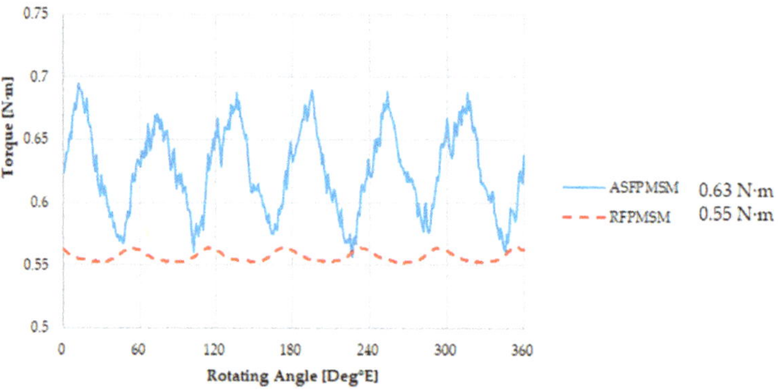

Figure 16. Load torque comparison between RFPMSM and ASFPMSM.

In the case of an axial flux motor, the main requirements are small size, lightweight, high torque density, and high power density. Therefore, torque ripple is not a main characteristic. Vibration and noise caused by torque ripple are commonly buried in external noise or other structural device noise in the applications. Even considering that the current torque ripple is 10% to 20%, it is not a high level, and there is no problem with the motor performance.

To verify the validity of the design, the magnetic flux line and magnetic flux density saturation of ASFPMSM were checked via FEA as shown in Figures 17 and 18.

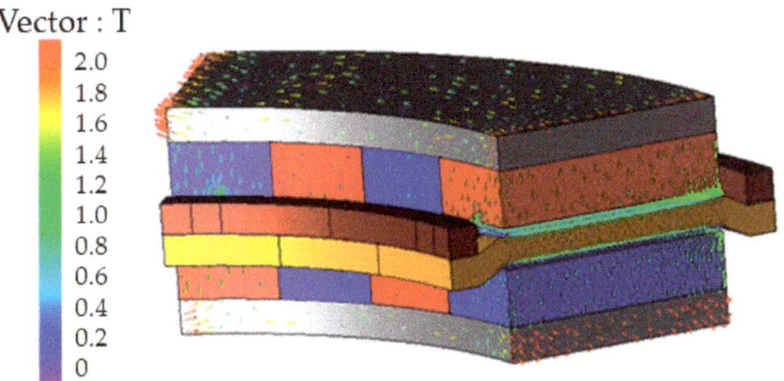

Figure 17. Vector plot of optimally designed ASFPMSM.

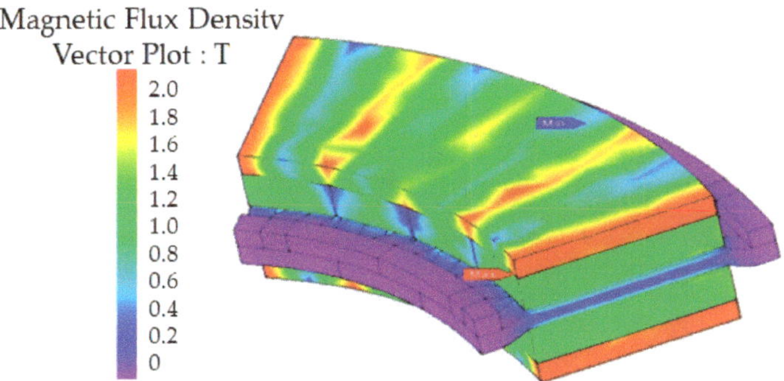

Figure 18. The magnetic flux density of optimally designed ASFPMSM.

As a result of analyzing the vector diagram in Figure 17, it is confirmed that magnetic flux is formed as a closed-loop without a leakage flux problem. In addition, it is also confirmed that there was no problem with the saturation of the core at the rating power. Therefore, it was confirmed that the torque density and output power density were improved compared to the same size of the target model at the rated 3500 RPM.

5. Conclusions

In this paper, the high torque density of ASFPMSM using 3D printing technology for collaborative robots was proposed to replace the RFPMSM used in collaborative robot joints. For accurate performance comparison, the same inner diameter, outer diameter size, volume, winding size, and current density as the target model were used. The back EMF of the ASFPMSM may be smaller compared to the RFPMSM because the block coil increases the airgap. Therefore, it was necessary to analyze the design parameters to increase the performance while minimizing the leakage flux. The block winding thickness, and the total amount of magnet and pole/slot combinations were considered for optimization design. Then, the model with a core thickness of 2 mm, a magnet thickness of 3.3 mm, a coil thickness of 1.8 mm, and an air gap length of 0.3 mm was proposed. The model's 12% higher torque, 15% higher output power, and 8.3% higher efficiency than that of the RFPMSM were proved via FEA.

Author Contributions: Conceptualization, W.-H.K.; methodology, D.-Y.S.; software, D.-Y.S.; validation, D.-Y.S.; formal analysis, D.-Y.S.; investigation, D.-Y.S.; resources, D.-Y.S.; data curation, D.-Y.S. and K.-B.L.; writing—original draft preparation, D.-Y.S. and K.-B.L.; visualization, D.-Y.S., M.-J.J.; supervision, W.-H.K. and K.-D.L. All authors have read and agreed to the published version of the manuscript.

Funding: This work was supported by the National Research Foundation of Korea (NRF) grant funded by the Korea government (MSIT) (No. 2020R1A2C1013724), and this work was supported by the Gachon University research fund of 2019 (GCU-2019-0770).

Institutional Review Board Statement: Not applicable.

Informed Consent Statement: Not applicable.

Data Availability Statement: Not applicable.

Conflicts of Interest: The authors declare no conflict of interest.

References

1. Wu, S.-H.; Hong, X.-S. Integrating Computer Vision and Natural Language Instruction for Collaborative Robot Human-Robot Interaction. In Proceedings of the 2020 International Automatic Control Conference (CACS), Hsinchu, Taiwan, 4–7 November 2020; pp. 1–5. [CrossRef]
2. Gieras, J.F.; Wang, R.-J.; Kamper, M.J. *Axial Flux Permanent Magnet Brushless Machines*; Springer Science Business Media B. V.: Berlin, Germany, 2008. [CrossRef]
3. Hanselman, D. *Brushless Motors: Magnetic Design Performance and Control*; E-Man Press LLC: Amsterdam, NY, USA, 2012.
4. Messina, G.; De Bella, E.T.; Morici, L. HTS Axial Flux Permanent Magnets Electrical Machine Prototype: Design and Test Results. *IEEE Trans. Appl. Supercond.* **2019**, *29*, 1–5. [CrossRef]
5. Andriollo, M.; Bettanini, G.; Tortella, A. Design procedure of a small-size axial flux motor with Halbach-type permanent magnet rotor and SMC cores. In Proceedings of the 2013 International Electric Machines & Drives Conference, Chicago, IL, USA, 12–15 May 2013; pp. 775–780. [CrossRef]
6. Profumo, F.; Zhang, Z.; Tenconi, A. Axial flux machines drives: A new viable solution for electric cars. *IEEE Trans. Ind. Electron.* **2002**, *44*, 39–45. [CrossRef]
7. Polat, M.; Yildiz, A.; Akinci, R. Performance Analysis and Reduction of Torque Ripple of Axial Flux Permanent Magnet Synchronous Motor Manufactured for Electric Vehicles. *IEEE Trans. Magn.* **2021**, *57*, 1–9. [CrossRef]
8. Ueda, Y.; Takahashi, H. Cogging Torque Reduction on Transverse-Flux Motor with Multilevel Skew Configuration of Toothed Cores. *IEEE Trans. Magn.* **2019**, *55*, 1–5. [CrossRef]
9. Jia, L.; Lin, M.; Le, W.; Li, N.; Kong, Y. Dual-Skew Magnet for Cogging Torque Minimization of Axial Flux PMSM With Segmented Stator. *IEEE Trans. Magn.* **2020**, *56*, 1–6. [CrossRef]
10. Polat, M.; Akinci, R. Design and analysis of axial flux double rotor permanent magnet synchronous motor for electric vehicles. *Fırat Univ. J. Eng. Sci.* **2020**, *32*, 345–358.
11. Aydin, M.; Gulec, M.; Demir, Y. Design and validation of a 24-pole coreless axial flux permanent magnet motor for a solar powered vehicle. In Proceedings of the XXII International Conference on Electrical Machines (ICEM), Lausanne, Switzerland, 4–7 September 2016.
12. Aydin, M.; Gulec, M. Reduction of Cogging Torque in Double-Rotor Axial-Flux Permanent-Magnet Disk Motors: A Review of Cost-Effective Magnet-Skewing Techniques with Experimental Verification. *IEEE Trans. Ind. Electron.* **2013**, *61*, 5025–5034. [CrossRef]
13. Neethu, S.; Nikam, S.P.; Singh, S. High-speed coreless axial-flux permanent-magnet motor with printed circuit board winding. *IEEE Trans.* **2019**, *55*, 1954–1962.
14. Wang, X.; Li, C.; Lou, F. Geometry Optimize of Printed Circuit Board Stator Winding in Coreless Axial Field Permanent Magnet Motor. In Proceedings of the IEEE Vehicle Power and Propulsion Conference (VPPC), Hangzhou, China, 17–20 October 2016; pp. 1–6. [CrossRef]
15. Jang-Young, C.; Sung-Ho, L.; Kyoung-Jin, K.; Seok-Myeong, J. Improved analytical model for electromagnetic analysis of axial flux machines with double-sided permanent magnet rotor and coreless stator windings. *IEEE Trans. Magnet.* **2011**, *47*, 2760–2763.
16. Gonzalez-Parada, A.; Trillaud, F.; Guzman-Cabrera, R.; Abatal, M. Torque Ripple Reduction in an Axial Flux High Temperature Superconducting Motor. *IEEE Trans. Appl. Supercond.* **2014**, *25*, 1–5. [CrossRef]
17. Bruzinga, G.R.; Filho, A.J.S.; Pelizari, A. Analysis and Design of 3 kW Axial Flux Permanent Magnet Synchronous Motor for Electric Car. *IEEE Lat. Am. Trans.* **2022**, *20*, 855–863. [CrossRef]
18. Zhao, F.; Kim, M.-S.; Kwon, B.-I.; Baek, J.-H. A Small Axial-Flux Vernier Machine with Ring-Type Magnets for the Auto-Focusing Lens Drive System. *IEEE Trans. Magn.* **2016**, *52*, 1–4. [CrossRef]

Article

Research on a Limit Analytical Method for a Low-Speed Micro Permanent Magnet Torque Motor with Back Winding

Shuangshuang Guo [1], Bo Zhao [1,*], Cunshan Zhang [1], Binglin Lu [1], Yukang Chu [2] and Peng Yang [2]

[1] School of Electrical and Electronic Engineering, Shandong University of Technology, Zibo 255000, China; gss@sdut.edu.cn (S.G.); zcs@sdut.edu.cn (C.Z.); lubinglin@sdut.edu.cn (B.L.)

[2] Shandong Shanbo Electric Machine Group Co., Ltd., Zibo 255000, China; cyk864145006@163.com (Y.C.); dengshan111@163.com (P.Y.)

* Correspondence: zhaobo@sdut.edu.cn; Tel.: +86-188-4515-1552

Abstract: The conventional permanent magnet torque motor (PMTM) is slotted on the inner surface of the stator core. When the size of the stator core is small, the winding coils are difficult to embed into the slots. To solve the problem, a novel PMTM is presented, which is slotted on the outer surface of the stator core. As a result, the winding coils can be conveniently embedded into the slots from the outer surface of stator core. The novel structure of the PMTM with back winding (BWPMTM) is introduced, and the advantage of the novel structure is analyzed. Furthermore, this paper proposes a limit analytical method to solve the optimal parameters of the motor which comprehensively considers four constraints: no-load back electromotive force (EMF), torque, temperature and slot space factor. The optimal parameters of the motor are directly solved to maximize torque density within the constrained range. This method avoids repeated iterative processes and greatly reduces the amount of calculation for PMTM design. Electromagnetic performance and thermal performance are analyzed based on the finite element model (FEM). Finally, the building of a prototype and the experimental results obtained with it are discussed. The rationality of the novel structure and the limit analytical method are verified.

Keywords: PMTM; back-winding structure; limit analytical method; FEM

Citation: Guo, S.; Zhao, B.; Zhang, C.; Lu, B.; Chu, Y.; Yang, P. Research on a Limit Analytical Method for a Low-Speed Micro Permanent Magnet Torque Motor with Back Winding. *Energies* **2022**, *15*, 4662. https://doi.org/10.3390/en15134662

Academic Editors: Federico Barrero, Quntao An, Bing Tian and Xinghe Fu

Received: 27 April 2022
Accepted: 23 June 2022
Published: 25 June 2022

Publisher's Note: MDPI stays neutral with regard to jurisdictional claims in published maps and institutional affiliations.

1. Introduction

The PMTM can be directly connected to a load without a gearbox and can operate stably at low-speed or in locked-rotor conditions [1–3]. It has the merits of low-speed, high torque density, fast response speed, high precision control and small torque ripple and is widely used in many fields, such as aerospace, micro electric platforms and micro-robot joint and weapon equipment of artillery follow-up systems [4–6]. Therefore, many researchers have conducted in-depth studies on low-speed and large torque PMTMs.

With the development of industry, the demand for high-performance PMTMs has increased rapidly [7]. Research on low-speed and large-torque PMTMs focuses on torque ripple and torque density. Cogging torque is the important part of torque ripple. Due to the interaction between the permanent magnet and the stator teeth, cogging torque is inevitably present which can affect the running performance of a PMTM. At present, researchers have carried out a lot of work on weakening cogging torque, mainly including the following measures: magnet skewing [8,9], slot skewing [10], asymmetric windings [11], magnetic wedges filling stator slots, etc. Cogging torque reduction by optimizing the shape of permanent magnets was presented in [12,13]. The study reported in [14] applied 1J22 soft magnetic alloy to a PMTM and proposed a multi-objective optimization method based on a non-dominant sorting genetic algorithm II to improve motor performance, which was applied to the optimal design of surface mount permanent magnet synchronous motors. The simulation results showed the effectiveness of this method. In [15], a low-speed and high-torque dual-stator permanent magnet direct drive motor was proposed, which could improve the torque density at the same the motor volume. In [16], various cogging torque suppression methods were proposed, the relationship between cogging torque and structural parameters was derived by formulas and the torque distortion caused by armature reaction force was predicted. In [17], a permanent magnet slotting machine with a special tooth tip structure was introduced, which could effectively reduce magnetic leakage and armature reaction to improve motor performance, while [18] proposed a technique to increase torque density by changing the shape of the permanent magnet with the minimum magnet volume. Increasing the PM thickness more than the width increased torque density. In [19], a multi-objective optimization method to determine the optimization target by analyzing the factors that affect torque ripple was proposed. The existing analysis methods improve torque based on the baseline motor or by determining motor size. The design process for the size of the prototype was not described. Hence, this paper proposes a limit analytical method to maximize the torque density of the motor by the stack length based on the torque and other indicators. The design process for solving motor parameters is given in this paper.

In this study, a PMTM is proposed which is slotted on the outer surface of the stator core. It solves the problem of embedding winding when the stator inner diameter is small. In addition, this paper proposes a limit analysis method which can quickly find the optimal parameters that meet all the constraints. Finite element software was used to build the model. The simulation results verify the rationality of the novel structure and the limit analysis method.

2. Structure and Advantage Analysis of the BWPMTM

2.1. Motor Structure

There are two common embedding winding methods: automatic embedding by an embedded robot and artificial embedding. The conventional PMTM is slotted on the inner surface of the stator core, as shown in Figure 1a. It is difficult to pass the embedded robot through the inner hole of the stator when the stator inner diameter is less than 40 mm. Artificial embedding is also extremely difficult. In view of this problem, a novel structure of PMTM is proposed, which is slotted on the outer surface of the stator core. With the novel structure, winding coils can be simply embedded into the slot from the outer surface of the stator core by the above two methods, with high productivity and efficiency and low production cost. The main magnetic flux path of the conventional PMTM is: permanent magnet, air gap, stator tooth, stator yoke. On account of slotting on the outer surface of stator, the novel stator structure has no yoke. In order to close the magnetic flux lines to form a loop, the shell is made of a high-permeability material as a part of the magnetic circuit. So, the main magnetic flow path of the BWPMTM is permanent magnet, air gap, stator tooth, shell. The eddy loss in the shell is tiny and can be ignored under low-speed or locked-rotor conditions. Therefore, the novel structure is suitable for micro PMTM, with an inner diameter for the stator less than 40 mm, working under low-speed or rocked-rotor conditions. Fractional slot concentrated winding was used in this study because of its advantages of high torque density and high slot filling coefficient [20].

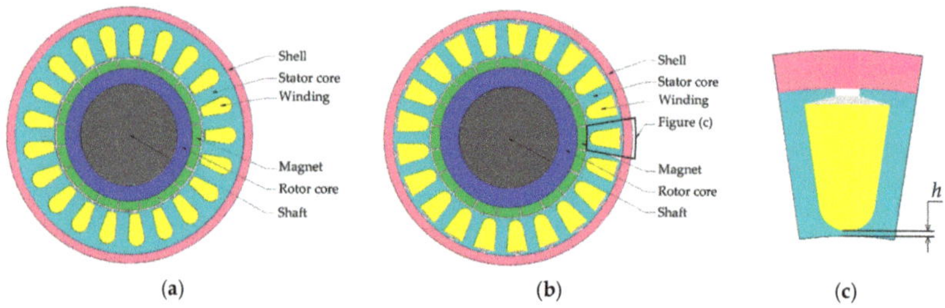

Figure 1. Cross-sectional view: (**a**) the conventional PMTM; (**b**) the novel BWPMTM; (**c**) the local structure.

The slot type of the BWPMTM is equivalent to the closed slot, which can effectively reduce the torque ripple and improve the running performance [21,22]. As shown in Figure 1c, there is a magnetic barrier (the connection between the bottom of the slot and the inner surface of the stator) to maintain the integrity of the stator. The magnetic barrier will increase the magnetic leakage. Therefore, the mechanical strength of the magnetic barrier and the electromagnetic performance of the motor should be comprehensively considered for the magnetic barrier thickness h in the design.

2.2. Fractional Slot Concentrated Winding

The slot number of per phase per pole of fractional slot concentrated winding is a fractional number. Compared with the integer slot permanent magnet motor, the fractional slot concentrated winding permanent magnet motor has the following advantages: greatly shortened end length, effectively reduced copper weight and copper loss, improved efficiency and power density and the winding coefficient of the fundamental wave of the magnetomotive force is high. The winding coefficient can be more than 0.93 by selecting the appropriate pole–slot combination; improvement of the slot space factor helps improve the power density; and the cogging torque can be effectively restrained by selecting a reasonable pole–slot combination [23,24]. Based on the above advantages, the pole–slot combination of 20 poles and 22 slots is adopted in this paper. The spatial distribution diagram of the three-phase winding motor is shown in Figure 2:

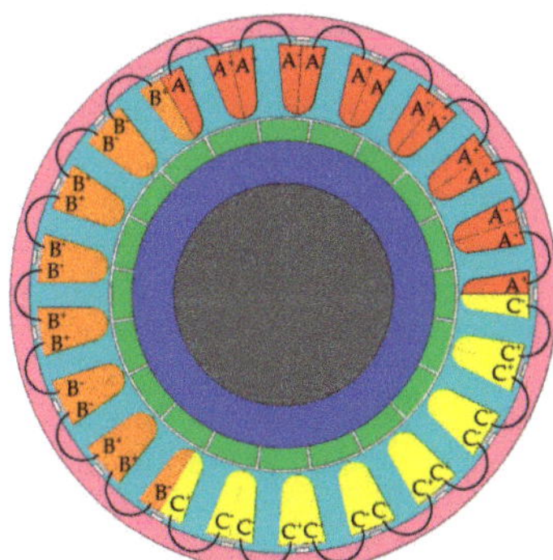

Figure 2. Spatial distribution diagram of the three−phase winding motor.

As can be seen from the figure above, the three phases of the motor are distributed centrally in space with differences of 120 degrees. Since the pitch of the motor is 1, there is no need to consider the starting position when winding. In addition, there is no part for the three-phase winding to cross, so the winding can be made into a type in advance and directly embedded in the stator slot.

2.3. Advantage in Torque Density

The general equation of AC motor electromagnetic torque can be expressed as:

$$T_e = \frac{\pi}{2}p^2 F\Phi$$
$$F = \frac{\sqrt{2}mN_aK_w}{\pi p}I \tag{1}$$

where p is the number of pole pairs; F is the amplitude of magnetomotive force; Φ is air gap flux, $\Phi = 2B\tau L/\pi$; B is the amplitude of air gap flux density; τ is pole pitch; L is the stack length; N_a is the number of series turns per phase; K_w is the winding factor; and I is the phase current.

The number of ampere conductors per unit of the armature circumference is called the electric load, A.

$$A = \frac{2mN_aI}{\pi D} \tag{2}$$

where D is the stator inner diameter.

Substitution of I of Equation (2) in Equation (1) yields the torque equation, shown as:

$$T_e = \frac{\pi K_w A B D^2 L}{2\sqrt{2}} \tag{3}$$

It can be seen that the torque is proportional to the product of the stator inner diameter squared and the stack length. The BWPMTM is slotted on the outer surface of the stator without the stator yoke. A larger stator inner diameter can be obtained in the novel structure. Based on the same shell outer diameter, the larger the stator inner diameter, the shorter the stack length, the lighter the weight of the motor and the higher the torque density. Hence, a higher torque density can be obtained theoretically.

3. Limit Analytical Method

The design flow chart of the traditional analytical method is shown in Figure 3a. Firstly, the traditional design of the motor determines the number of pole pairs p and number of slots Q and prefetches the stator inner diameter D and the stack length L according to experience. Then, the number of series turns per phase are obtained according to the no-load back EMF as well as the section area of single wire according to the slot space factor. Finally, the motor parameters are obtained by judging whether the torque and temperature constraints are met, otherwise the parameters should be re-calculated. However, the motor parameters obtained by the traditional method may not be optimal to satisfy all constraints. The traditional design method requires multiple iteration calculations, which take a large amount of calculation and consume a lot of manpower and time. This paper proposes a limit analytical method, as shown in Figure 3b, which can locate the optimal parameters accurately and avoid iterative calculation. The design is divided into four steps. In the first step, the minimum number of the number of series turns per phase N_{a1min} is obtained according to the no-load back EMF. In the second step, the minimum number of the number of series turns per phase N_{a2min} is obtained according to the maximum permissible temperature. In the third step, the minimum number of the section area of single wire S_{min} is obtained according to the locked-rotor torque. In the last step, the optimal parameters of the motor are obtained by integrating the relationship between constraints and judging by the slot space factor. This design method avoids the iterative calculation of parameters and models, which is beneficial in terms of saving labor and time costs.

The limit analytical method proposed in this paper is used to obtain the maximum torque density of the motor under the design specifications. The specifications of the BWPMTM are listed in Table 1.

Table 1. The specifications of the BWPMTM.

Parameter	Value
Maximum stack length	25 mm
Shell outer diameter	40 mm
Number of pole pairs	10
Number of slots	21
Maximum no-load speed	1200 rpm
Continuous locked-rotor voltage	14 V
Peak locked-rotor voltage	28 V
Continuous locked-rotor torque	0.3 Nm
Peak locked-rotor torque	0.6 Nm
Slot space factor	0.3
Maximum working temperature	180 °C
Initial temperature	20 °C

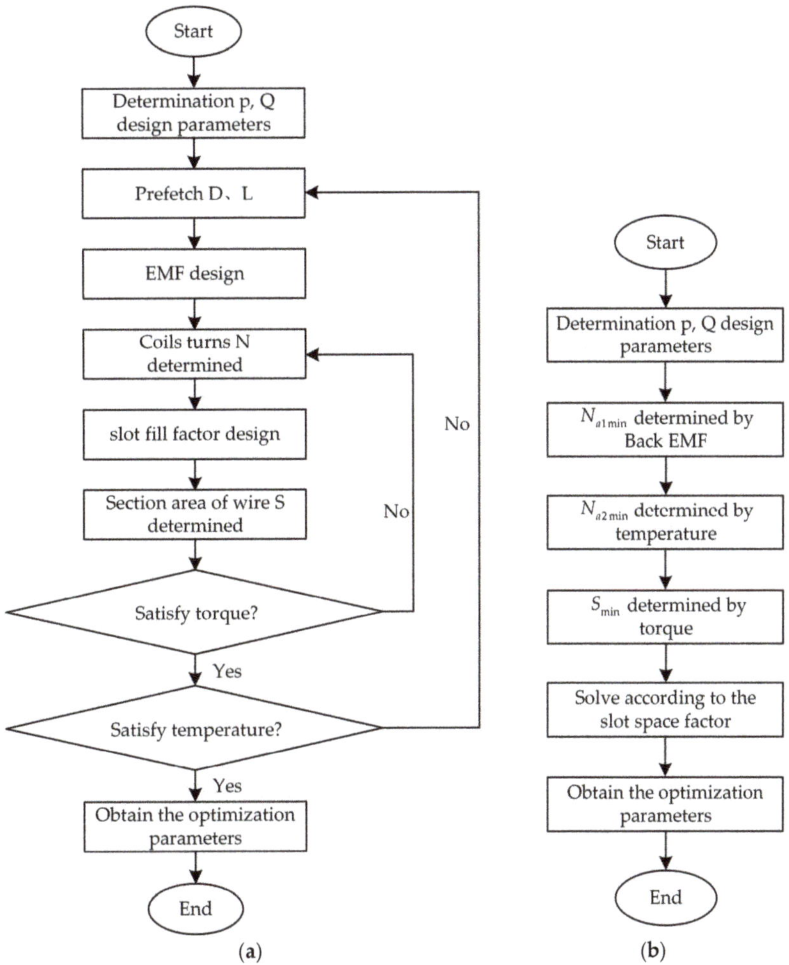

Figure 3. Design flow chart: (**a**) the traditional analytical method; (**b**) the proposed limit analytical method.

According to the theory of electromechanics, the magnetic field energy mainly exists in the air gap, and the electromechanical energy conversion is mainly carried out through the air gap. Practice shows that the stator inner diameter and the stack length near the air gap are the main parameters of the motor. The product of the square of the stator inner diameter and the stack length can be determined using Equation (4):

$$D^2L = \frac{2\sqrt{2}T_e}{\pi K_w AB} \tag{4}$$

The maximum torque density can be obtained at the minimum axial length based on constant shell outer diameter. As a result, taking the minimum stack length as the objective function to solve the optimal parameters of the motor is an effective method.

3.1. The Control Circuit of the Motor

Space vector pulse width modulation (SVPWM) technology is applied to the control system of the PMSM. Based on flux linkage tracking, SVPWM technology combines inverters and the PMSM [25]. The actual voltage space vector is synthesized by using different combinations of six bridge arms switching signals to generate eight basic space voltage vectors, and the rotation trajectory of the resultant vector forms the basic flux circle. Based on SVPWM technology, direct current U_{dc} can be converted to alternating current U_{AN}, U_{BN} and U_{CN} and applied to the three-phase winding motor. The control circuit of motor is shown in Figure 4.

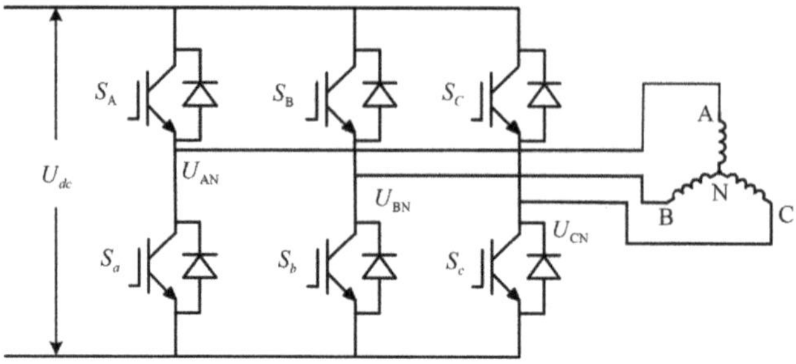

Figure 4. The control circuit of the motor.

3.2. Constraint of No-Load Back EMF

Assuming that the magnetic field is sinusoidally distributed in the air gap, the phase EMF can be expressed as:

$$E = \sqrt{2}\pi k_\sigma K_w f N_a \Phi \tag{5}$$

where f is the electric frequency, which is proportional to the speed.

It can be seen from Equation (5) that the EMF is proportional to the air gap magnetic flux. From the above, the magnetic barrier will increase the magnetic flux leakage, resulting in a decrease in air gap flux, which leads to a decrease in the EMF. Thus, it is necessary to multiply the coefficient $k_\sigma(k_\sigma < 1)$ in the process of calculating the EMF to reflect the influence of the magnetic flux leakage on the magnetic bridge.

The voltage balance equation of the motor on no-load can be written as:

$$U_{PN} = E + RI \tag{6}$$

where U_{PN} is the rated phase voltage, E is the phase EMF and R is the phase resistance.

When the motor is running under no-load condition, the current is nearly 0 A, so the no-load back EMF is approximately equal to the rated phase voltage. If the running speed reaches the maximum no-load speed and the no-load back EMF is less than the rated phase voltage, the motor speed will continue to increase until the two are equal. Meanwhile, the motor running speed exceeds the specified maximum speed, which will cause a great burden to the motor operation and even cause damage. Therefore, the no-load speed of the no-load back EMF at the rated phase voltage must be lower than the maximum allowable speed, i.e., $n \leq n_{max}$. Incorporating the inequality into Equation (5), the constraint of the number of series turns per phase can be expressed as:

$$N_a \geq N_{a1min} = \frac{60E}{\sqrt{2}\pi k_\sigma K_w BDLn_{max}} \tag{7}$$

3.3. Constraint of Slot Space Factor

The slot space factor is the ratio of the section area of bare copper of the winding wire to the stator slot section area, and a reasonable slot space factor is one of the specifications of the motor design. Generally, the slot space factor of the PMTM is less than 0.3, considering the manufacturing process. For the BWPMTM, the winding coil is embedded in the slot from the outer surface of the stator, simplifying the manufacturing process. Therefore, the slot space factor can be improved.

In this paper, in the stator with parallel teeth, the section area of the stator slot can be approximately expressed as:

$$S_{slot} = \frac{\pi\left(D_2{}^2 - D_1{}^2\right)}{4} - Qb_t(D_2 - D_1) \tag{8}$$

where S_{slot} is the slot section area, D_2 is the stator outer diameter, D_1 is the slot bottom diameter and b_t the width of the stator teeth.

The slot space factor is used to judge whether the area of winding wire is adequate. The section area of winding wire needs to meet the constraint:

$$\begin{aligned} S_w &\leq \lambda S_{slot} \\ S_w &= 2mN_aS \end{aligned} \tag{9}$$

where, λ is the slot space factor, S_w is the section area of bare copper of all winding wires and S is the section area of bare copper of a single wire.

3.4. Constraint of Temperature

To improving the utilization of motor material, higher electromagnetic load is often adopted, but most motors adopt natural cooling, which leads to poor heat dissipation conditions and makes the temperature rise problem for the motor more prominent. Excessive temperature of the motor can reduce the performance of motor material, e.g., leading to a decline in the magnetic properties of silicon steel sheets as well as the electrical conductivity of the winding. Hence, the temperature analysis is a significant part of motor design.

The instantaneous value of the three-phase current before the rotor of the motor is locked is as follows:

$$I_{A/B/C} = I_m \cos(\theta + \alpha) \tag{10}$$

where θ is the electrical angle; α is the initial position angle, corresponding to the A-phase, B-phase and C-phase, which are $0°$, $-120°$ and $120°$, respectively; and I_m is the amplitude of the phase current.

According to the voltage balance equation, the back EMF is 0 under the locked-rotor condition. Therefore, the winding current is the largest, the greater the loss of the motor, and the highest temperature of the motor. In addition, the motor designed in this study has the longest time under the locked-rotor condition, so this paper takes the locked-rotor condition as an example to design.

When the motor works under the locked-rotor condition, the speed is 0 rpm. Meanwhile, the motor has no core loss, only the winding copper loss.

$$\begin{aligned} P_{cu} &= \sum P_{cuA/B/C} \\ &= I_A{}^2R + I_B{}^2R + I_C{}^2R \\ &= I_m{}^2R\left(\cos\left(\theta\right)^2 + \cos\left(\theta - \tfrac{2\pi}{3}\right)^2 + \cos\left(\theta + \tfrac{2\pi}{3}\right)^2\right) \\ &= \tfrac{3}{2}I_m{}^2R \end{aligned} \tag{11}$$

The total copper loss is a constant value, which is independent of the locked-rotor position. However, the distribution of the copper loss is not constant. With the change of the locked-rotor position, the distribution of the copper loss is different, resulting in a difference in the temperature distribution and in the maximum temperature. The maximum temperature caused by the same copper loss is basically the same given that the stator outer diameter is constant, the stack length does not change much and the heat dissipation conditions are the same.

It can be seen form Figure 5 that the copper loss in the three-phase windings is a function of the square of the sine. The copper loss of the A-phase winding reaches the maximum value at $\theta = 0°$, and the copper loss of the B-phase winding reaches the maximum value at $\theta = 60°$. At these two positions, the winding copper loss distribution regularity is the same and only differs by $120°$ in the space position. In consequence, the maximum temperature of the motor is the same. The maximum temperature occurs cyclically with an electrical angle of $60°$. According to the symmetry of the loss distribution, the copper loss can be calculated using electromagnetic field simulation software at the four locked-rotor positions $\theta = 0°$, $\theta = 10°$, $\theta = 20°$ and $\theta = 30°$ and loaded into the temperature field as a heat source. As can be seen form the results of temperature field simulation, the motor has a maximum temperature in the A-phase winding at the locked-rotor position $\theta = 0°$, that is, the copper loss of the A-phase winding is the maximum. The temperature of the B-phase and the C-phase are the same according to symmetry when the temperature of the A-phase winding is the highest. The copper loss of the A-phase winding is four times that of the B-phase winding or the C-phase winding. According to Figure 2, when the A-phase winding loss is the largest, the heat source is mainly distributed on the A-phase winding in the temperature field simulation. Due to the concentrated distribution of the A-phase winding in space, the temperature of the motor at the position of the A-phase winding is higher than that of the B-phase winding and the C-phase winding. At this point, it is the maximum blocking position of the motor operation. As long as the maximum temperature meets the requirement at the locked-rotor position $\theta = 0°$, the maximum temperature meets the requirement, regardless of the locked rotor at any position.

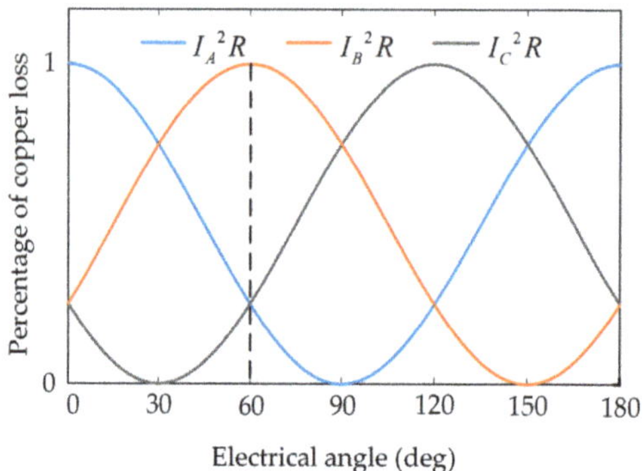

Figure 5. Copper loss curves of the three-phase winding motor.

At this point, switching tubes S_A, S_b and S_c are normally on. The voltage source acts directly on the three-phase winding motor. Therefore, this paper studies the limit design under the condition of the locked-rotor position $\theta = 0°$, which is not affected by the control method. The current flow path is shown in Figure 6.

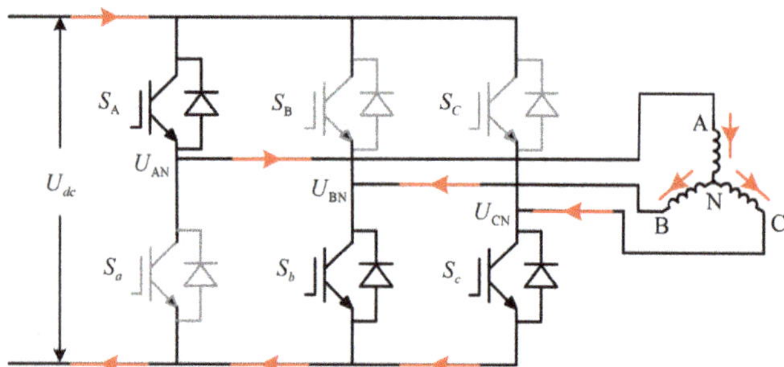

Figure 6. Circuit diagram at maximum A-phase current.

Ignoring the pressure drop on the switching tube, the winding current can be expressed as:

$$I_m = \frac{2U_{dc}}{3R} \tag{12}$$

where U_{dc} is the bus voltage; R is the phase resistance, which includes the winding resistance within the core, R_{ef}, and the end winding resistance, R_{end}.

$$R = R_{ef} + R_{end} = \frac{2\rho N_a L}{S} + \frac{\rho N_a L_{end}}{S} \tag{13}$$

where ρ is the copper resistivity and L_{end} is the end winding length, which can be approximated by the tooth pitch when the motor size is small and the concentrated winding is adopted.

Calculate the maximum copper loss at the locked-rotor position $\theta = 0°$ with the continuous locked-rotor voltage for 10 min and the peak locked-rotor voltage for 1 min while the motor reaches the maximum allowable temperature. That the copper loss is less than the maximum loss calculated above can ensure that the maximum temperature is within the allowable range.

$$P_{\max} \geq \frac{3}{2} I_m{}^2 R \tag{14}$$

Substitution of Equations (12) and (13) in Equation (15) yields the constraint of the number of series turns per phase:

$$N_a \geq N_{a2\min} = \frac{2U^2 S}{3\rho(2L + L_{end})P_{\max}} \tag{15}$$

3.5. Constraint of Torque

When the motor runs under the locked-rotor condition, the back EMF is zero, the winding inductance is zero and the armature current is a direct current. Hence, the bus voltage acts entirely on the winding resistance, resulting in an increase in copper loss and severe heating of the winding coil. As the resistance tends to increase as temperature increases and the torque tends to decrease as the resistance increases, the influence of temperature on the torque should be considered in the design process. It is necessary to ensure that the torque meets the design requirement at the maximum temperature. From the above analysis, the motor has the maximum temperature as the current of the A-phase winding reaches the maximum value (a case of the A-phase). Phases B and C have the same temperature distribution. Considering the effect of temperature, the phase resistance can be expressed as:

$$\begin{aligned} R_A &= \frac{T_A + 235}{T_0 + 235} R = k_1 R \\ R_{B/C} &= \frac{T_B + 235}{T_0 + 235} R = k_2 R \end{aligned} \tag{16}$$

where T_0 is the initial temperature and T_A and T_B are the temperatures of the A-phase winding and the B-phase (C-phase) windings, respectively.

Substitution of Equation (17) in Equation (1) yields the torque considering the temperature:

$$T_e = \frac{mK_w BDLUS}{\rho(2k_1 + k_2)(2L + L_{end})} \tag{17}$$

The torque considering the temperature should be greater than the rated torque, T_{eN}. The torque is proportional to the section area of wire. It can be determined that the cross-section of the wire needs to meet the following constraint:

$$S \geq S_{\min} = \frac{\rho(2k_1 + k_2)(2L + L_{end})}{mK_w BDLU} T_{eN} \tag{18}$$

3.6. Design of the Magnetic Barrier

When the motor works under the locked-rotor condition for a long time, the frequency of the motor is zero, so the influence of silicon steel sheet thickness on core loss can be negligible. A silicon steel sheet thickness with a specification of 0.5 mm was selected considering the cost. The design of the magnetic barrier must not only meet the requirement of mechanical strength but also consider electromagnetic performance.

1. The magnetic barrier should be as small as possible so that the magnetic flux at the slot bottom is in a state of oversaturation. This allows magnetic flux to pass through the stator teeth to the greatest extent possible, reducing magnetic leakage and improving the efficiency of the permanent magnet.

2. The thickness of the stator stamping die is generally larger than 0.15 mm–0.2 mm in the manufacturing process. In view of the wear of the die on the stamping die and the service life of the stamping die, $h = 0.2$ mm is selected. The maximum force that the slot bottom bears can be expressed as $F_{max} = PL_s h$, where P is the yield strength and L_s is the thickness of the silicon steel sheet. To ensure the safe operation of the motor, it is taken to be 50% of the yield strength of the silicon steel sheet for the purpose of calculation. The force that the magnetic barrier bears during processing is less than F_{max}.

3.7. The Solution of Optimal Parameters

From the above, the constraints of the no-load back EMF, temperature and torque are transformed into the constraints of the number of series turns per phase and the cross-section of the wire. According to the minimum number of series turns per phase and the minimum cross-section of the wire, the minimum section area of all winding wires can be obtained.

$$
\begin{aligned}
S_{w1} &= 2mN_{a1min}S_{min} \\
&= 2m\frac{60E}{\sqrt{2}\pi k_\sigma K_w BDL n_{max}}\frac{\rho(2k_1+k_2)(2L+L_{end})}{mK_w BDLU}T_{eN} \\
&= \frac{60\sqrt{2}\rho E T_{eN}(2k_1+k_2)(2L+\frac{\pi D}{Q})}{\pi k_\sigma K_w{}^2 B^2 D^2 L^2 U n_{max}}
\end{aligned}
\tag{19}
$$

$$
\begin{aligned}
S_{w2} &= 2mN_{a2min}S_{min} \\
&= 2m\frac{2U^2}{3\rho P_{max}(2L+L_{end})}S_{min}{}^2 \\
&= \frac{4\rho T_{eN}{}^2(2k_1+k_2)^2(2L+L_{end})}{3mP_{max}K_w{}^2 B^2 D^2 L^2}
\end{aligned}
\tag{20}
$$

where S_{w1} is the section area of all winding wires satisfying the constraints of the no-load back EMF and temperature increase and S_{w2} is the winding section satisfying the constraints of the torque and temperature increase.

Then, the optimal parameters of the motor can be obtained by judging whether the winding wires can be reasonably placed in the stator slot under the constant of shell outer diameter $D_s = 40$mm.

$$
\lambda S_{slot} - S_{w1} \geq 0
\tag{21}
$$

$$
\lambda S_{slot} - S_{w2} \geq 0
\tag{22}
$$

As can be seen from Figure 7, both $\lambda S_{slot} - S_{w1}$ and $\lambda S_{slot} - S_{w2}$ increase with increase in L, which first increases then decreases with increase in D. The minimum value of L that satisfies the Equation (22) is 17.1 mm, and D ranges from 23.2 mm to 24.7 mm. The minimum value of L that satisfies Equation (23) is 17.6 mm, and D ranges from 23.2 mm to 24.8 mm. Leaving a certain margin for the stator inner diameter considering the wire size, take $D = 24.6$mm and $L = 18$mm. The value range of N and S can be obtained by Equations (7), (16) and (19). Within the range, the maximum resistance is preferred to minimize the temperature increase.

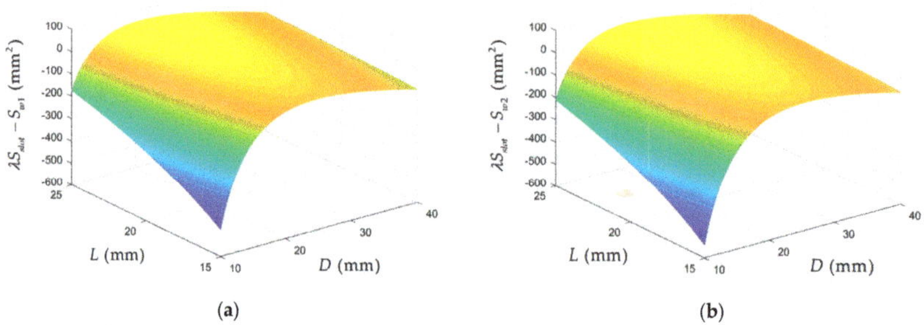

Figure 7. (**a**) Solutions of Equation (20). (**b**) Solutions of Equation (21).

The parameters of the BWPMTM are shown in Table 2.

Table 2. Main parameters of the BWPMTM.

Parameters	Value	Unit
Stack length	18	mm
Shell outer diameter	40	mm
Stator outer diameter	37	mm
Stator inner diameter	24.6	mm
Rotor outer diameter	22	mm
Permanent magnet thickness	2	mm
Air gap length	0.2	mm
Magnetic barrier thickness	0.2	mm
Section area of single wire	0.056	mm^2
Number of series turns per phase	322	-
Phase resistance	4.8	Ω
Continuous locked-rotor current	2	A
Peak locked-rotor current	4	A

4. Simulation Results and Discussion

4.1. Finite Element Model

The finite element numerical calculation of the electromagnetic field has the advantages of accurate calculation, flexible meshing and can better deal with nonlinear problems, which make it suitable for solving the problem of complex geometric shapes, such as a motor has. It is widely used in the analysis of and calculations for various motors.

The process of using finite element software to establish a motor model can be divided into the following steps: according to the size of the motor, draw the motor model; determine the material properties of the motor; determine the boundary conditions; determine the meshing; then set the load and solve the matrix.

Firstly, with the parameters in Table 2, the model of the BWPMTM was drawn based on the finite element software. Secondly, the default boundary condition was automatically added to the outer boundary of the object during the solution process. With two objects in contact, the tangential component of the magnetic field intensity H and the normal component of the magnetic induction intensity B remain continuous at the contact surface. The vector potential boundary condition is mainly applied to the edge of the solution domain or the calculation model, and all points on the edge can be defined to satisfy the following two formulas:

$$A_z = Const \text{ or } rA_\theta = Const \tag{23}$$

where the former applies to the Cartesian coordinate system and the latter to the cylindrical coordinate system, Const is a given constant, and A_z and A_θ are the vector magnetic potential in the Z direction in the Cartesian coordinate system and the vector magnetic potential in the θ direction in the cylindrical coordinate system, respectively.

Thirdly, the material in the finite element software can be divided into air, magnetic material, conductive material and permanent magnet material. In modeling, corresponding materials are added to the different parts. They can be selected from the material library or according to the material properties of their own inputs. The material properties of the stator rotor are shown in Table 3 and added to the corresponding motor parts. The stator core and rotor core are added with silicon steel sheet material. The shell is added with Steel c101.

In addition, the meshing is the basis of finite element solution. Manual division is needed to ensure the accuracy of the calculation. The mesh is finer in the part where the magnetic field is strong or the magnetic field varies greatly. The mesh within a pitch is copied and rotated to ensure that the stator under different poles has the same mesh. The finite element mesh generation result within one pitch is shown in Figure 8.

Table 3. The material properties.

Permanent Magnet		Silicon Steel Sheet		Steel C101	
Density/kg·m^{-3}	7400	Density/kg·m^{-3}	7650	Density/kg·m^{-3}	7900
Remanence/T	1.05	Thickness/mm	0.5	Resistivity/Ωm	6.098×10^{-7}
Coercive force/kA·m^{-1}	764	Saturation flux density/T	2.2	Saturation flux density/T	1.8

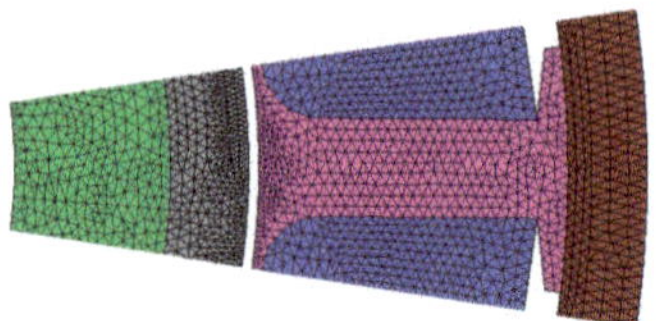

Figure 8. Result of mesh generation.

Finally, according to the design requirements, to set different excitations and loads, solve the matrix and solve the analysis.

4.2. Electromagnetic Characteristic

With the parameters in Table 2, the model of the BWPMTM was established based on finite element software. According to the simulation results, the no-load air gap flux density, no-load back EMF, temperature and torque were analyzed. Figure 9 presents the magnetic flux density distribution diagram of the BWPMT on no-load. It is easy to see that the magnetic flux lines are closed by the permanent magnet, air gap, stator teeth and shell to form a loop. The maximum magnetic flux density was 1.85 T in the magnetic barrier, as shown in Figure 9b, which is in an oversaturation state. The maximum magnetic flux density of the stator tooth is 1.2 T.

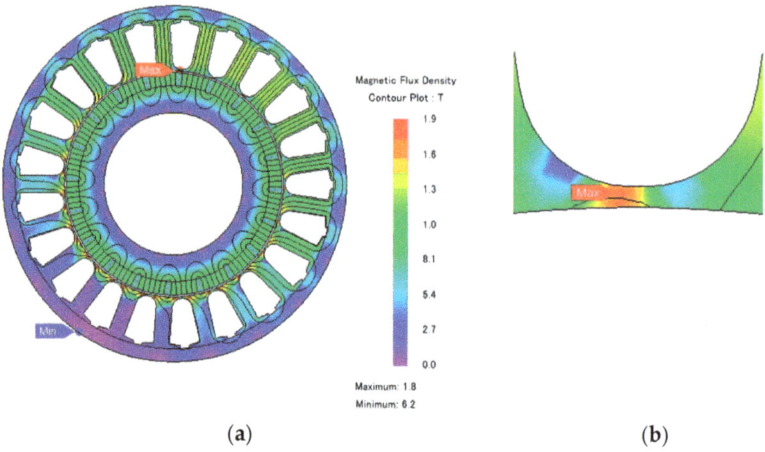

(a)

(b)

Figure 9. The magnetic flux density distributions on no-load: (**a**) the motor; (**b**) the magnetic barrier.

The air gap flux density waveform and the corresponding harmonics of the BWPMTM in the no-load condition are shown in Figure 10. Figure 10a shows the air gap flux density waveform. Figure 10b depicts the spectral analysis of the air gap flux density. The amplitude of the air gap flux density is 0.95 T, and the amplitude of the fundamental component is 1.14 T.

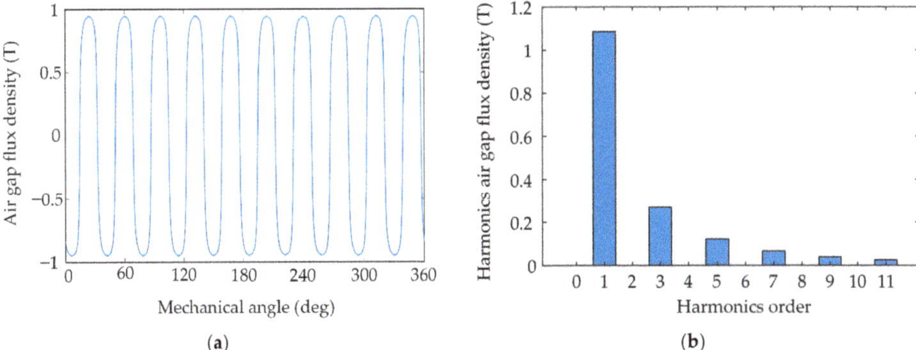

Figure 10. (**a**) Air gap flux density waveform. (**b**) Harmonics of air gap flux density.

The no-load back EMF waveform and the corresponding harmonics are shown in Figure 11 at a rotation speed of 1200 rpm. The phase back EMF waveform approximates the sinusoidal waveform and is symmetric. Therefore, there are no even harmonics in the back EMF waveform. The third harmonic content is small, and the remaining harmonics are almost zero. The amplitude of no-load phase back EMF is 17.2 V. The amplitude of the fundamental component is 18 V. The back EMF coefficient is defined as:

$$k = \frac{E_{p-p}}{f} \tag{24}$$

where E_{p-p} is the peak-to-peak value of back EMF.

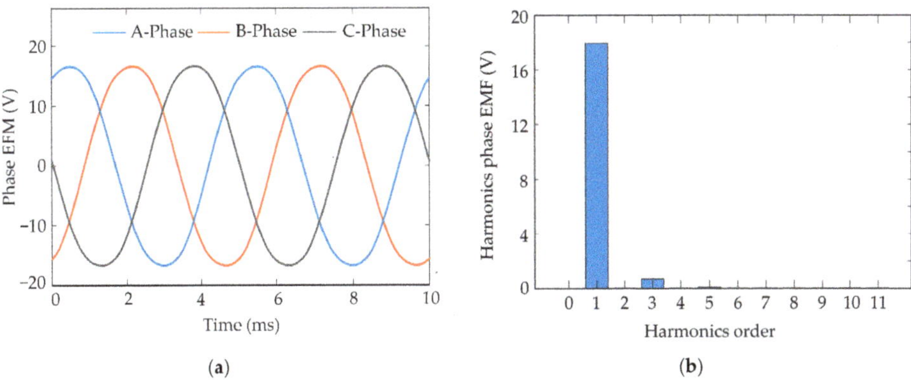

Figure 11. (**a**) The phase back EMF under the no-load condition. (**b**) Harmonics of phase back EMF.

It can be inferred from the above equation that the phase back EMF coefficient is 0.172 V/Hz. According to the relationship between line EMF and phase EMF, the line back EMF coefficient is 0.298 V/Hz.

Figure 12 shows the cogging torque waveforms of the conventional PMTM and the BWPMTM. It can be observed that the BWPMTM can effectively reduce the cogging torque compared with the conventional PMTM. The maximum cogging torques of the BWPMTM and the conventional PMTM are 14 μNm and 31 μNm, respectively. The closed slot is an effective measure for weakening the cogging torque. Due to the characteristics of the slots on the outer surface of the stator of the BWPMTM, no slotting of the inner surface of the stator will not result in variation in air-gap reluctance. So, the structure of the BWPMTM can effectively reduce the cogging torque.

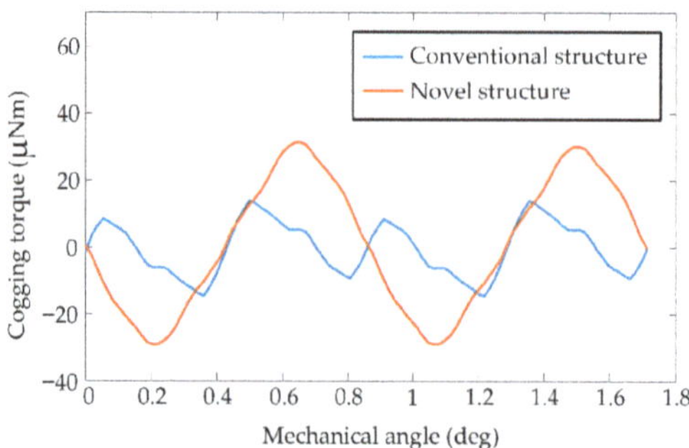

Figure 12. The cogging torque waveforms for two different configurations.

4.3. Temperature Distribution

The core loss is small and can be ignored when the motor works under the locked-rotor condition. Therefore, the winding copper loss is only considered in the temperature field analysis. The model of the BWPMTM was established by three-dimensional finite element temperature field simulation software, with the initial temperature set at 20 °C. The influence of heat radiation on the temperature must be considered when the motor adopts the natural cooling method. As copper loss decreases with the increase in resistance and resistance increases with the increase in temperature, copper loss decreases with the increase in temperature. In this section, the temperature distribution of the motor is analyzed under the two working conditions of the continuous locked-rotor condition for 10 min and the peak locked-rotor condition for 1 min.

Figure 13 shows a diagram of the temperature field and the maximum temperature curves of the winding, shell and permanent magnet when the motor works under the continuous locked-rotor condition for 10 min at the locked-rotor position $\theta = 0°$. As the heat source is generated by the winding, the winding temperature rises rapidly at the beginning, and after about 1 min, the increase drops off slightly. With the effect of heat conduction and radiation, the temperature of the shell and permanent magnet increases rapidly with the temperature of the winding. At 10 min, the maximum temperature of the motor is 136.2 °C at the A-phase winding. The maximum temperature of the B-phase winding is equal to that of the C-phase winding and is 127.3 °C. This is consistent with the above theoretical analysis. The maximum temperature of the shell is slightly lower than that of the winding at 130.5 °C, and the maximum temperature of the permanent magnet is 94.8 °C.

Figure 14 shows a diagram of the temperature field and the maximum temperature curves when the motor works under the peak locked-rotor condition for 1 min at the locked-rotor position $\theta = 0°$. The temperature distribution of the motor and the maximum temperature curves of the winding and shell are similar to that of the motor under the continuous locked-rotor condition. The temperature difference of each part is greater. At 1 min, the maximum temperature of the motor is 160.8 °C at the A-phase winding. The maximum temperature of the B-phase winding is equal to that of the C-phase winding and is 124.2 °C. The maximum temperature of the shell is 127.9 °C. As the inner surface of the stator and the outer surface of the rotor dissipate and absorb heat through thermal radiation, less heat is transferred to the rotor through the air gap per minute. The temperature of the rotor is lower. The maximum temperature of the permanent magnet is 48.6 °C, which is lower than the maximum temperature of the permanent magnet under the continuous locked-rotor condition for 10 min. It can be seen that the temperature has not reached a steady state under the two locked-rotor conditions, that is, the motor is at the highest temperature for only a short time, which is conducive to its safe operation. Meanwhile, the maximum temperatures under the two conditions all meet the design requirements and retain a certain safety margin.

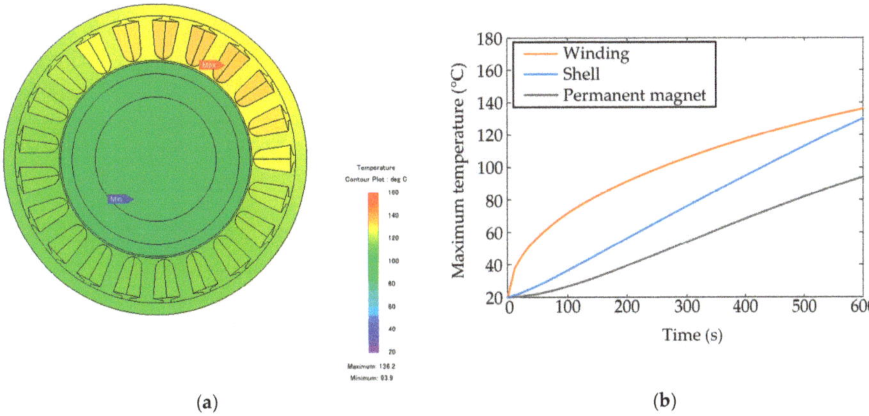

(a)   (b)

Figure 13. Temperature under the continuous locked-rotor condition for 10 min: (**a**) diagram of the temperature field; (**b**) maximum temperature curves of winding, shell and permanent magnet.

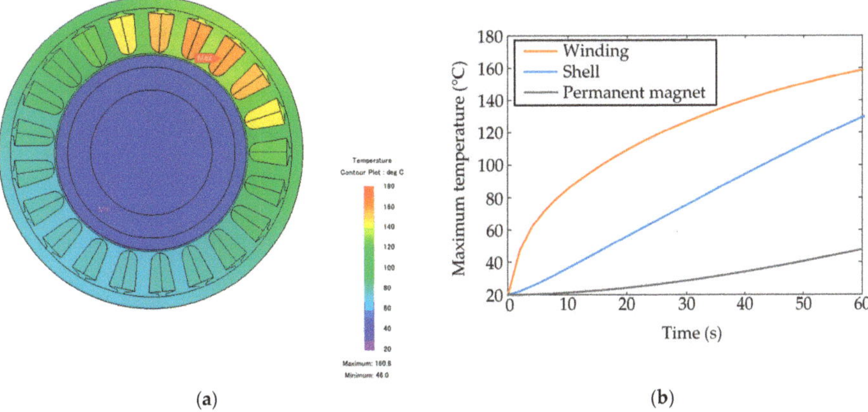

(a)   (b)

Figure 14. The temperature under the peak locked-rotor condition for 1 min: (**a**) diagram of the temperature field; (**b**) maximum temperature curves of winding, shell and permanent magnet.

4.4. Impact of Temperature on Torque

From the simulation results for the temperature field, the temperature of the motor increases rapidly, which increases the winding resistance and decreases the torque. Therefore, the torque calculation considering temperature is necessary. It can be seen from Figures 13 and 14 that the temperature difference between the different slots is large, but the temperature difference in a slot is small. To simplify the calculation, it is reasonable to take the average temperature in the slot as the winding temperature. Figure 15 shows the torque curves considering temperature corresponding to the conditions of the continuous locked-rotor condition for 10 min and the peak locked-rotor condition for 1 min. As the motor has the highest temperature and the largest resistance at the last moment, the locked-rotor torque is the smallest. At the last moment, the continuous locked-rotor torque is 0.3 Nm. The peak locked-rotor torque is 0.6 Nm. The torques meet the design specifications under the two working conditions. The continuous locked-rotor torque density is 2 Nm/kg, wherein the weight is 0.151 kg. The torque density is increased by 30% compared with the maxon motor of specification 758031 with torque density of 1.54 Nm/kg, wherein continuous locked-rotor torque is 0.416 Nm and its weight is 0.27 kg.

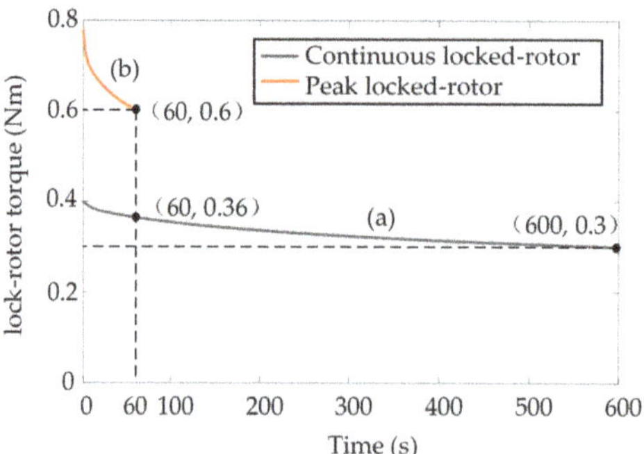

Figure 15. The torque curves: (**a**) continuous locked-rotor condition for 10 min; (**b**) peak locked-rotor condition for 1 min considering temperature.

5. Prototype Motor and Experiments

The manufacture of the prototype and the experimental results for the proposed BEPMTM are presented in this section. The designed motor and the experiment bench are presented in Figure 16. In order to verify the performance of the BWPMTM, no-load and lock-load experiments were carried out. All the performances of the BWPMTM prototype were verified successfully.

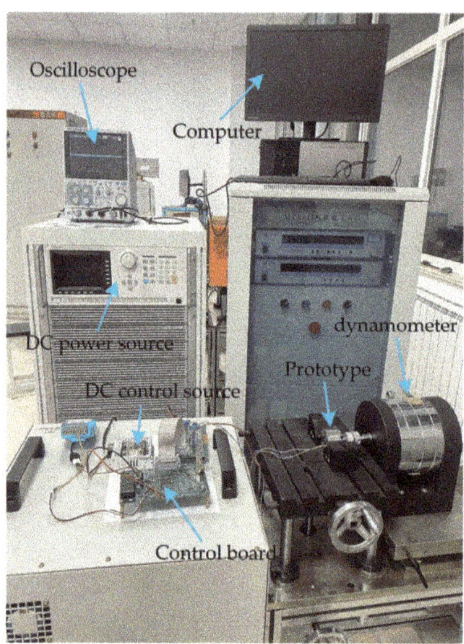

Figure 16. Prototype and experiment platform of the BWPMTM prototype.

5.1. No-Load Experiment

Figure 17 shows the back EMF measurement of the BWPMTM under the no-load condition. The experimental line back EMF coefficient was 0.318 V/Hz. By comparing this with the simulated line back EMF coefficient of 0.298 V/Hz, it was found that the experimental result fits well with the FEM result. A small difference was observed due to the accuracy of the experimental system. The FEM result is slightly lower than the experimental ones, with a maximum error of 6.3%, showing the general agreement between the FEM and experiment.

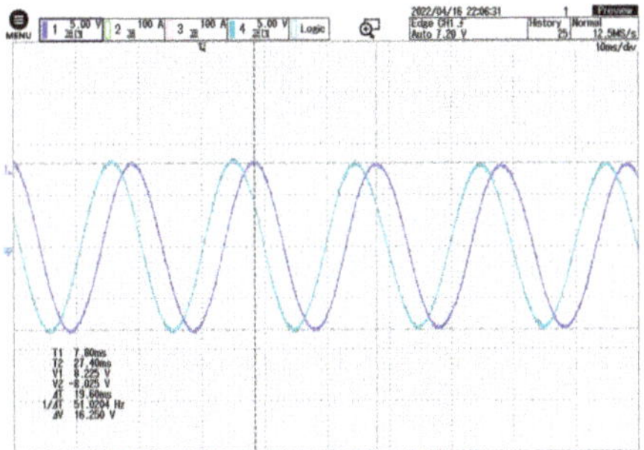

Figure 17. No-load back EMF.

5.2. Temperature Experiment

The temperature of the prototype under the two experimental locked-rotor conditions was measured using a noncontact infrared thermometer. The measured temperature curves versus time are shown in Figures 18 and 19, respectively. The initial temperature was 22.1 °C. As shown, the experimental temperature distribution of the BWPMTM model shows great agreement with the FEM. The maximum temperature of the shell was 123.7 °C under the continuous locked-rotor condition after 1 min. The maximum error rate between the experiment and the FEM was less than 6%. The maximum temperature was 136.9 °C under the peak locked-rotor condition after 10 min. The maximum error rate between the experiment and the FEM was less than 7%. From an engineering perspective, these errors are acceptable.

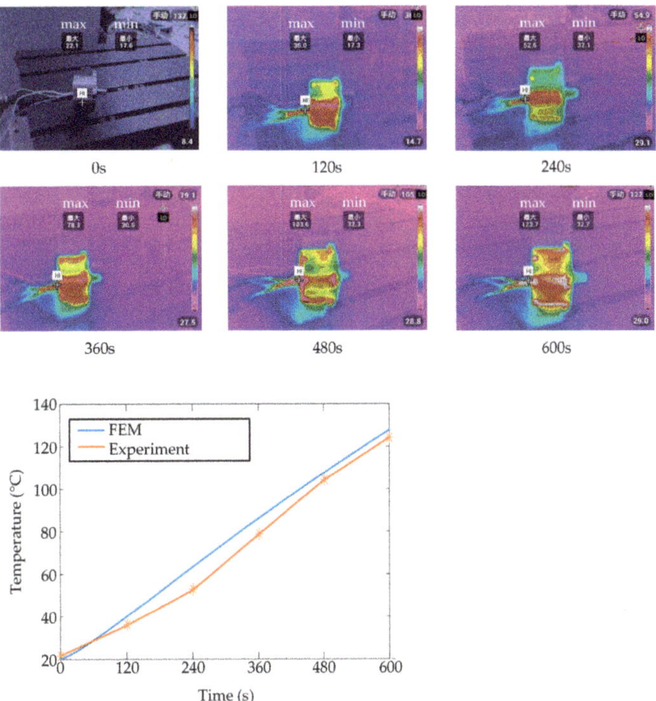

Figure 18. Temperature for the experiment under the continuous locked-rotor condition and the maximum temperature curve of the shell.

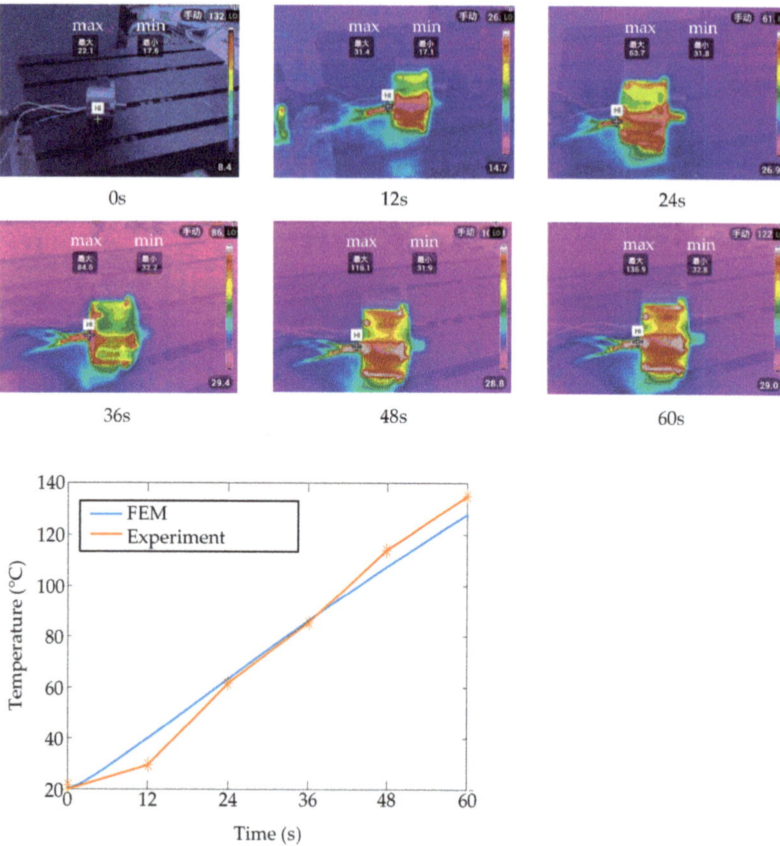

Figure 19. The temperature experiment under the peak locked-rotor condition and the maximum temperature curve of the shell.

6. Discussion

In the BWPMTM proposed in this paper, the outer surface of the stator is slotted. From the point of view of processing technics, the winding coils are embedded in the slots from the outer diameter of the stator, which makes processing more convenient and simpler. From the point of view of performance, the novel structure does not have the stator yoke and the size of the motor can be made smaller, which could increase motor power. Moreover, due to the simpler process, the slot space factor can be improved to a certain extent, making the motor performance superior. This paper proposes a limit analytical method, which takes minimum stack length as the objective function to determine the optimal parameters of the motor, considering four constraints of no-load back EMF, torque, temperature and slot space factor. In this design method, the optimal parameters of the motor are solved directly by known constraints, avoiding the calculation of large-scale searches through scanning and iteration. With this design method, the temperature and the slot space factor have reached the limit. This indicates that the motor parameters have been minimized and that motor materials have been fully used.

The electromagnetic performance was investigated with finite element electromagnetic simulation. Finally, experimental tests on the BWPMTM prototype were undertaken. The experimental results were in good agreement with the results of the FEM, which successfully verified the performance of the motor.

The FEM and experimental results verified the feasibility of the proposed structure and the rationality of the design scheme. The proposed limit analytical method can also serve as a reference for the design of other types of motor.

Author Contributions: Conceptualization, B.Z.; investigation, Y.C. and P.Y.; writing—original draft preparation, S.G.; software, S.G.; writing—review and editing, B.L.; supervision, C.Z. All authors have read and agreed to the published version of the manuscript.

Funding: This research was funded by the National Natural Science Foundation of China (52075307) and the Natural Science Foundation of Shandong Province (ZR2020QE217).

Conflicts of Interest: The authors declare no conflict of interest.

References

1. Onsal, M.; Cumhur, B.; Demir, Y.; Yolacan, E. Rotor design optimization of a new flux-assisted consequent pole spoke-type permanent magnet torque motor for low-speed applications. *IEEE Trans. Magn.* **2018**, *54*, 8206005. [CrossRef]
2. Jo, I.H.; Lee, J.; Lee, H.W.; Lee, J.B.; Lim, J.H.; Kim, S.H.; Park, C.B. A Study on MG-PMSM for High Torque Density of 45 kW–Class Tram Driving System. *Energies* **2022**, *15*, 1749. [CrossRef]
3. Gao, C.; Gao, M.; Si, J.; Hu, Y.; Gan, C. A novel direct-drive permanent magnet synchronous motor with toroidal windings. *Energies* **2019**, *12*, 432. [CrossRef]
4. Zhang, Z.; Yu, S.; Zhang, F.; Jin, S.; Wang, X. Electromagnetic and Structural Design of a Novel Low-Speed High-Torque Motor with Dual-Stator and PM-Reluctance Rotor. *IEEE Trans. Appl. Supercon.* **2020**, *30*, 5203605. [CrossRef]
5. Bacco, G.; Bianchi, N.; Luise, F. High-torque low-speed permanent magnet assisted synchronous reluctance motor design. In Proceedings of the 2019 IEEE International Electric Machines and Drives Conference (IEMDC), San Diego, CA, USA, 12–15 May 2019; pp. 644–649.
6. Wu, S.; Zhao, X.; Jiao, Z.; Luk, P.C.-K.; Jiu, C. Multi-objective optimal design of a toroidally wound radial-flux Halbach permanent magnet array limited angle torque motor. *IEEE Trans. Ind. Electron.* **2016**, *64*, 2962–2971. [CrossRef]
7. Chai, W.; Lipo, T.A.; Kwon, B.I. Design and optimization of a novel wound field synchronous machine for torque performance enhancement. *Energies* **2018**, *11*, 2111. [CrossRef]
8. Chu, W.Q.; Zhu, Z.Q. Reduction of on-load torque ripples in permanent magnet synchronous machines by improved skewing. *IEEE Trans. Magn.* **2013**, *49*, 3822–3825. [CrossRef]
9. Zhao, W.; Lipo, T.A.; Kwon, B.I. Torque pulsation minimization in spoke-type interior permanent magnet motors with skewing and sinusoidal permanent magnet configurations. *IEEE Trans. Magn.* **2015**, *51*, 8110804. [CrossRef]
10. Zhu, Z.Q.; Howe, D. Influence of design parameters on cogging torque in permanent magnet machines. *IEEE Trans. Energy Convers.* **2000**, *15*, 407–412. [CrossRef]
11. Demir, Y.; Aydin, M. A novel dual three-phase permanent magnet synchronous motor with asymmetric stator winding. *IEEE Trans. Magn.* **2016**, *52*, 8105005. [CrossRef]
12. Chen, N.; Ho, S.L.; Fu, W.N. Optimization of permanent magnet surface shapes of electric motors for minimization of cogging torque using FEM. *IEEE Trans. Magn.* **2010**, *46*, 2478–2481. [CrossRef]
13. Vicente, S.-S.; Auxiliadora, S.-G.; Manuel, B.-P.; José-Ramón, C.-B. Optimisation of Magnet Shape for Cogging Torque Reduction in Axial-Flux Permanent-Magnet Motors. *IEEE Trans. Energy Convers.* **2021**, *36*, 2825–2838.
14. Yu, H.; Yu, G.; Xu, Y.; Zou, J. Torque performance improvement for slotted limited-angle torque motors by combined SMA application and GA optimization. *IEEE Trans. Magn.* **2020**, *57*, 8200305. [CrossRef]
15. Zhang, J.; Zhang, B.; Feng, G.; Gan, B. Design and Analysis of a Low-Speed and High-Torque Dual-Stator Permanent Magnet Motor With Inner Enhanced Torque. *IEEE Access* **2020**, *8*, 182984–182995. [CrossRef]
16. Ma, P.; Wang, Q.; Li, Y.; Jiang, S.; Zhao, M. Research on Torque Ripple Suppression of the Slotted Limited Angle Torque Motor. *IEEE Trans. Magn.* **2020**, *57*, 8200106. [CrossRef]
17. Zou, J.B.; Yu, G.D.; Xu, Y.X.; Li, J.L.; Wang, Q. Design and analysis of a permanent magnet slotted limited-angle torque motor with special tooth-tip structure for torque performance improvement. In Proceedings of the 2015 IEEE International Conference on Applied Superconductivity and Electromagnetic Devices (ASEMD), Shanghai, China, 20–23 November 2015; pp. 246–247.
18. Kim, M.; Cho, S.; Lee, K.; Lee, J.; Han, J.; Jeong, T.; Kim, W.; Koo, D.; Lee, J. Torque density elevation in concentrated winding interior PM synchronous motor with minimized magnet volume. *IEEE Trans. Magn.* **2013**, *49*, 3334–3337. [CrossRef]
19. Chai, J.; Zhao, T.; Gui, X. Multi-Objective Optimization Design of Permanent Magnet Torque Motor. *World Electr. Veh. J.* **2021**, *12*, 131. [CrossRef]
20. Chung, S.U.; Kim, J.W.; Chun, Y.D.; Woo, B.C.; Hong, D.K. Fractional slot concentrated winding PMSM with consequent pole rotor for a low-speed direct drive: Reduction of rare earth permanent magnet. *IEEE Trans. Energy Convers.* **2015**, *30*, 103–109. [CrossRef]
21. Li, Y.; Zhu, Z.Q.; Li, G.J. Influence of stator topologies on average torque and torque ripple of fractional-slot SPM machines with fully closed slots. *IEEE Trans. Ind. Appl.* **2018**, *54*, 2151–2164. [CrossRef]

22. Hu, Y.; Zhu, S.; Xu, L.; Jiang, B. Reduction of Torque Ripple and Rotor Eddy Current Losses by Closed Slots Design in a High-speed PMSM for EHA Applications. *IEEE Trans. Magn.* **2022**, *58*, 8102206. [CrossRef]
23. Chen, H.; Qu, R.; Li, J.; Li, D. Demagnetization performance of a 7 MW interior permanent magnet wind generator with fractional-slot concentrated windings. *IEEE Trans. Magn.* **2015**, *51*, 2442263. [CrossRef]
24. De Donato, G.; Capponi, F.G.; Rivellini, G.A.; Caricchi, F. Integral-slot versus fractional-slot concentrated-winding axial-flux permanent-magnet machines: Comparative design, FEA, and experimental tests. *IEEE Trans. Ind. Appl.* **2012**, *48*, 1487–1495. [CrossRef]
25. Jia, G.X.; Zhao, H.C.; Shao, H.J. Simulation research on PMSM vector control system based on SVPWM. In Proceedings of the 2010 International Conference on Electrical and Control Engineering, Wuhan, China, 25–27 June 2010; pp. 1936–1940.

Article

Observer Based Improved Position Estimation in Field-Oriented Controlled PMSM with Misplaced Hall-Effect Sensors

Mengji Zhao, Quntao An *, Changqing Chen, Fuqiang Cao and Siwen Li

School of Electrical Engineering and Automation, Harbin Institute of Technology, Harbin 150001, China
* Correspondence: anquntao@hit.edu.cn

Abstract: Low resolution Hall-effect sensors have been commonly applied in PMSM drives for the reason of cost and volume. Generally, rotor speed and position are estimated inaccurately due to the installation error of the sensors. The inaccurate position degrades the performance of current control and also increases torque ripples, which aggravates mechanical vibration and noise. An improved dual observer is proposed in this paper to suppress the impact of misplaced Hall-effect sensors and improve estimation accuracy. By a cascading dual Luenberger observer and combining feedback decoupling control, the low-order noises produced by the deviation of Hall signals are effectively suppressed. The effectiveness of the proposed method is verified by experimental results.

Keywords: permanent magnet synchronous motor; Hall-effect sensors; speed and position estimation; dual observer

Citation: Zhao, M.; An, Q.; Chen, C.; Cao, F.; Li, S. Observer Based Improved Position Estimation in Field-Oriented Controlled PMSM with Misplaced Hall-Effect Sensors. *Energies* **2022**, *15*, 5985. https://doi.org/10.3390/en15165985

Academic Editor: Mario Marchesoni

Received: 11 July 2022
Accepted: 14 August 2022
Published: 18 August 2022

Publisher's Note: MDPI stays neutral with regard to jurisdictional claims in published maps and institutional affiliations.

1. Introduction

The permanent magnet synchronous motor (PMSM) has the advantages of high- power density, compact volume and excellent dynamic performance. It is increasingly used in various high-performance applications, such as aerospace, military, automotive, industrial and household products. An accurate rotor position is necessary for the field-oriented control (FOC) which is the main control algorithm for PMSM. It is usually measured by high-precision sensors such as a rotary transformer or encoder. However, the use of such sensors will increase the cost, weight and volume of the system, which is in conflict with the high-power density of the motor system. In order to improve the power density, the ultimate goal is to realize the high performance sensorless control of PMSM. However, there are still some disadvantages in the high dynamic motor drive systems. The sensorless control scheme is used in pump or fan systems, or as a redundant alternative in case of position sensor failure in some high reliability motor systems.

The Hall-effect sensor has the advantages of small size, low cost and strong anti-interference ability, which can also provide the speed and position information. In some cost sensitive fields, low resolution Hall-effect sensors have been widely used. It is a good choice that enables both the high-power density and the high reliability to be considered. The Hall-effect sensor- based speed and position measurements can be used in field-oriented controlled PMSM drives, which is a sinusoidal current drive scheme of PMSM with low resolution sensors. As the position directly provided by the low-resolution Hall-effect sensors cannot meet the accuracy requirement in the FOC, it is necessary to estimate the rotor position accurately according to the Hall signals. Due to processing and installation errors of the Hall sensors, the asymmetric output signals will lead to increasement of speed and position estimation errors and affect the performance of the system. It is equally important to weaken the impact of misplaced Hall-effect sensors.

Generally, the PMSM integrates three Hall-effect sensors, which can only provide six different states in an electrical cycle. The six states can be used to calculate or observe rotor speed and position. At present, the interpolation method, the filter method and the observer method, which are reviewed in [1], are usually employed to estimate the rotor speed and position based on low resolution position sensors. In [2–4], estimation strategies based on average speed, average acceleration and least squares method were proposed, respectively. The interpolation algorithm is employed in these strategies, which requires the use of multiple different pieces of Hall state information. A filter designed in the synchronous rotating coordinate system of the motor was proposed in [5], which has a better frequency selection characteristic under the variable frequency [6,7]. However, when the rotor speed changes greatly or the Hall signals deviate from the ideal ones, the filtering effect will decrease obviously. As non-model methods, the interpolation method and the filter one have advantages of simple algorithm and easy implementation, but the estimation results have serious delays. In [8], an extended Kalman filter was used to extract the fundamental frequency component of Hall signals. However, the algorithm is complex and not practical in low-cost motor systems. In [9], an observer combined with Hall signals was proposed to obtain the back electromotive force (EMF) by using the stator voltages and currents, and then the high-precision position can be achieved. On the basis of [9], some improvements were made in [10–14] to enhance performance of the observers. In [15], a vector-tracking observer was presented, and detailed analyses were made and some improvement measures were presented in [16–19]. In [20], a dual observer was proposed to suppress the low-order harmonics introduced by the misplaced Hall-effect sensors. The first employed observer is used for the speed estimation and the second one is for the rotor position. In [21], a design scheme of the Hall signal processing module using an FPGA was proposed, which provided a new solution for the low-cost motor drive with sinusoidal currents. In [22,23], a Hall-effect sensor-based fault diagnosis and protection strategy were also considered. In [24], a position observer combined with non-smooth feedback phase locked loop was employed to obtain continuous rotor position and speed. An angle representation method with the Hall-effect sensors based on the first-order Taylor series approximation was proposed in [25]. In [26], available techniques for motion control, monitoring and diagnosis of PMSM using Hall-effect sensors were summarized. In [27,28], the rotor position estimation error caused by misplaced Hall sensors was analyzed, and the least square estimation method (LSM) as well as back EMF vector tracking method were compared in the steady and dynamic situations, respectively.

In this paper, an improved dual Luenberger observer is proposed for PMSM drive system based on the low-resolution Hall-effect sensors. The 5th, 7th, 11th and 13th harmonics are used in the first observer by means of feedback and then can be eliminated, and the obtained result is directly input into the second observer. By the feedback decoupling and the dual observer, the low-order harmonics introduced by the misplaced Hall-effect sensors are suppressed effectively, and the position estimation accuracy is improved. Finally, the estimation method is verified by the simulations and experiments.

2. Field-Oriented Control of SPMSM Based on Hall-Effect Sensors

In the field-oriented control of PMSM, the direct and quadrature axis currents can be obtained and controlled independently by using the measured three-phase currents and coordinate transformation, and excellent control performance can be achieved. The surface mounted PMSM (SPMSM) has the characteristic of equal d and q axis inductances due to the symmetrical magnetic circuit. When the $i_d = 0$ control is adopted, all the current can be controlled to generate electromagnetic torque by controlling the d-axis current to zero, and the maximum torque per ampere control of SPMSM can be realized. In this control scheme, it is necessary to have a precise angle of rotor flux, which can be obtained by detecting the position of the rotor. Generally, the position sensors are a rotary transformer and a photoelectric encoder. In volume-limited systems, the Hall-effect sensors are commonly used to detect the rotor position and speed, and the field-oriented control of the PMSM drive based on the Hall-effect sensors is shown in Figure 1.

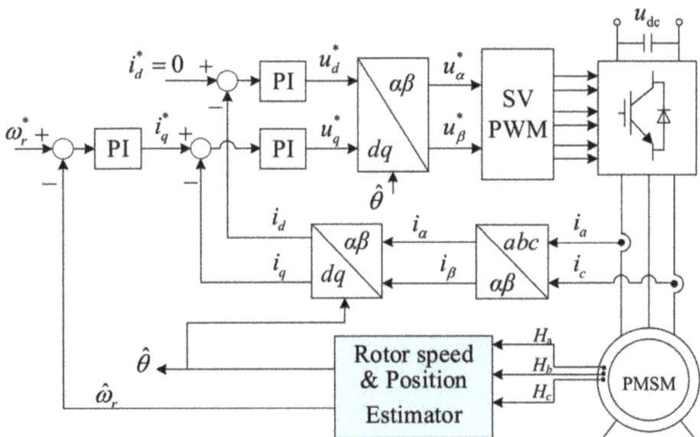

Figure 1. Block diagram of PMSM FOC based on Hall-effect sensors.

Inside the PMSM, three Hall-effect sensors are placed on the stator at intervals of 120 electrical degrees. When the rotor rotates, the Hall-effect sensors output three square wave signals along with the change of magnetic field, and the frequency of Hall signals is positively correlated with the speed. With the help of certain estimation methods, the rotor position and speed can be obtained according to the Hall signals and used in the FOC consequently.

3. Conventional Rotor Position Estimation Methods

3.1. Average Speed Method

The interpolation is a common method in the function fitting of discrete data points. All the data for the whole interval can be obtained through the known data points. According to the different fitting functions, the interpolation methods used for Hall signal processing are usually divided into average speed method, average acceleration method and least squares method. In practical applications, the most widely used one is the average speed method for easy implementation, in which the motor speed is assumed to be a constant during one Hall state.

As mentioned above, three Hall-effect sensors are usually installed inside the stator at intervals of 120 electrical degrees. The sensors output square wave signals according to the polarity of the rotating magnets, and these signals are recorded as H_a, H_b and H_c, as shown in Figure 2, respectively. When the south magnet pole approaches, the output signal switches to low level, and otherwise it turns to high level.

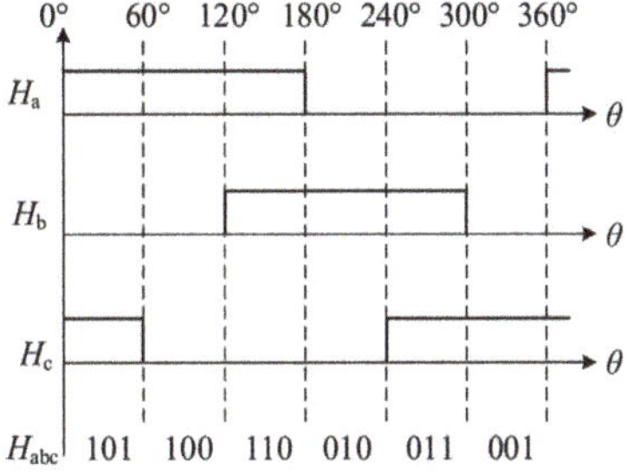

Figure 2. The waveforms of three Hall signals.

Due to the inertia of the rotor, the speed usually cannot change violently during a Hall state. Therefore, in the average speed method, the speed is assumed to be constant in one Hall state, and the average speed of the previous Hall state is used to calculate the rotor position inside the current Hall state by interpolation. The average speed in one Hall state can be calculated by

$$\omega_n = \frac{\pi}{3T_n} \tag{1}$$

where ω_n is the average speed, and T_n is the time interval of one Hall state.

The edge of each Hall signal corresponds to an exact position. According to the average speed, Hall edge position, the rotor position at T_r can be calculated by

$$\theta_{(t)} = \theta_n + \omega_{(n-1)} \times T_r \tag{2}$$

where θ_n is the position on the edge, and T_r is the running time from the Hall edge.

The position estimation strategy based on the average speed performs well in the high-speed range. However, under low-speed or dynamic situations, delay of the calculated speed will reduce accuracy of the position estimation.

3.2. Luenberger Observer

The Luenberger observer has superior dynamic characteristics and high accuracy in estimation, and also has an intrinsic zero-delay tracking capability. The structure of the Luenberger observer is shown in Figure 3, and it constructs an observed model by using the known or measured system parameters. The output error between the observed model and the real system can converge to zero by the feedback matrix, and then model can track the real system.

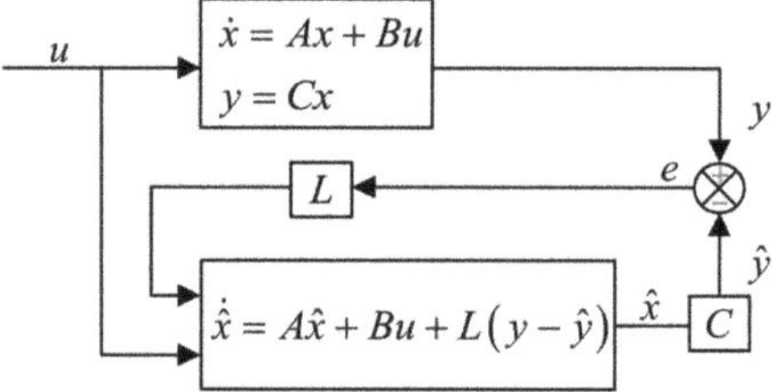

Figure 3. Luenberger Observer.

The mechanical equation of PMSM can be given by

$$\begin{cases} T_e - T_L = J\frac{d\omega_m}{dt} + B\omega_m \\ \frac{d\theta_m}{dt} = \omega_m = \theta/P_n \end{cases} \tag{3}$$

where T_e and T_L are the electromagnetic and load torque, respectively. J and B are the inertia and friction coefficient of the motor, respectively. ω_m and θ_m are the mechanical rotor speed and position, respectively. θ is the electrical position, and P_n is the pole pairs.

Generally, B is very small and can be ignored. Thus, the mechanical equation can be rewritten as

$$\begin{cases} \frac{d}{dt}\begin{bmatrix} \theta \\ \omega_m \\ T_L \end{bmatrix} = \begin{bmatrix} 0 & P_n & 0 \\ 0 & 0 & -1/J \\ 0 & 0 & 0 \end{bmatrix}\begin{bmatrix} \theta \\ \omega_m \\ T_L \end{bmatrix} + \begin{bmatrix} 0 \\ 1/J \\ 0 \end{bmatrix}T_e \\ \theta = \begin{bmatrix} 1 & 0 & 0 \end{bmatrix}\begin{bmatrix} \theta \\ \omega_m \\ T_L \end{bmatrix} \end{cases} \tag{4}$$

By replacing the J of the actual system with the off-line measured value $\hat{J}$, the observer equation can be obtained as

$$
\begin{bmatrix} \dot{\hat{\theta}} \\ \dot{\hat{\omega}}_m \\ \dot{\hat{T}}_L \end{bmatrix} = \begin{bmatrix} 0 & P_n & 0 \\ 0 & 0 & -1/\hat{J} \\ 0 & 0 & 0 \end{bmatrix} \begin{bmatrix} \hat{\theta} \\ \hat{\omega}_m \\ \hat{T}_L \end{bmatrix} + \begin{bmatrix} 0 \\ 1/\hat{J} \\ 0 \end{bmatrix} T_e + \begin{bmatrix} l_1 \\ l_2 \\ l_3 \end{bmatrix} (\theta - \hat{\theta})
\tag{5}
$$

where $L = \begin{bmatrix} l_1 & l_2 & l_3 \end{bmatrix}^T$ is the state feedback gain matrix, and ^ stands for the estimated values of these variables.

The characteristic equation of the observer system matrix is

$$
s^3 + l_1 s^2 + l_2 P_n s - l_3 P_n / \hat{J} = 0
\tag{6}
$$

In order to ensure that the observed state variables can converge to the real value, it is necessary to let the roots of the characteristic equation be located on the left half plane. Supposing the expected poles are α, β, γ, respectively, which are located on the negative real axis, they meet $\alpha < 0$, $\beta < 0$, $\gamma < 0$. Then the expected characteristic equation can be rewritten as

$$
s^3 - (\alpha + \beta + \gamma)s^2 + (\alpha\beta + \beta\gamma + \gamma\alpha)s - \alpha\beta\gamma = 0
\tag{7}
$$

The expression of state feedback gain matrix L is

$$
\begin{cases}
l_1 = -(\alpha + \beta + \gamma) \\
l_2 = (\alpha\beta + \beta\gamma + \gamma\alpha)/P_n \\
l_3 = \alpha\beta\gamma\hat{J}/P_n
\end{cases}
\tag{8}
$$

For convenience of calculation and analysis, α is set as a triple root ($\alpha = \beta = \gamma < 0$), and the matrix L can be rewritten as

$$
\begin{cases}
l_1 = -3\alpha \\
l_2 = 3\alpha^2/P_n \\
l_3 = \hat{J}\alpha^3/P_n
\end{cases}
\tag{9}
$$

Thus, expression of the state observer is

$$
\begin{cases}
d\hat{\theta}/dt = P_n\hat{\omega}_m + l_1(\theta - \hat{\theta}) \\
d\hat{\omega}_m/dt = (T_e - \hat{T}_L)/\hat{J} + l_2(\theta - \hat{\theta}) \\
d\hat{T}_L/dt = l_3(\theta - \hat{\theta})
\end{cases}
\tag{10}
$$

According to Equation (10), the block diagram of the Luenberger observer can be obtained, as shown in Figure 4.

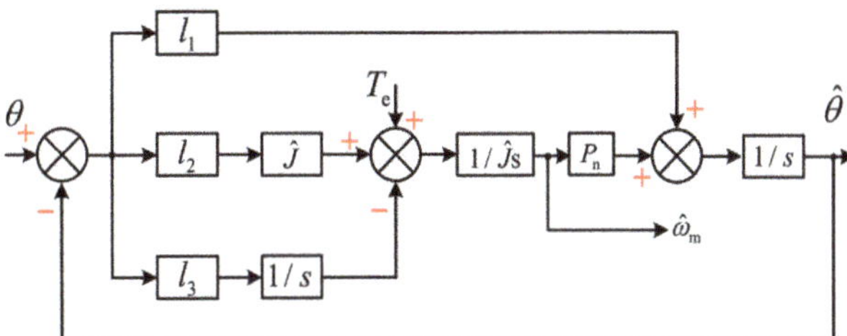

Figure 4. Block diagram of Luenberger observer.

The $\hat{\theta}$ and $\hat{\omega}_m$ can be written as Equations (11) and (12).

$$
\hat{\theta} = \frac{\hat{J}l_1 s^2 + \hat{J}P_n l_2 s - P_n l_3}{\hat{J}s^3 + \hat{J}l_1 s^2 + \hat{J}P_n l_2 s - P_n l_3}\theta + \frac{P_n s}{\hat{J}s^3 + \hat{J}l_1 s^2 + \hat{J}P_n l_2 s - P_n l_3}T_e
\tag{11}
$$

$$\hat{\omega}_{\mathrm{m}} = \frac{\hat{J}\hat{l}_2 s^2 - l_3 s}{\hat{J}s^3 + \hat{J}l_1 s^2 + \hat{J}P_{\mathrm{n}}l_2 s - P_{\mathrm{n}}l_3}\theta + \frac{s^2 + l_1 s}{\hat{J}s^3 + \hat{J}_1 s^2 + \hat{J}P_{\mathrm{n}}l_2 s - P_{\mathrm{n}}l_3}T_{\mathrm{e}} \tag{12}$$

when $J \approx \hat{J}$, the gain matrix L is taken according to Equations (10)–(12) can be rewritten as

$$\hat{\theta} = \frac{-3\alpha s^2 + 3\alpha^2 s - \alpha^3}{(s-\alpha)^3}\theta + \frac{P_{\mathrm{n}}s}{\hat{J}(s-\alpha)^3}T_{\mathrm{e}} \tag{13}$$

$$\hat{\omega}_{\mathrm{m}} = \frac{3\alpha^2 s - \alpha^3}{P_{\mathrm{n}}(s-\alpha)^3}\omega_{\mathrm{m}} + \frac{s^2 - 3\alpha s}{\hat{J}(s-\alpha)^3}T_{\mathrm{e}} \tag{14}$$

where, α can be used to adjust the observer gain. When α takes value of 50, 100 and 150, respectively, the Bode diagram of the observer are shown in Figure 5. It can be seen from the Bode diagram that the observer has the function of low-pass filtering, which can suppress the high-order harmonics in the input signal. The bandwidth increases with the increasing of α, and the dynamic response of the system improves as well. However, the ability to suppress high-frequency noises will decline. The bandwidth of the observer changes with α accordingly. In practical applications, α can be reasonably selected according to the speed range. When the bandwidth of the observer is 2 to 5 times the input signal frequency, the system can achieve both good tracking performance and suppression ability to high-frequency noises.

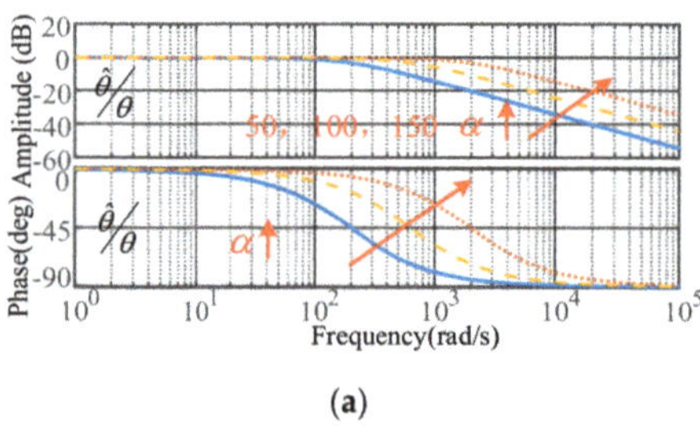

(a)

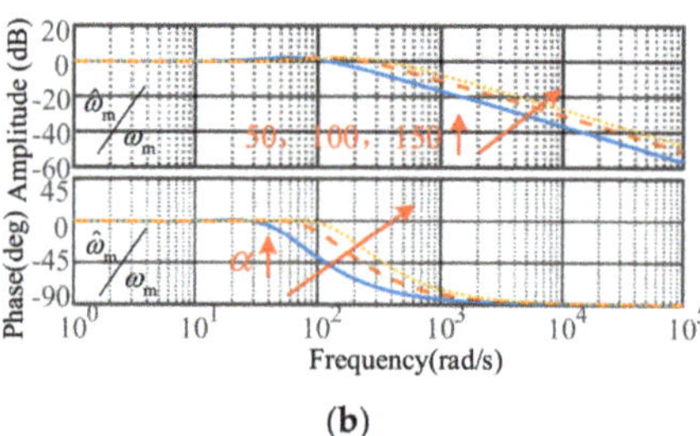

(b)

Figure 5. (**a**) Bode diagram of the observed position with different gain α; (**b**) Bode diagram of the observed speed with different gain α.

3.3. Luenberger Observer with Feedback Decoupling

Hall position and torque are required as the inputs in the Luenberger observer in order to obtain the position and speed of the motor. The torque can be calculated according to the three-phase currents, but the discrete Hall position as the input will lead to the fluctuation of observation.

The Hall position signals can be regarded as a discrete position vector rotating in space with an electrical angle increment of 60 degrees, as shown in Figure 6. If the Hall position vector is decomposed into two orthogonal components, it can be written as follows:

$$\vec{H} = e^{j\Delta\theta} = H_\alpha + jH_\beta \,(\Delta\theta = k\pi/3,\ k = 0,1,2\ldots) \tag{15}$$

where $\vec{H}$ is the hall position vector, H_α and H_β are two orthogonal components. The Hall vector transformation is implemented by

$$\begin{bmatrix} H_\alpha \\ H_\beta \end{bmatrix} = \begin{bmatrix} 1 & -\frac{1}{2} & -\frac{1}{2} \\ 0 & -\frac{\sqrt{3}}{2} & \frac{\sqrt{3}}{2} \end{bmatrix} \begin{bmatrix} H_{\mathrm{a}} \\ H_{\mathrm{b}} \\ H_{\mathrm{c}} \end{bmatrix} \tag{16}$$

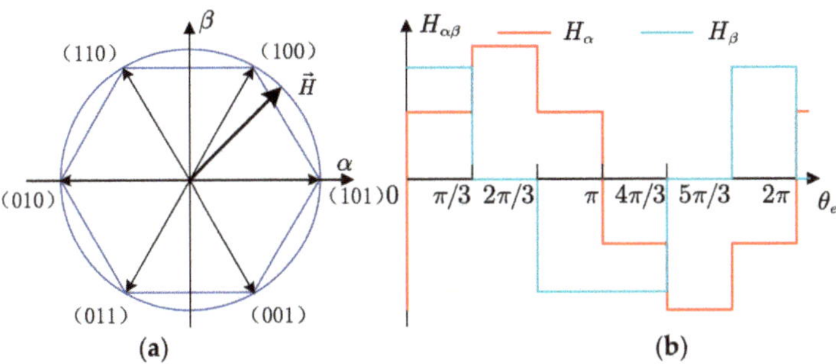

Figure 6. (**a**) Discrete Hall position vector; (**b**) Two orthogonal components.

The Fourier transform of H_α and H_β can be expressed as

$$
\begin{aligned}
H_\alpha &= \sum_{n=1}^{\infty} \frac{1}{n\pi}\left[1 + \cos\left(\frac{n\pi}{3}\right) - \cos\left(\frac{2n\pi}{3}\right) - \cos(n\pi)\right]\sin(n\omega t) \\
&= \frac{3}{\pi}\sin(\omega t) + \frac{3}{5\pi}\sin(5\omega t) + \frac{3}{7\pi}\sin(7\omega t) + \frac{3}{11\pi}\sin(11\omega t) + \dots
\end{aligned}
\tag{17}
$$

$$
\begin{aligned}
H_\beta &= \sum_{n=1}^{\infty} \frac{\sqrt{3}}{n\pi}\left[\sin\left(\frac{n\pi}{3}\right) + \sin\left(\frac{2n\pi}{3}\right)\right]\cos(n\omega t) \\
&= \frac{3}{\pi}\cos(\omega t) - \frac{3}{5\pi}\cos(5\omega t) + \frac{3}{7\pi}\cos(7\omega t) - \frac{3}{11\pi}\cos(11\omega t) + \dots
\end{aligned}
\tag{18}
$$

Thus, their fundamental components can be regarded as two-phase orthogonal sine and cosine signals, as shown in Figure 7.

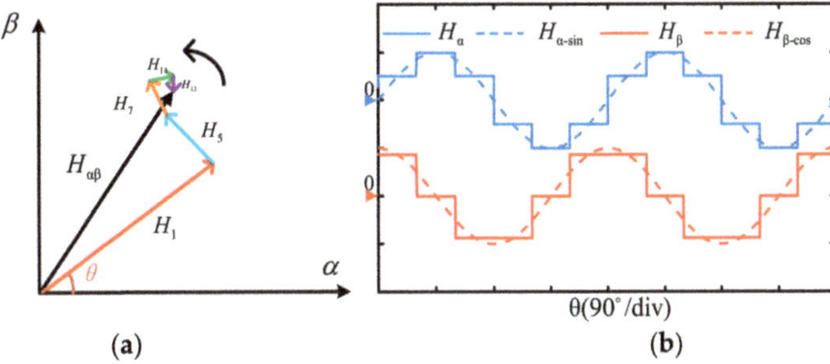

Figure 7. (**a**) The Hall position vector decomposition; (**b**) The fundamental components of H_α, H_β.

When the components of the input are known, the low-order harmonics can be reduced by using harmonic feedback decoupling. The improved observer structure is shown in Figure 8. According to the phase angle at the moment, the 5th, 7th, 11th and 13th harmonics are directly calculated and fed back to the input. The low-order harmonics are subtracted from the input, and the remaining high-order harmonics can be well suppressed by the low-pass filtering characteristic of the observer. The function of feedback decoupling is to further increase the gain of the observer, reduce the low-frequency noises of the input, and obtain a better dynamic performance while reducing the error of rotor position estimation [29].

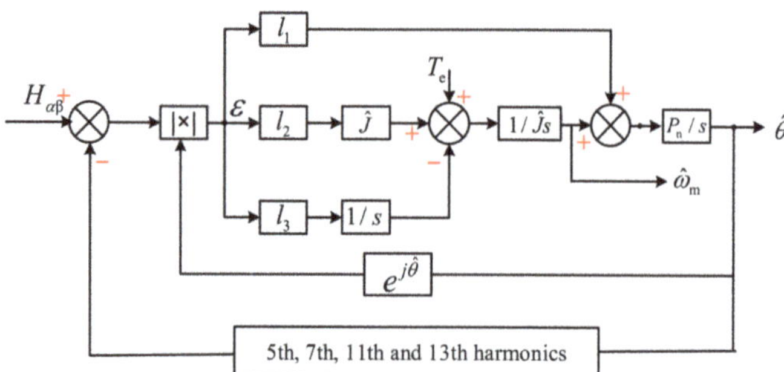

Figure 8. Block diagram of Luenberger observer with feedback decoupling.

4. Improved Dual Observer

4.1. Misplaced Position Sensors

Due to the mechanical errors either in the manufacturing or assembling process, each Hall-effect sensor may be misplaced by a certain mechanical angle regarding its corresponding phase winding. There is often a certain deviation between the output signal of Hall-effect sensors and the ideal signal.

Figure 9 shows the output signals of three Hall-effect sensors when considering the deviation. In this case, the Hall states are no longer a 60° electrical angle, which will produce position and speed estimation errors. Especially when the rotor position deviates, the sinusoidity of the motor current will be seriously damaged, which will aggravate the speed fluctuation and reduce the dynamic performance.

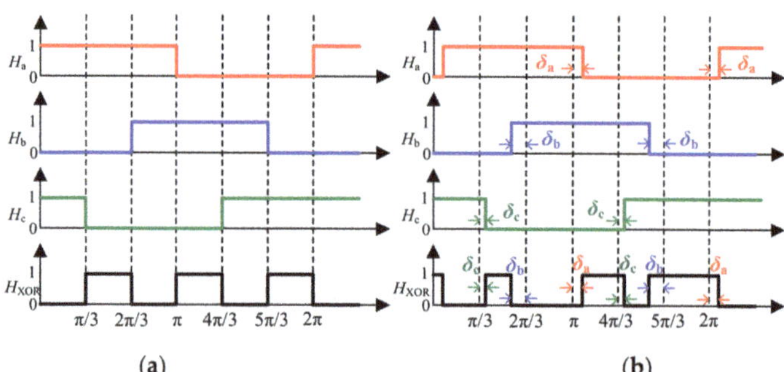

Figure 9. (**a**) Ideal Hall signals; (**b**) Hall signals with deviation.

4.2. The Proposed Dual Observer

The observer with low-pass filtering property can suppress noises of the input rotor position angle. When there is a deviation in the Hall signal, the harmonics feedback decoupling can not effectively suppress the low harmonics, resulting in an error of rotor position estimation. To solve this problem, a novel position estimation method with a cascaded dual observer is proposed. By taking the rotor position obtained by the first observer as the input of the second observer, the low-order harmonics can be further suppressed, and the influence of deviation will be reduced significantly. The block diagram of the dual observer is shown in Figure 10, where $F = \begin{bmatrix} f_1 & f_2 & f_3 \end{bmatrix}^T$ is the state feedback gain matrix of the second observer.

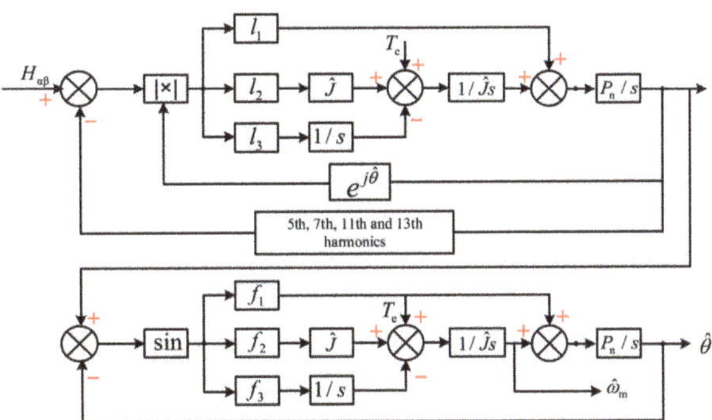

Figure 10. Block diagram of the proposed dual observer.

In the first observer, the 5th, 7th, 11th and 13th harmonics are fed back to the input, and the obtained position is used as input of the second observer. Through the feedback decoupling and double observers, the noises and observation errors can be reduced. The two observers have similar structure, and the same principle can be followed in parameter selection. When $L = F$, the dual observer can improve the observation accuracy of speed and rotor position, rather than affecting the anti-disturbance performance of the system since the same gain is chosen in the two observers.

In order to evaluate the performance of the improved observer, the observer in Figure 8 and the proposed dual observer are compared by the simulation, in which the parameters are shown as Table 1.

Table 1. Parameters of the simulation.

Parameters	Value (Unit)
Polar pairs P_N	5
Flux ψ_f	0.022 (Wb)
Stator resistance R	0.18 (Ω)
Stator inductance L	0.35 (mH)
Rated speed n_N	2000 (rpm)
Rated current i_N	7 (A)
Inertia J	0.0001 (kg.m^2)
Hall average deviation	2 (deg)

Figure 11 shows the estimated waveforms of rotor speed and position of the conventional Luenberger observer with feedback decoupling and the improved dual observer, respectively, when the motor runs at 1200 rpm. From Figure 11a,b, it can be seen that the observation accuracy of the Luenberger observer is affected by the Hall signal deviation. The speed estimation error is 28 rpm, and the maximum error of rotor position is 5.5° of electrical angle. The waveforms of the dual observer ($\alpha = 250$) are shown in Figure 11c,d. The error in the estimated speed is significantly reduced, to about 12 rpm, and the rotor position error is reduced to within 3°.

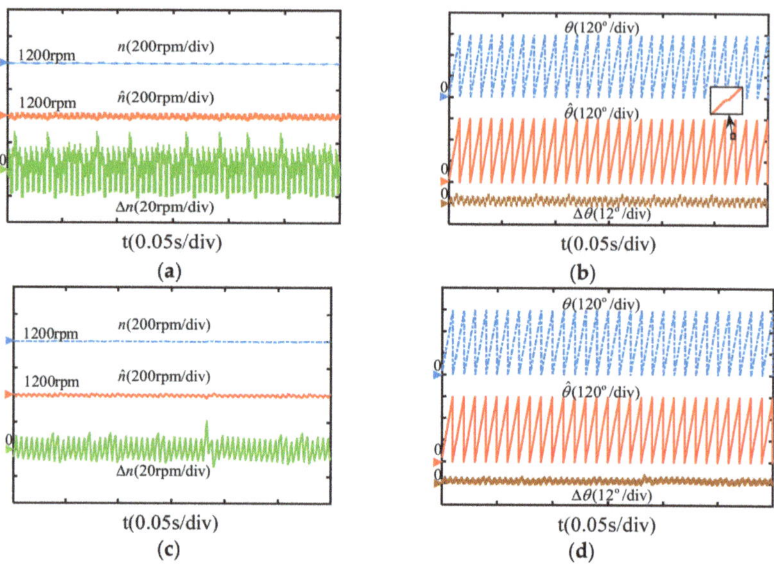

Figure 11. (**a**) Speed waveforms of Luenberger observer; (**b**) Position waveforms of Luenberger observer; (**c**) Speed waveforms of the dual observer; (**d**) Position waveforms of the dual observer.

5. Experimental Verification

Experimental platform of the PMSM drive with Hall-effect sensors is shown in Figure 12. It consists of the power part, the control part and the measurement part. The power drive module is realized by a three-phase half bridge circuit composed of Infineon power MOSFET. The motor control algorithm is carried out in the TMS320F28375D chip of TI. In order to realize the accurate measurement of speed and rotor position, a 2500-line encoder and a FPGA-based measurement board are used.

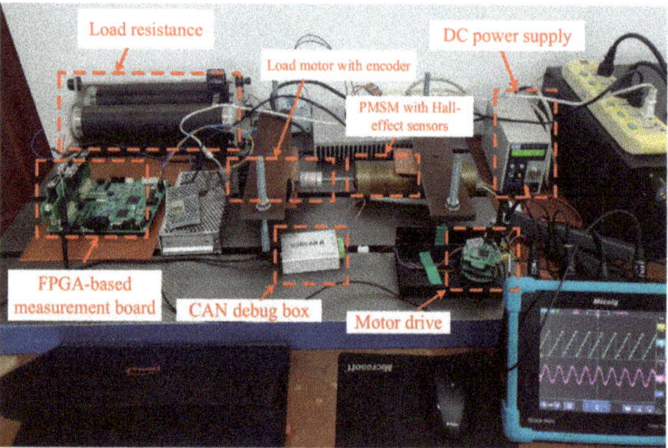

Figure 12. The PMSM drive experimental platform.

Firstly, the deviation of the Hall signals of the motor is tested. When the motor rotates at an approximately constant speed and the states of Hall signals are measured as shown in Figure 13. Theoretically, the phase angle of each Hall state is 60° electrical angle, but they are unequal obviously for the test motor.

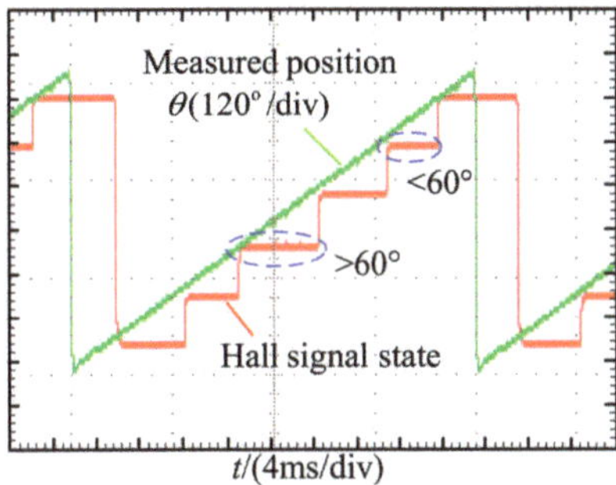

Figure 13. The deviation of the Hall signal state.

The position and motor current waveforms of the average speed method are shown in Figure 14. It can be clearly seen from Figure 14a that the estimated rotor position is unsmooth due to the misplaced Hall-effect sensors. The fluctuant rotor position will cause current distortion and increase lower harmonic components, as shown in Figure 14b.

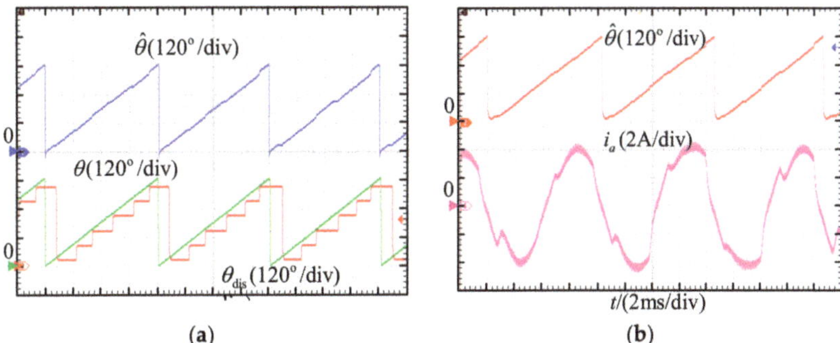

Figure 14. Experimental results of the average speed method. (**a**) Rotor position; (**b**) Phase current.

Figure 15 shows the estimated rotor position, one phase current and its FFT analysis obtained by using the Luenberger observer with feedback decoupling and the dual observer, respectively. According to Figure 15a, when the motor runs at 750 rpm, the error of rotor position is obvious, and the current is distorted due to the error. From the FFT analysis in Figure 15b, it can be seen that the phase current contains low-order harmonics. The results obtained by the dual observer are shown in Figure 15c,d. The phase current is effectively improved and the current THD is decreased significantly.

To evaluate the dynamic performance of the proposed observer, speed step and load step responses are tested respectively. In the acceleration and deceleration experiments, the motor speed steps from 750 rpm to 1500 rpm with the 0.5 N·m load, and then steps back to 750 rpm after 0.5 s. The actual, estimated speed and their error obtained by the average speed method and the dual observer are shown in Figure 16, respectively. It can be seen that the estimated speed of the dual observer has better steady-state performance and dynamic response compared with the average speed method.

Figure 17 shows the speed and current waveforms with the load step changes. When the motor runs at 1500 rpm, the load torque suddenly decreases from 0.5 N·m to zero and then back to 0.5 N·m again. During the load step changes, the estimated speed of dual observer closely follows the measured speed. Therefore, the improved dual observer does not affect the dynamic performance of the system and can well improve accuracy of the position and speed estimations.

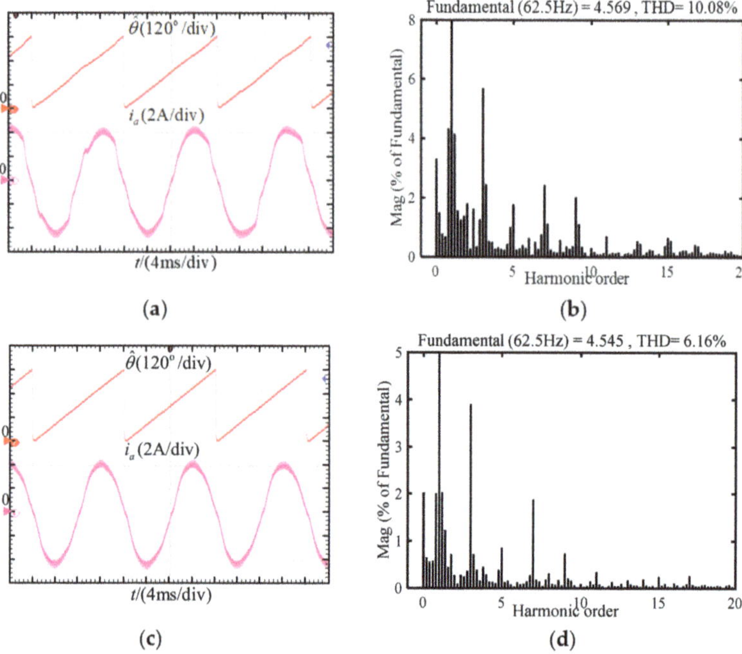

Figure 15. (**a**) Position and current with Luenberger observer; (**b**) FFT analysis of phase current with Luenberger observer; (**c**) Position and current with the dual observer; (**d**) FFT analysis of phase current with the dual observer.

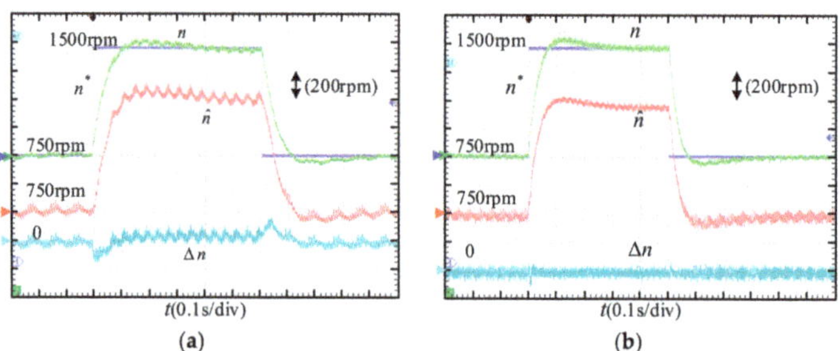

Figure 16. The experimental results under step speed. (**a**) Average speed method; (**b**) Dual observer method.

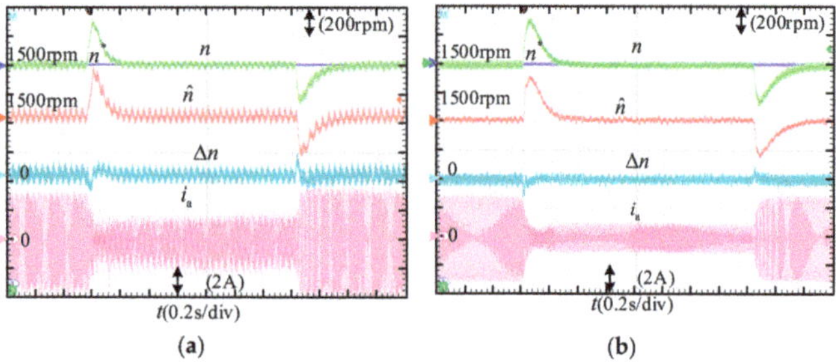

Figure 17. The experimental results under step load. (**a**) Average speed method; (**b**) Dual observer method.

6. Conclusions

This paper analyzes the PMSM rotor position estimation strategy and implementation method based on the low-resolution Hall-effect sensors. In order to improve rotor position estimation accuracy, an improved dual observer scheme is proposed in this paper. By cascading two observers, the proposed scheme can effectively suppress the low-order noises introduced by the deviation of Hall signal, and consequently the estimation errors of speed and position can be reduced significantly. The experimental results show that the rotor position estimation and the current sinusoidal characteristic can be improved by the proposed scheme compared with the conventional average speed method and the Luenberger observer with feedback decoupling, and also the proposed scheme has an unattenuated dynamic performance. Along with decrease of the speed, bandwidth of the two observers can be adjusted in order to suppress lower noises and improve estimation performance.

The rotor position estimation technology based on low-resolution Hall-effect sensors is a low-cost and high-reliability drive scheme for PMSMs. The observer method with closed-loop characteristic has better tolerance to Hall sensor installation errors. It is necessary to consider how to improve estimation performance under low speed and/or speed dynamics in the following research.

Author Contributions: Data curation, M.Z.; Investigation, M.Z. and F.C.; Project administration, Q.A.; Software, M.Z. and S.L.; Supervision, Q.A.; Validation, C.C. All authors have read and agreed to the published version of the manuscript.

Funding: This research received no external funding.

Institutional Review Board Statement: Not applicable.

Informed Consent Statement: Not applicable.

Data Availability Statement: Not applicable.

Conflicts of Interest: The authors declare no conflict of interest.

References

1. Wang, J.; Jiang, Q.; Xiong, D. Review of Rotor Position and Speed Estimation Method of PMSM with Hall Sensor. In Proceedings of the 2021 IEEE 16th Conference on Industrial Electronics and Applications (ICIEA), Chengdu, China, 1–4 August 2021; pp. 1832–1837.
2. Morimoto, S.; Sanada, M.; Takeda, Y. Sinusoidal Current Drive System of Permanent Magnet Synchronous Motor with Low Resolution Position Sensor. In Proceedings of the IAS '96, Conference Record of the 1996 IEEE Industry Applications Conference Thirty-First IAS Annual Meeting, San Diego, CA, USA, 6–10 October 1996; Volume 1, pp. 9–14.
3. Capponi, F.G.; De Donato, G.; Del Ferraro, L. Brushless AC Drive Using an Axial Flux Synchronous PM Motor with Low Resolution Position Sensors. In Proceedings of the 2004 IEEE 35th Annual Power Electronics Specialists Conference (IEEE Cat. No.04CH37551), Aachen, Germany, 20–25 June 2004; Volume 3, pp. 2287–2292.
4. Brown, R.H.; Schneider, S.C.; Mulligan, M.G. Analysis of Algorithms for Velocity Estimation from Discrete Position versus Time Data. *IEEE Trans. Ind. Electron.* **1992**, *39*, 11–19. [CrossRef]
5. Dalala, Z.M.; Cho, Y.; Lai, J.-S. Enhanced Vector Tracking Observer for Rotor Position Estimation for PMSM Drives with Low Resolution Hall-Effect Position Sensors. In Proceedings of the 2013 International Electric Machines & Drives Conference, Chicago, IL, USA, 12–15 May 2013; pp. 484–491.
6. Liu, G.; Chen, B.; Song, X. High-Precision Speed and Position Estimation Based on Hall Vector Frequency Tracking for PMSM With Bipolar Hall-Effect Sensors. *IEEE Sens. J.* **2019**, *19*, 2347–2355. [CrossRef]
7. Lim, J.-H.; Park, S.M.; Hyon, B.J.; Park, J.S.; Kim, J.-H.; Choi, J.-H. Study on PMSM Rotor Position Compensation Method for Hall-Effect Sensor Installation Error Using WBG- Based Inverter. In Proceedings of the 2021 24th International Conference on Electrical Machines and Systems (ICEMS), Gyeongju, Korea, 31 October–3 November 2021; pp. 1844–1848.
8. Bolognani, S.; Tubiana, L.; Zigliotto, M. Extended Kalman Filter Tuning in Sensorless PMSM Drives. *IEEE Trans. Ind. Appl.* **2003**, *39*, 1741–1747. [CrossRef]
9. Batzel, T.D.; Lee, K.Y. Commutation Torque Ripple Minimization for Permanent Magnet Synchronous Machines with Hall Effect Position Feedback. *IEEE Trans. Energy Convers.* **1998**, *13*, 257–262. [CrossRef]

10. Zaim, S.; Martin, J.P.; Nahid-Mobarakeh, B.; Meibody-Tabar, F. High Performance Low Cost Control of a Permanent Magnet Wheel Motor Using a Hall Effect Position Sensor. In Proceedings of the 2011 IEEE Vehicle Power and Propulsion Conference, Chicago, IL, USA, 6–9 September 2011; pp. 1–6.

11. Kim, K.S.; Lee, J.H.; Lee, J.H.; Won, C.Y. Control Algorithm PMSM Using Rectangular 2 Hall Sensors Compensated by Sensorless Control Method. In Proceedings of the 2011 International Conference on Electrical Machines and Systems, Beijing, China, 20–23August 2011; pp. 1–6.

12. Kim, S.-Y.; Choi, C.; Lee, K.; Lee, W. An Improved Rotor Position Estimation With Vector-Tracking Observer in PMSM Drives With Low-Resolution Hall-Effect Sensors. *IEEE Trans. Ind. Electron.* **2011**, *58*, 4078–4086. [CrossRef]

13. Zhang, P.; Zhang, Q. A Controller of PMSM for Elctrical Bicycle with Hall Effect Sensors. In Proceedings of the 2016 IEEE 11th Conference on Industrial Electronics and Applications (ICIEA), Hefei, China, 5–7 June 2016; pp. 619–623.

14. Yao, X.; Huang, S.; Wang, J.; Zhang, F.; Wang, Y.; Ma, H. Pseudo Sensorless Deadbeat Predictive Current Control for PMSM Drives With Hall-Effect Sensors. In Proceedings of the 2021 IEEE International Electric Machines & Drives Conference (IEMDC), Virtual Event. 16–18 May 2021; pp. 1–6.

15. Giulii Capponi, F.; De Donato, G.; Del Ferraro, L.; Honorati, O.; Harke, M.C.; Lorenz, R.D. AC Brushless Drive with Low-Resolution Hall-Effect Sensors for Surface-Mounted PM Machines. *IEEE Trans. Ind. Appl.* **2006**, *42*, 526–535. [CrossRef]

16. Harke, M.C.; De Donato, G.; Giulii Capponi, F.; Tesch, T.R.; Lorenz, R.D. Implementation Issues and Performance Evaluation of Sinusoidal, Surface-Mounted PM Machine Drives With Hall-Effect Position Sensors and a Vector-Tracking Observer. *IEEE Trans. Ind. Appl.* **2008**, *44*, 161–173. [CrossRef]

17. Ahn, H.-J.; Lee, D.-M. A New Bumpless Rotor-Flux Position Estimation Scheme for Vector-Controlled Washing Machine. *IEEE Trans. Ind. Inf.* **2016**, *12*, 466–473. [CrossRef]

18. Ni, O.; Yang, M.; Odhano, S.A.; Zanchetta, P.; Liu, X.; Xu, D. Analysis and Design of Position and Velocity Estimation Scheme for PM Servo Motor Drive with Binary Hall Sensors. In Proceedings of the 2018 IEEE Energy Conversion Congress and Exposition (ECCE), Portland, OR, USA, 23–27 September 2018; pp. 6967–6974.

19. Ni, Q.; Yang, M.; Odhano, S.A.; Tang, M.; Zanchetta, P.; Liu, X.; Xu, D. A New Position and Speed Estimation Scheme for Position Control of PMSM Drives Using Low-Resolution Position Sensors. *IEEE Trans. Ind. Appl.* **2019**, *55*, 3747–3758. [CrossRef]

20. Yoo, A.; Sul, S.-K.; Lee, D.-C.; Jun, C.-S. Novel Speed and Rotor Position Estimation Strategy Using a Dual Observer for Low-Resolution Position Sensors. *IEEE Trans. Power Electron.* **2009**, *24*, 2897–2906. [CrossRef]

21. Zhao-yong, Z.; Zheng, X.; Tie-cai, L. FPGA Implementation of a New Hybrid Rotor Position Estimation Scheme Based on Three Symmetrical Locked Hall Effect Position Sensors. In Proceedings of the The 4th International Power Electronics and Motion Control Conference (IPEMC), IPEMC 2004, Xi'an, China, 14–16 August 2004; Volume 3, pp. 1592–1596.

22. Zhao, Y.; Huang, W.; Yang, J. Fault Diagnosis of Low-Cost Hall-Effect Sensors Used in Controlling Permanent Magnet Synchronous Motor. In Proceedings of the 2016 19th International Conference on Electrical Machines and Systems (ICEMS), Chiba, Japan, 13–16 November; pp. 1–5.

23. Park, Y.; Yang, C.; Lee, S.B.; Lee, D.-M.; Fernandez, D.; Reigosa, D.; Briz, F. Online Detection and Classification of Rotor and Load Defects in PMSMs Based on Hall Sensor Measurements. *IEEE Trans. Ind. Applicat.* **2019**, *55*, 3803–3812. [CrossRef]

24. Wang, Z.; Wang, K.; Zhang, J.; Liu, C.; Cao, R. Improved Rotor Position Estimation for Permanent Magnet Synchronous Machines Based on Hall-Effect Sensors. In Proceedings of the 2016 IEEE International Conference on Aircraft Utility Systems (AUS), Beijing, China, 10–12 October 2016; pp. 911–916.

25. Miguel-Espinar, C.; Heredero-Peris, D.; Igor-Gross, G.; Llonch-Masachs, M.; Montesinos-Miracle, D. An Enhanced Electrical Angle Representation in PMSM Control with Misplaced Hall-Effect Switch Sensors. In Proceedings of the 2020 23rd International Conference on Electrical Machines and Systems (ICEMS), Hamamatsu, Japan, 24–27 November 2020; pp. 1454–1459.

26. Fernandez, D.; Reigosa, D.; Park, Y.; Lee, S.; Briz, F. Hall-Effect Sensors as Multipurpose Devices to Control, Monitor and Diagnose AC Permanent Magnet Synchronous Machines. In Proceedings of the 2021 IEEE Energy Conversion Congress and Exposition (ECCE), Vancouver, BC, Canada, 10–14 October 2021; pp. 4967–4972.

27. Wu, Z.; Zuo, S.; Huang, Z.; Hu, X.; Chen, S.; Liu, C.; Zhuang, H. Effect of Hall Errors on Electromagnetic Vibration and Noise of Integer-Slot Inset Permanent Magnet Synchronous Motors. *IEEE Trans. Transp. Electrif.* **2022**, 1. [CrossRef]

28. Muley, N.; Saxena, A.; Chaudhary, P. Comparative Evaluation of Methods for Continuous Rotor Position Estimation Using Low Resolution Hall Sensors. In Proceedings of the 2021 National Power Electronics Conference (NPEC), Bhubaneswar, India, 15 December 2021; pp. 1–6.

29. An, Q.; Chen, C.; Zhao, M.; Ma, T.; Ge, K. Research on Rotor Position Estimation of PMSM Based on Hall Position Sensor. In Proceedings of the 2021 IEEE 16th Conference on Industrial Electronics and Applications (ICIEA), Chengdu, China, 23–26 April 2021; pp. 2088–2094.

Article

Variable Weighting Coefficient of EMF-Based Enhanced Sliding Mode Observer for Sensorless PMSM Drives

Fuqiang Cao, Quntao An *, Jianqiu Zhang, Mengji Zhao and Siwen Li

School of Electrical Engineering and Automation, Harbin Institute of Technology, Harbin 150001, China
* Correspondence: anquntao@hit.edu.cn

Abstract: In the field of permanent magnet synchronous motor (PMSM) control, the sliding mode observer (SMO)-based sensorless control is widely used; however, the actual control input of the current observation function is asymmetric. It can lead to different velocities of the estimated currents approaching to the actual currents and will make the current and back EMF fluctuations more severe, and result in more skewed angle and speed estimates, especially at a lower carrier ratio. In response to the above problems, this paper proposes a variable weighting coefficient of an EMF-based sliding mode observer (VWC-SMO). Unlike the traditional sliding mode observers, the weighted sliding mode switching variables and their bandpass-filtered values are used as the input of the current observer in the VWC-SMO. Thereby, the asymmetry of the control input in the current observation function can be well-suppressed, and almost the same approaching velocity on the two sides of the sliding surface can be obtained. Therefore, chattering near the sliding surface can also be suppressed. The method is verified on a motor controller experimental platform, and the comparative results shows that the VWC-SMO can reduce chattering of the observed currents and mitigate back EMFs fluctuations and improve the dynamic and steady-state performance.

Keywords: permanent magnet synchronous motor (PMSM); sensorless control; sliding mode observer (SMO); approaching velocity; low carrier ratio

Citation: Cao, F.; An, Q.; Zhang, J.; Zhao, M.; Li, S. Variable Weighting Coefficient of EMF-Based Enhanced Sliding Mode Observer for Sensorless PMSM Drives. *Energies* **2022**, *15*, 6001. https://doi.org/10.3390/en15166001

Academic Editors: Adolfo Dannier and Federico Barrero

Received: 11 July 2022
Accepted: 10 August 2022
Published: 18 August 2022

Publisher's Note: MDPI stays neutral with regard to jurisdictional claims in published maps and institutional affiliations.

1. Introduction

Due to the continuous progress and perfection of the utilization of electric energy, the permanent magnet synchronous motors (PMSMs) with many excellent properties have been widely applied in many areas. With more application scenarios and higher demands, much time and effort has been devoted to research in the field of sensorless control since the 1980s [1,2]. Mainstream sensorless control approaches can be divided into two categories: high-frequency injection methods based on detecting the response to this signal [3–5] and back electromotive force (EMF) methods based on various machine models [6–29].

The first method can only work well in a low-speed range, especially for the interior PMSM (IPMSM), but it has degraded performance for non-salient surface-mounted PMSM (SPMSM). On the contrary, the back EMF method is mainly applied for angle estimation in the technologies of high-speed motor control. At present, the most widely used back EMF methods mainly include the following: model reference adaptive system (MRAS) [6,7], Luenberger observer [8], sliding mode observer (SMO) [9–22], linear extended state observer (LESO) [23], disturbance observer [24], high-order sliding mode approach and nonlinear observer [25], extended Kalman filter (EKF) [26,27], etc. Among the above methods, the sliding mode observer is widely used in the field of sensorless control due to its good dynamic and static characteristics, good anti-disturbance, and other advantages.

In [9,10], the death compensation strategy is raised to reduce the harm of the inverter nonlinearity, so that the error of position estimation can be decreased. In addition, the flux spatial harmonics also affect the accuracy of position estimates [11–17]. The fifth and seventh harmonics are generally included in the back EMF observation, so that there will be sixth harmonics in the speed and angle estimates. Therefore, the adaptive notch filter (ANF) [11], the second-order generalized integrator (SOGI) [12], the recursive-least-square (RLS) adaptive filter [13], the bilinear recursive least squares (BRLS) adaptive filter [14], the adaptive hybrid generalized integrator (AHGI) [15], the adaptive synchronous filter (ASF) [16], and the frequency-adaptive complex-coefficient filter (FACCF) [17] are proposed, respectively, to extract the back EMF and enhance the accuracy of angle and speed estimation at high speed. For the sensorless control of interior PMSMs, an extended back EMF model is raised in [18], and the SMO is used to calculate the speed and angle in IPMSM drives [18,19].

In addition, the system performance will degrade because of the inherent characteristics of high-frequency vibration of the SMO; therefore, it is indispensable to reduce chattering to enhance the speed estimation accuracy. In order to improve the vibration of the traditional SMO, the saturation and sigmoid function are applied and compared in [20], as well as the super twisting algorithm in [21], which are proposed as the sliding mode switching functions to take the place of the conventional switching function. Additionally, an adaptive adjusting method for the sliding mode gain is proposed in [22]. The chattering can be improved by using such strategies to a certain degree. In recent years, the SMO has been applied in high-power and high-speed occasions, and because of the lower carrier ratio and greater delay of control, the chattering becomes severe and the current harmonics are more intense. A quasi-proportional resonant (QPR) controller-based adaptive observer is proposed in [28], and the discrete-time domain design is proposed in [29] which can enhance the accuracy of angle and speed estimation at a lower carrier ratio.

To suppress the chattering of the traditional sliding mode observer and improve the accuracy of the angle and velocity estimations in the sensorless PMSM drive at a lower carrier ratio, this paper proposes a variable weighting coefficient of an EMF-based sliding mode observer (VWC-SMO). The proposed observer can reduce the asymmetry of the control input in the current observation function by weighting the sliding mode switching variables and their bandpass-filtered values and can obtain almost the same approaching velocity on the two sides of the sliding surface. In this way, the chattering amplitude near the sliding surface can be effectively reduced. Comparative experiments of the proposed VWC-SMO and the traditional SMO under different PWM carrier ratios were carried out in a motor control experimental platform, and the results showed that the VWC-SMO can reduce chattering of the observed currents and back EMFs, so that it can improve the static and dynamic properties. Moreover, the simulation and experimental results showed good dynamic and steady-state performance of the VVC-SMO. In addition, compared with the methods mentioned in [20], the VVC-SMO method is easier to implement and has more advantages in the application of sensorless control.

2. SMO-Based Sensorless Control of PMSM

2.1. Field-Oriented Control of Position of Sensorless SPMSM

Figure 1 shows the control structure of sensorless motor control. The VWC-SMO replaces the traditional SMO in this system and uses the same inputs with the traditional SMO. The angle and speed are acquired by the normalized phase locked loop (PLL) based on the back EMFs observed by the VWC-SMO.

2.2. SMO-Based Position Estimation

Based on the theory of sliding mode variable structure control, the state of a control system as described in (1) can track the sliding surface as shown in (2) when the control input meets the condition of (3).

$$\dot{x} = f(x, u, t) \tag{1}$$

$$s(x) = 0 \tag{2}$$

$$u(x,t) = \begin{cases} u^+(x,t) & s(x) > 0 \\ u^-(x,t) & s(x) < 0 \end{cases} \tag{3}$$

where x and u are the system state variable and the control variable, respectively. The symbol "·" means derivative of the variables. u^+ and u^- indicate the control functions of both sides of the sliding surface, respectively. Under the action of the two control functions, the system can approach the sliding surface from both sides. That is to say, the system reaches the sliding mode control surface as:

$$s^\mathrm{T}\dot{s} < 0 \tag{4}$$

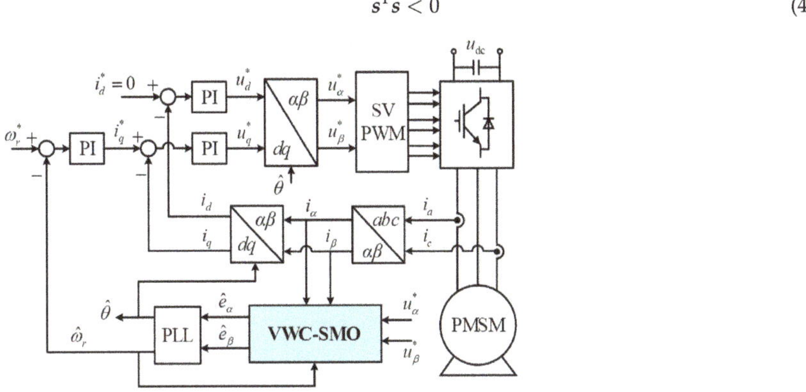

Figure 1. Position of sensorless field-oriented control of PMSM drive.

As for a SPMSM, the current state function in the α-β frame can be written as:

$$\dot{i} = -\frac{R}{L}i + \frac{1}{L}(u - e) \tag{5}$$

where $i = [i_\alpha \; i_\beta]^\mathrm{T}$ is current, $u = [u_\alpha \; u_\beta]^\mathrm{T}$ is voltage, $e = [e_\alpha \; e_\beta]^\mathrm{T}$ are EMF vectors, and, i_α, i_β are currents of α- and β-axis, u_α, u_β are voltages, and e_α, e_β are back EMFs. R is the stator resistance and L is the inductance.

The back EMFs can be expressed as:

$$e = \begin{bmatrix} e_\alpha \\ e_\beta \end{bmatrix} = \omega\psi_f \begin{bmatrix} -\sin\theta \\ \cos\theta \end{bmatrix} \tag{6}$$

where ω is the electric angular speed, ψ_f is the flux linkage, and θ is the rotor angle.

Based on (5), the SMO can be written as:

$$\dot{\hat{i}} = -\frac{R}{L}\hat{i} + \frac{1}{L}(u - u_c) \tag{7}$$

where the sign "^" represents the estimated current values. $u_c = [u_{c\alpha} \; u_{c\beta}]^\mathrm{T}$, and $u_{c\alpha}$, $u_{c\beta}$ are the functions of sliding mode control of α and β axes, respectively.

According to (5) and (7), the error function can be acquired as:

$$\Delta\dot{i} = -\frac{R}{L}\Delta i + \frac{1}{L}(e - u_c) \tag{8}$$

where $\Delta i = \hat{i} - i = [\hat{i}_\alpha - i_\alpha \; \hat{i}_\beta - i_\beta]^\mathrm{T} = [\Delta i_\alpha \; \Delta i_\beta]^\mathrm{T}$ indicates the current error, and it is usually chosen to be the sliding surface function, i.e.,

$$s = \hat{i} - i = \begin{bmatrix} \hat{i}_\alpha - i_\alpha \\ \hat{i}_\beta - i_\beta \end{bmatrix} = 0 \tag{9}$$

The sliding mode control function can have various control laws. Typically, the sign function as shown below is used as the sliding mode control function:

$$
\boldsymbol{u}_{\mathrm{c}} = \boldsymbol{z} = \begin{bmatrix} z_\alpha \\ z_\beta \end{bmatrix} = k_1 \begin{bmatrix} \mathrm{sgn}(\hat{i}_\alpha - i_\alpha) \\ \mathrm{sgn}(\hat{i}_\beta - i_\beta) \end{bmatrix} \tag{10}
$$

where sgn() stands for the sign function. k_1 is the ratio, and there is $k_1 > \max\{|e_\alpha|, |e_\beta|\}$ according to the sliding mode arrival condition as in (11) based on Lyapunov stability theory:

$$
\boldsymbol{s}^{\mathrm{T}} \dot{\boldsymbol{s}} = \Delta i_\alpha \Delta \dot{i}_\alpha + \Delta i_\beta \Delta \dot{i}_\beta < 0, \; \forall t > 0 \tag{11}
$$

Under the action of the control variables, i.e., z_α and z_β, the current errors could approach to zero, and there are position and speed information in z_α and z_β. Once the system states achieve the sliding surface, it used to be that the current errors were zero, so z_α and z_β can be applied to obtain the value of the back EMFS through a low-pass filter, as:

$$
\begin{cases} \hat{e}_\alpha = \mathrm{LPF}(z_\alpha) = \dfrac{\omega_c}{s + \omega_c} z_\alpha \\ \hat{e}_\beta = \mathrm{LPF}(z_\beta) = \dfrac{\omega_c}{s + \omega_c} z_\beta \end{cases} \tag{12}
$$

where ω_c is the cutoff frequency of the LPF. The angle and position information can then be obtained from the back EMFs using the normalized PLL, as shown in Figure 2.

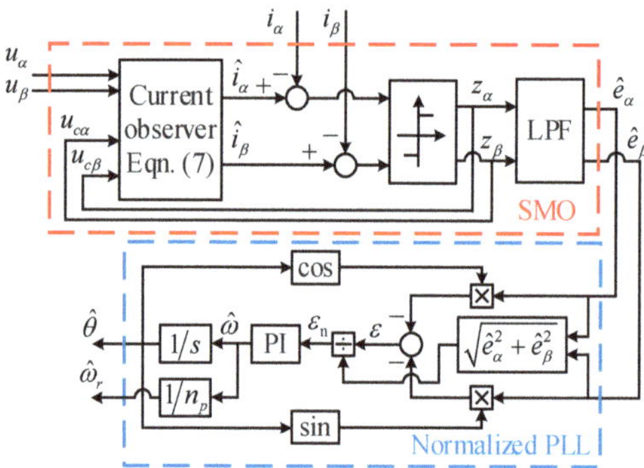

Figure 2. SMO and normalized PLL-based angle and speed estimation.

2.3. Asymmetric Approaching Velocity of SMO

According to (8), the sliding surface function can be written as:

$$
\boldsymbol{s} = \Delta \boldsymbol{i} = \frac{\boldsymbol{e} - \boldsymbol{u}_{\mathrm{c}}}{R + pL} = 0 \tag{13}
$$

where p stands for the differential operator. The current error is with a first-order inertia of the difference between back EMF and the switching function. Therefore, the actual control input is:

$$
\boldsymbol{u}(\boldsymbol{x}, t) = \boldsymbol{e} - \boldsymbol{u}_{\mathrm{c}} = \boldsymbol{e} - k_1 \mathrm{sgn}(\hat{\boldsymbol{i}} - \boldsymbol{i}) \tag{14}
$$

The sliding mode control function $\boldsymbol{u}_{\mathrm{c}}$ has two symmetric values, k_1 or $-k_1$, but the actual control variable $\boldsymbol{u}(\boldsymbol{x}, t)$ has asymmetric values since the back EMF varies when the motor runs. Therefore, the effects on the current evolution are different, which leads to the asymmetric approaching velocity on two sides of the sliding surface. As shown in Figure 3, due to the discrete sampling and control in the actual digital system, the control variable $\boldsymbol{u}_{\mathrm{c}}$ will remain constant for the whole period. Thus, when the back EMF has a large positive amplitude, the positive control input $(e_{\alpha,\beta} + k_1)$ is much larger than the negative one $(e_{\alpha,\beta} - k_1)$. Then, the current error could increase in the control period of $(e_{\alpha,\beta} + k_1)$

much more than the decrease in the period of $(e_{\alpha,\beta} - k_1)$; that is, the current error could deviate even further from the sliding surface. Moreover, the difference of approaching velocity is even greater near the peak point of back EMF, since the actual control variable is severely asymmetric, which leads to more serious chattering at that time.

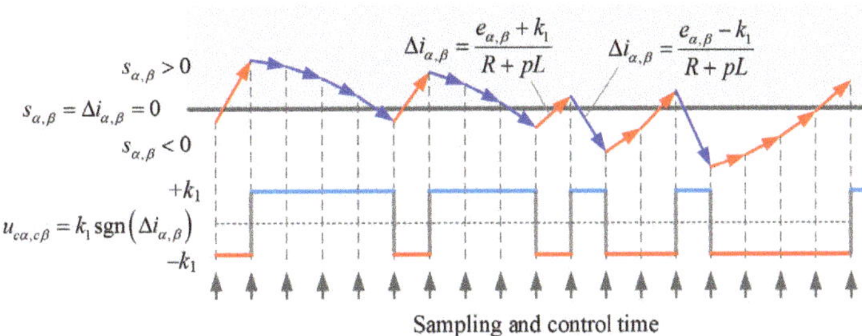

Figure 3. Asymmetric approaching velocity on both sides of the sliding surface in the traditional SMO.

Figure 4 presents the simulation analysis results of the actual currents, the observed currents, and the practical sliding control variable under 5 and 1 kHz switching frequencies, respectively.

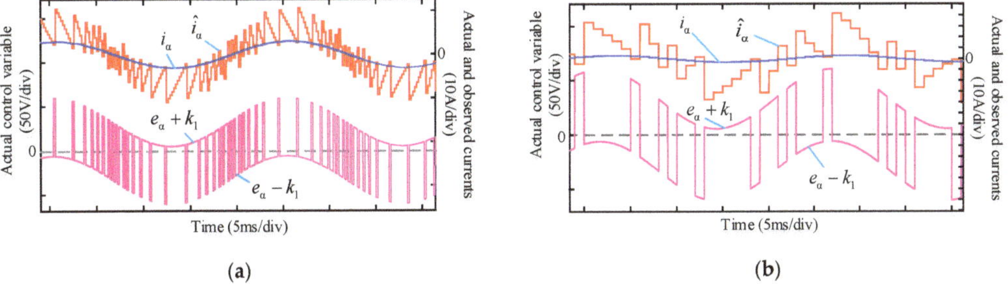

Figure 4. Simulation analysis results of SMO at switching frequencies of (**a**) 5 and (**b**) 1 kHz.

The speed is 600 r/m in Figure 4a. The fundamental frequency is 40 Hz, and the carrier ratio is 125 at this time. The speed is also 600 r/m in Figure 4b. The fundamental frequency is 40 Hz, and the carrier ratio is 25 at this time.

Near the peak of back EMF, the actual control input is far from zero at one side of the sliding surface, so the observed current can cross the actual current in one control cycle, but it cannot recover in a couple of coming control cycles since the control input is near zero at the other side of the sliding surface. The asymmetric approaching velocity increases the chattering of the observed current, and it is more deteriorating when the carrier ratio is reduced, as shown in Figure 4b. Then, the observed current chattering can lead to increased position and speed estimation errors.

3. Improved SMO Based on Variable Weighting Coefficient of Back EMFS

3.1. VWC-SMO

For improving the observed current chattering, the approaching velocity asymmetry on two sides of the sliding surface should be improved. Therefore, the actual control variable $u(x, t)$ is assumed to be another symmetric sign function with gain k_2. Thus, the sliding switching function can be written as (15) according to (14):

$$u_c = e - u(x, t) = e + k_2 \operatorname{sgn}(\hat{i} - i) \tag{15}$$

In (15), the actual back EMF is unknown, and it can be replaced by the estimated value. To reduce the phase shift and amplitude attenuation of the estimated back EMF, a bandpass filter (BPF) as in (17) was employed instead of the LPF. Thus, the sliding mode switching function can be written as the weighted sum of the expected $u(x, t)$ and the bandpass-filtered z; that is,

$$u_c = k_2\mathrm{sgn}(\hat{\pmb{i}} - \pmb{i}) + \mathrm{BPF}\left[k_1\mathrm{sgn}(\hat{\pmb{i}} - \pmb{i})\right] = \frac{k_2}{k_1}z + z_F \tag{16}$$

where z_F stands for the bandpass-filtered z. The control model of VWC-SMO is presented in Figure 5. The BPF can be written as:

$$G_{\mathrm{BPF}}(s) = \frac{2k_{\mathrm{BPF}}\omega_0 s}{s^2 + 2k_{\mathrm{BPF}}\omega_0 s + \omega_0^2} \tag{17}$$

where K_{BPF} is the adjustable coefficient of bandwidth and ω_0 is the center frequency.

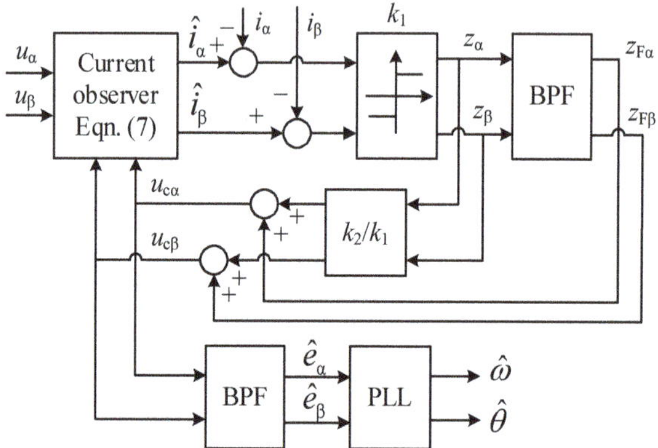

Figure 5. Block diagram of the proposed VWC-SMO.

3.2. Parameter Design

In the steady-state condition, the fundamental value can be considered constant before and after filtering, and the control variable u_c is equal to the back EMF. According to (16) and Figure 5, there is:

$$z_F = \frac{k_1}{k_1 + k_2}e \tag{18}$$

Combining (14), (16), and (18), the actual sliding mode control function and the arrival condition are described as (19) and (20), respectively.

$$u(x, t) = e - u_c = \frac{k_2}{k_1 + k_2}e - k_2\mathrm{sgn}(\hat{\pmb{i}} - \pmb{i}) \tag{19}$$

$$\left| \frac{1}{k_1 + k_2}\max\{|e_\alpha|, |e_\beta|\} \right| < 1 \tag{20}$$

Under dynamic conditions, considering that there is a back EMF error e_{err}, (18) can be rewritten as (21) and the sliding arrival condition can be expressed as (22):

$$z_F = \frac{k_1}{k_1 + k_2}e + e_{\mathrm{err}} \tag{21}$$

$$\left| \frac{1}{k_1 + k_2}\max\{|e_\alpha|, |e_\beta|\} + \frac{1}{k_2}\max\{|e_{\mathrm{err}\alpha}|, |e_{\mathrm{err}\beta}|\} \right| < 1 \tag{22}$$

The same as the traditional SMO, k_1 is generally a constant bigger than the maximum of the back EMF. Similarly, k_2 is designed as:

$$k_2 = k_{\text{smo}}\hat{\omega}\psi_f \tag{23}$$

where k_{smo} is an adjustable coefficient. Along with the increase of k_{smo}, the robustness can be improved, but the chattering increases. To reduce the chattering, k_{smo} should be properly chosen.

The back EMF error is mainly caused by the BPF. When there is an error on the center frequency of the BPF, the back EMF error can be expressed as:

$$e_{\text{err}} = \omega\psi_f[\sin\theta - A_{\text{BPF}}\sin(\theta - \varphi_{\text{BPF}})] \tag{24}$$

where A_{BPF} and φ_{BPF} are the filter gain and phase shift at the real speed, respectively.

If the maximum back EMF error coefficient is defined as:

$$k_{\text{err}} = \frac{\max(e_{\text{err}\alpha}, e_{\text{err}\beta})}{\omega\psi_f} \tag{25}$$

the relationship between the back EMF error coefficient K_{err} and the BPF bandwidth coefficient K_{BPF}, and speed error ratio η, can be presented in Figure 6. The diagram clearly shows that the error coefficient k_{err} is 0.707 if k_{BPF} is selected the same as the speed error η. Therefore, k_{BPF} should be selected as larger than the speed error η. Assuming that the speed error is $\pm 2\%$ and k_{BPF} is 0.1, the back EMF error coefficient k_{err} is 0.198 according to Figure 6, and the parameter k_{smo} should be selected as 0.3.

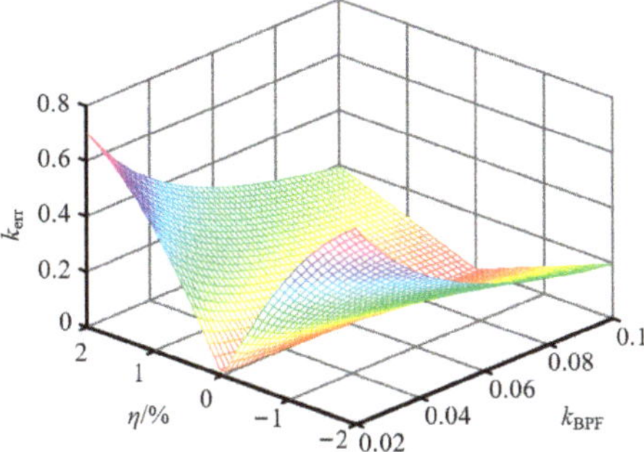

Figure 6. The relationship between k_{err} and η, and k_{BPF}.

The simulation results of the VWC-SMO are shown in Figure 7, and parameters of the SMO are set as above. In the proposed SMO system, the actual control input of axis α state observation function is $(e_\alpha - z_{F\alpha} \pm k_2)$ instead of $(e_\alpha \pm k_1)$, and its asymmetry is reduced. Therefore, the observed current chattering can be suppressed.

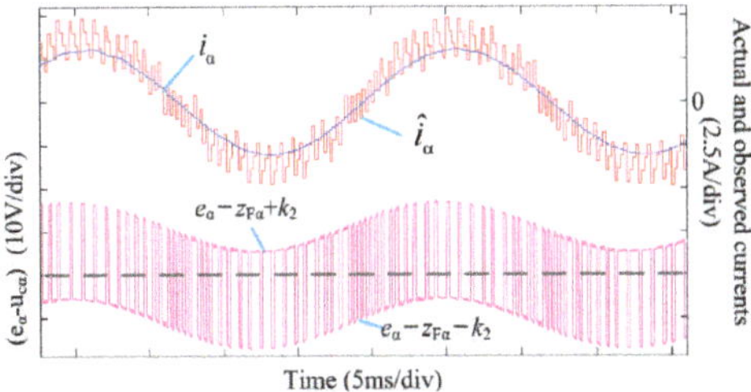

Figure 7. Simulation results of the VWC-SMO at a 5 kHz switching frequency.

4. Experimental Results

The test platform of the proposed VWC-SMO is shown in Figure 8. The specific parameters of the motor applied in the experimental platform are presented in Table 1. The currents were sampled using the current sensor on the driver, and the rotor angle and speed information was sampled by the resolver installed on the motor and decoded by the decoding chip. The proposed VWC-SMO and the traditional SMO have carried out comparative experiments under high and low carrier ratios, and the dynamic and static performance of the VWC-SMO was also validated by the experiments.

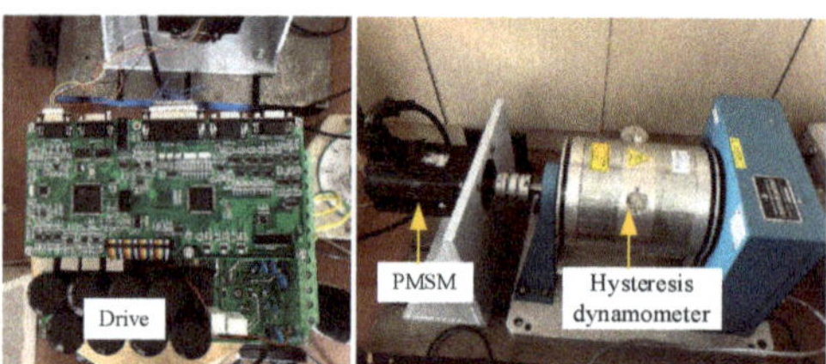

Figure 8. Test platform.

Table 1. Specific parameters of the experimental platform.

Parameter	Symbol	Value
Rated power	P_N (kW)	3.0
Rated current	I_N (A)	17.8
Rated torque	T_{eN} (N·m)	14.3
Rated speed	n_N (r/min)	2000
Winding resistance	R (Ω)	0.1
d-axis inductance	L_d (mH)	1.5
q-axis inductance	L_q (mH)	1.5
PM flux linkage	ψ_f (Wb)	0.11
Number of pole pairs	n_p	4
Rotor inertia	J (kg·m^2)	0.00223
DC link voltage	U_{dc} (V)	300
Dead time	T_d (μs)	3

4.1. Steady-State Performance Comparison

Comparative experiments of the VWC-SMO with the traditional SMO were conducted when the speed of the motor was 600 r/min and the torque was 2 N·m. The switching frequencies are set to 5 kHz and 600 Hz, respectively, and the corresponding carrier ratios are 125 and 15, respectively. The actual effects of the traditional SMO are presented in Figures 9 and 10, and the cutoff frequency of the low-pass filter was set to 2 times the motor angular speed. The results of the VWC-SMO are presented in Figures 11 and 12.

The experiment of the traditional SMO at a 5 kHz switching frequency is presented in Figure 9. At this time, the base frequency was 40 Hz, and the carrier ratio was 25 at 600 rpm. Although the currents shown in Figure 9a have harmonics, the rotor information can be well-resolved from the back EMFs in the $\alpha\beta$ frame, as shown in Figure 9b. The maximum error between the actual rotor angle and the estimated one was 6.1° and the maximum error of the speed value of the motor was 5.6 r/min, and all the waveforms can be seen in Figure 9c.

The experimental effects of the traditional SMO at a 600 Hz switching frequency are presented in Figure 10. At this time, the carrier ratio was 15. It can be seen from Figure 10a,b that the currents and back EMFs have more severe chattering and harmonics because of the lower carrier ratio. The maximum errors of the rotor angle and the speed reached 12.1° and 32 r/min, respectively, which can be seen in Figure 10c. This larger angle error and speed error have a serious adverse impact during the operation of the motor servo system, which will affect the system performance.

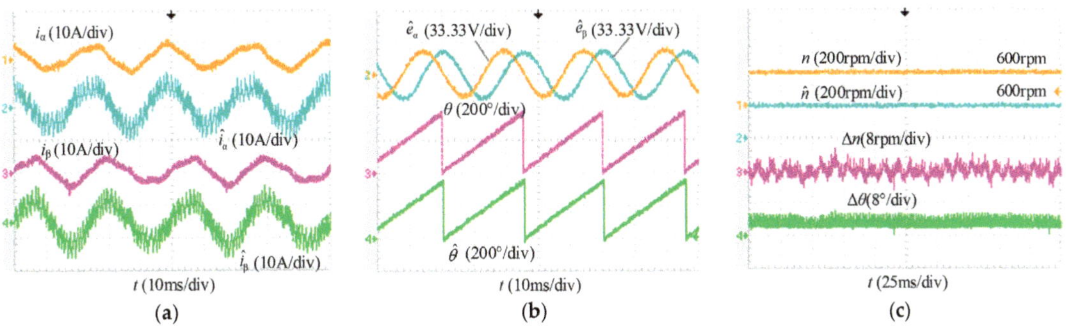

Figure 9. Experimental waveforms of the traditional SMO under a 5 kHz switching frequency. (**a**) Currents. (**b**) Estimated back EMFs and angle. (**c**) Speed and angle.

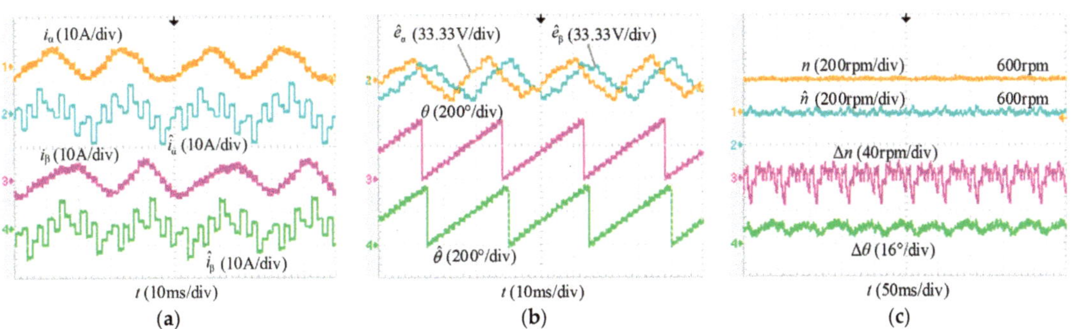

Figure 10. Experimental waveforms of the traditional SMO under a 600 Hz switching frequency. (**a**) Currents. (**b**) Estimated back EMFs and angle. (**c**) Speed and angle.

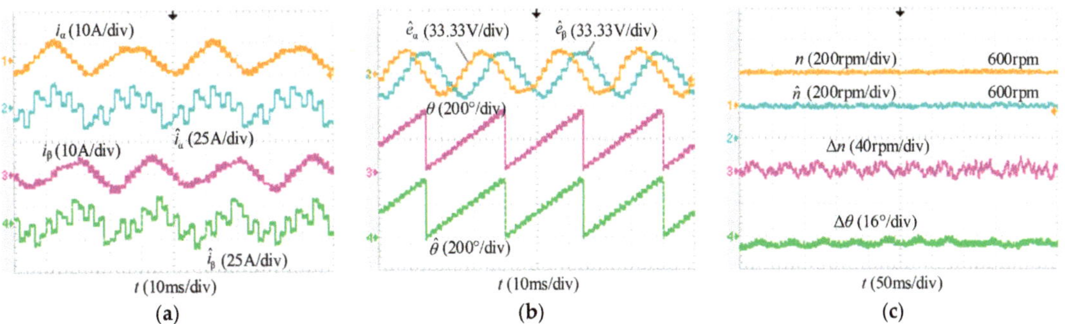

Figure 11. Experimental waveforms of the VWC-SMO under a 5 kHz switching frequency. (**a**) Currents. (**b**) Estimated back EMFs and angle. (**c**) Speed and angle.

The experiment of the VWC-SMO proposed in this paper with a 5 kHz switching frequency is shown in Figure 11, and the VWC-SMO had an improved performance compared with the traditional SMO. The maximum angle and speed errors of the rotor reached 3.2° and 5.2 r/min, respectively.

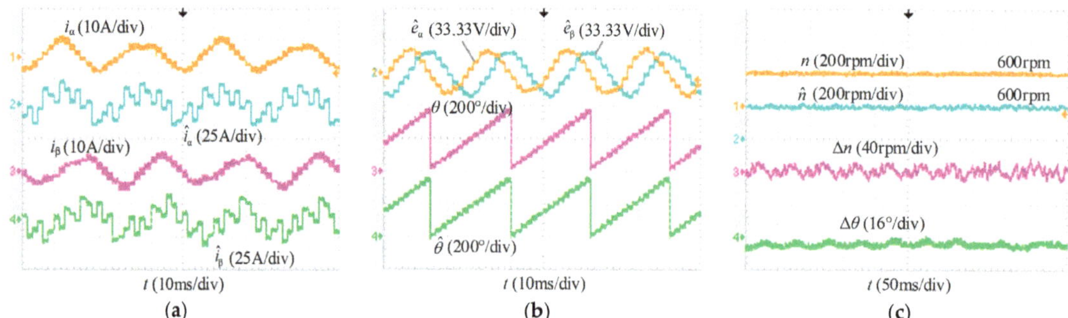

Figure 12. Experimental waveforms of the VWC-SMO at a 600 Hz switching frequency. (**a**) Currents. (**b**) Estimated back EMFs and angle. (**c**) Speed and angle.

The experimental results in Figure 12 show the VWC-SMO under a 600 Hz switching frequency. The chattering of estimated currents and back EMFs was reduced by using the VWC-SMO, as shown in Figure 12a,b. From Figure 12c, because of the lower carrier ratio, the maximum errors of the angle and speed expanded to 6.4° and 11.2 r/min, respectively. From 5 kHz to 600 Hz, it can be seen that although the experiments showed that the lower the carrier ratio, the larger the deviation of the observer estimate, the proposed observer still had good performance.

Figure 13 shows, in detail, the measured and observed currents with the traditional SMO and the VWC-SMO, respectively. In the traditional SMO, the observed current ripples severely near the peak of the current, which is caused by the different approaching velocity on the two sides of the sliding surface. However, the asymmetry of the sliding mode control variables can be reduced by the VWC-SMO and the current ripple can be mitigated, as shown in Figure 13b.

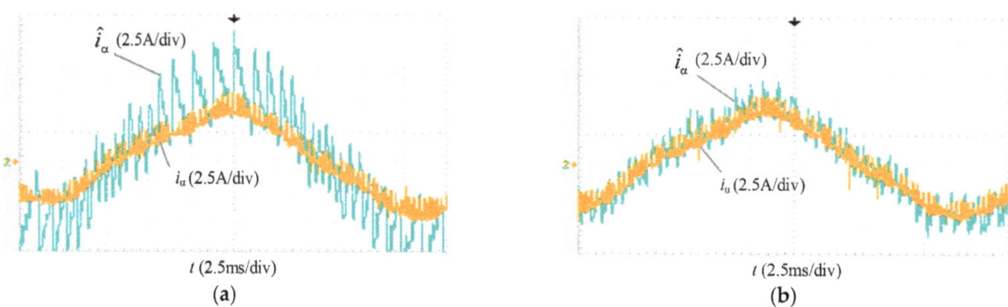

Figure 13. Actual and observed currents at a 5 kHz switching frequency. (**a**) The traditional SMO. (**b**) The VWC-SMO.

4.2. Dynamic Performance of the Proposed VWC-SMO

To verify the dynamic effect of the proposed VWC-SMO for angle and speed estimations, acceleration, deceleration, and changing load experiments in various states were conducted. Figure 14 shows the results. At a 5 kHz switching frequency, the motor speed was ramped up from 200 to 2000 r/min and then back down to 200 r/min. Figure 14a shows the real speed, estimated speed, and the errors of speed and angle. Obviously, VVC-SMO can track the real-time angle and speed very well when the motor is running. The speed error was less than 10 r/min, and the angle error was no more than 9°.

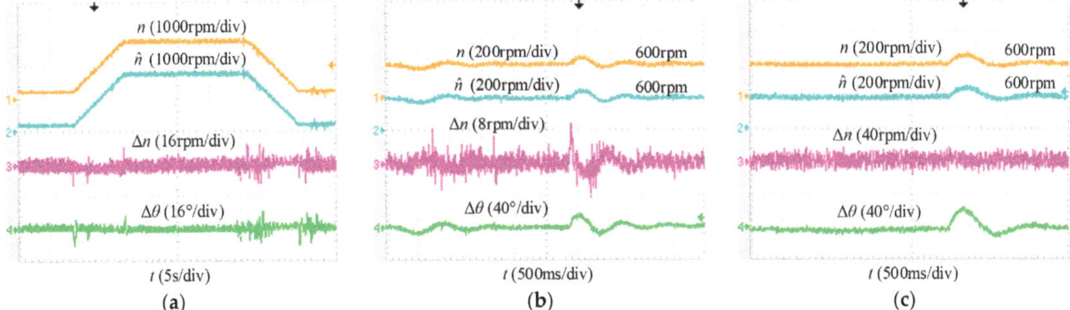

Figure 14. Dynamic experimental waveforms. (**a**) Speed up and down under a 5 kHz switching frequency. (**b**) Speed and angle errors in addition and subtraction load under a 5 kHz switching frequency. (**c**) Speed and angle errors in addition and subtraction load under a 600 Hz switching frequency.

Figure 14b,c present the results of adding and subtracting the load experiment of the VWC-SMO under 5 kHz and 600 Hz switching frequencies, respectively. The motor first ran smoothly at 600 r/min, and then the load of the motor suddenly changed from 0 to 6 N·m; them, 3 s later, the load was back to 0 N·m again. When the motor was operating with sudden load changes, the speed and angle errors were, respectively, 9.6 r/min and 17° under a 5 kHz switching frequency. Under a 600 Hz switching frequency, the estimated speed and angle were 18 r/min and 22°, respectively. The experimental results for the sudden load of the motor during operation showed that the proposed VWC-SMO can effectively resist torque disturbance.

5. Conclusions

In applications with lower carrier ratios, the performance degradation of SMO makes it impossible to meet engineering needs. For improving the performance of SMO and reducing the estimation errors of angle and speed, this paper proposed a VWC-SMO. The new VWC-SMO can suppress the asymmetry of the sliding mode control variable on the two sides of the sliding surface, and then reduce distortion of the observed current. The experimental results on the experimental platform showed that the proposed method can effectively improve the angle and position estimation accuracy, especially under a low carrier ratio, and can virtually enhance the performance of the PMSM drives.

Author Contributions: Conceptualization, Q.A.; methodology, F.C. and J.Z.; software, J.Z.; validation, F.C.; investigation, M.Z. and S.L.; writing—original draft preparation, F.C.; writing—review and editing, F.C.; funding acquisition, Q.A. All authors have read and agreed to the published version of the manuscript.

Funding: This research received no external funding.

Institutional Review Board Statement: Not applicable.

Informed Consent Statement: Not applicable.

Data Availability Statement: Not applicable.

References

1. Wang, G.; Valla, M.; Solsona, J. Position Sensorless Permanent Magnet Synchronous Machine Drives—A Review. *IEEE Trans. Ind. Electron.* **2020**, *67*, 5830–5842. [CrossRef]
2. Stanica, D.M.; Bizon, N.; Arva, M.C. A brief review of sensorless AC motors control. In Proceedings of the 2021 13th International Conference on Electronics, Computers and Artificial Intelligence (ECAI), Pitesti, Romania, 1–3 July 2021; pp. 1–7.

3. Lu, Q.; Wang, Y.; Mo, L.; Zhang, T. Pulsating High Frequency Voltage Injection Strategy for Sensorless Permanent Magnet Synchronous Motor Drives. *IEEE Trans. Appl. Supercond.* **2021**, *31*, 1–4. [CrossRef]
4. Wang, G.; Kuang, J.; Zhao, N.; Zhang, G.; Xu, D. Rotor Position Estimation of PMSM in Low-Speed Region and Standstill Using Zero-Voltage Vector Injection. *IEEE Trans. Power Electron.* **2018**, *33*, 7948–7958. [CrossRef]
5. Zhang, G.; Wang, G.; Wang, H.; Xiao, D.; Li, L.; Xu, D. Pseudorandom-Frequency Sinusoidal Injection Based Sensorless IPMSM Drives with Tolerance for System Delays. *IEEE Trans. Power Electron.* **2019**, *34*, 3623–3632. [CrossRef]
6. Kivanc, O.C.; Ozturk, S.B. Sensorless PMSM Drive Based on Stator Feedforward Voltage Estimation Improved with MRAS Multiparameter Estimation. *IEEE/ASME Trans. Mechatron.* **2018**, *23*, 1326–1337. [CrossRef]
7. Ni, Y.; Shao, D. Research of Improved MRAS Based Sensorless Control of Permanent Magnet Synchronous Motor Considering Parameter Sensitivity. In Proceedings of the 2021 IEEE 4th Advanced Information Management, Communicates, Electronic and Automation Control Conference (IMCEC), Chongqing, China, 18–20 June 2021; pp. 633–638.
8. Andersson, A.; Thiringer, T. Motion Sensorless IPMSM Control Using Linear Moving Horizon Estimation with Luenberger Observer State Feedback. *IEEE Trans. Transp. Electrif.* **2018**, *4*, 464–473. [CrossRef]
9. Zhao, Y.; Qiao, W.; Wu, L. Dead-Time Effect Analysis and Compensation for a Sliding-Mode Position Observer-Based Sensorless IPMSM Control System. *IEEE Trans. Ind. Appl.* **2015**, *51*, 2528–2535. [CrossRef]
10. Wang, Y.; Xu, Y.; Zou, J. Sliding-Mode Sensorless Control of PMSM With Inverter Nonlinearity Compensation. *IEEE Trans. Power Electron.* **2019**, *34*, 10206–10220. [CrossRef]
11. Wang, G.; Zhan, H.; Zhang, G.; Gui, X.; Xu, D. Adaptive Compensation Method of Position Estimation Harmonic Error for EMF-Based Observer in Sensorless IPMSM Drives. *IEEE Trans. Power Electron.* **2014**, *29*, 3055–3064. [CrossRef]
12. Sun, S.; Cheng, H.; Wang, W.; Liu, H.; Mi, S.; Zhou, X. Sensorless DPCC of PMLSM Using SOGI-PLL Based High-Order SMO with Cogging Force Feedforward Compensation. In Proceedings of the 2021 13th International Symposium on Linear Drives for Industry Applications (LDIA), Wuhan, China, 1–3 July 2021; pp. 1–7.
13. Wang, G.; Li, T.; Zhang, G.; Gui, X.; Xu, D. Position Estimation Error Reduction Using Recursive-Least-Square Adaptive Filter for Model-Based Sensorless Interior Permanent-Magnet Synchronous Motor Drives. *IEEE Trans. Ind. Electron.* **2014**, *61*, 5115–5125. [CrossRef]
14. Wu, X.; Huang, S.; Liu, K.; Lu, K.; Hu, Y.; Pan, W.; Peng, X. Enhanced Position Sensorless Control Using Bilinear Recursive Least Squares Adaptive Filter for Interior Permanent Magnet Synchronous Motor. *IEEE Trans. Power Electron.* **2020**, *35*, 681–698. [CrossRef]
15. Kashif, M.; Murshid, S.; Singh, B. Adaptive Hybrid Generalized Integrator Based SMO for Solar PV Array Fed Encoderless PMSM Driven Water Pump. *IEEE Trans. Sustain. Energy* **2021**, *12*, 1651–1661. [CrossRef]
16. Jian, H.; Song, W. A sliding mode observer of IPMSM combining adaptive synchronous filter and back EMF estimator. In Proceedings of the 2020 7th International Forum on Electrical Engineering and Automation (IFEEA), Hefei, China, 25–27 September 2020; pp. 121–126.
17. An, Q.; Zhang, J.; An, Q.; Liu, X.; Shamekov, A.; Bi, K. Frequency-Adaptive Complex-Coefficient Filter-Based Enhanced Sliding Mode Observer for Sensorless Control of Permanent Magnet Synchronous Motor Drives. *IEEE Trans. Ind. Appl.* **2020**, *56*, 335–343. [CrossRef]
18. Chen, Z.; Zhang, H.; Zhang, Z. EEMF-based Sensorless control for IPMSM Drives with an Optimized Asymmetric Space Vector Modulation. In Proceedings of the 2019 22nd International Conference on Electrical Machines and Systems (ICEMS), Harbin, China, 11–14 August 2019; pp. 1–5.
19. Yin, Z.; Zhang, Y.; Cao, X.; Yuan, D.; Liu, J. Estimated Position Error Suppression Using Novel PLL for IPMSM Sensorless Drives Based on Full-Order SMO. *IEEE Trans. Power Electron.* **2022**, *37*, 4463–4474. [CrossRef]
20. Petro, V.; Kyslan, K. A Comparative Study of Different SMO Switching Functions for Sensorless PMSM Control. In Proceedings of the 2021 International Conference on Electrical Drives & Power Electronics (EDPE), Dubrovnik, Croatia, 22–24 September 2021; pp. 102–107.
21. Liang, D.; Li, J.; Qu, R. Sensorless Control of Permanent Magnet Synchronous Machine Based on Second-Order Sliding-Mode Observer with Online Resistance Estimation. *IEEE Trans. Ind. Appl.* **2017**, *53*, 3672–3682. [CrossRef]
22. Yang, C.; Ma, T.; Che, Z.; Zhou, L. An Adaptive-Gain Sliding Mode Observer for Sensorless Control of Permanent Magnet Linear Synchronous Motors. *IEEE Access* **2018**, *6*, 3469–3478. [CrossRef]
23. Yu, B.; Shen, A.; Chen, B.; Luo, X.; Tang, Q.; Xu, J.; Zhu, M. A Compensation Strategy of Flux Linkage Observer in SPMSM Sensorless Drives Based on Linear Extended State Observer. *IEEE Trans. Energy Convers.* **2022**, *37*, 824–831. [CrossRef]
24. Indriawati, K.; Widjiantoro, B.L.; Rachman, N.R.i. Disturbance Observer-Based Speed Estimator for Controlling Speed Sensorless Induction Motor. In Proceedings of the 2020 3rd International Seminar on Research of Information Technology and Intelligent Systems (ISRITI), Yogyakarta, Indonesia, 10 December 2020; pp. 301–305.
25. Belkhier, Y.; Shaw, R.N.; Bures, M.; Islam, M.R.; Bajaj, M.; Albalawi, F.; Alqurashi, A.; Ghoneim, S.S. Robust interconnection and damping assignment energy-based control for a permanent magnet synchronous motor using high order sliding mode approach and nonlinear observer. *Energy Rep.* **2022**, *8*, 1731–1740. [CrossRef]
26. Gopinath, G.; Das, S.P. An extended Kalman filter based sensorless permanent magnet synchronous motor drive with improved dynamic performance. In Proceedings of the 2018 IEEE International Conference on Power Electronics, Drives and Energy Systems (PEDES), Chennai, India, 18–21 December 2018; pp. 1–6.

27. Tondpoor, K.; Saghaiannezhad, S.M.; Rashidi, A. Sensorless Control of PMSM Using Simplified Model Based on Extended Kalman Filter. In Proceedings of the 2020 11th Power Electronics, Drive Systems, and Technologies Conference (PEDSTC), Tehran, Iran, 4–6 February 2020; pp. 1–5.
28. An, Q.; Zhang, J.; An, Q.; Shamekov, A. Quasi-Proportional-Resonant Controller Based Adaptive Position Observer for Sensorless Control of PMSM Drives Under Low Carrier Ratio. *IEEE Trans. Ind. Electron.* **2020**, *67*, 2564–2573. [CrossRef]
29. Zhang, G.; Wang, G.; Xu, D.; Yu, Y. Discrete-Time Low-Frequency-Ratio Synchronous-Frame Full-Order Observer for Position Sensorless IPMSM Drives. *IEEE J. Emerg. Sel. Top. Power Electron.* **2017**, *5*, 870–879. [CrossRef]

Article

Analysis and Error Separation of Capacitive Potential in the Inductosyn

Chengjun Liu, Jianfei Sun, Yifei Zhang and Jing Shang *

Institute of Special Motor and Control, Harbin Institute of Technology, Harbin 150001, China
* Correspondence: shangjing@hit.edu.cn; Tel.: +86-0451-8641-3613

Abstract: High-precision rotor position information is usually needed in permanent-magnet synchronous motors, which are critical to high-performance motor control based on vector algorithm. Therefore, inductosyn is the best choice for the permanent-magnet synchronous motor position sensor. Capacitive potential is an important component of the ineffective potential in the inductosyn angle measuring system. When the effective potential of the inductosyn approaches zero, the proportion of capacitive potential in the output potential will be greatly amplified. As a result, the zero-position accuracy will be seriously affected. Error potential and effective potential always exist at the same time, so it is difficult to measure and study quantitatively. In this paper, the capacitance network model of inductosyn was established and the analytical calculation method was proposed. The factors affecting the capacitive potential and the suppression strategy were studied through the combination of theoretical analysis and the finite element method. In addition, the error separation method of capacitive potential was also proposed in this paper, which realized the accurate measurement of this part of error. The accuracy of the theoretical calculation and finite element analysis was verified by the experimental results.

Keywords: inductosyn; capacitive potential; error separation; suppression strategy

Citation: Liu, C.; Sun, J.; Zhang, Y.; Shang, J. Analysis and Error Separation of Capacitive Potential in the Inductosyn. *Energies* **2022**, *15*, 6910. https://doi.org/10.3390/en15196910

Academic Editor: Andrea Mariscotti

Received: 9 August 2022
Accepted: 16 September 2022
Published: 21 September 2022

Publisher's Note: MDPI stays neutral with regard to jurisdictional claims in published maps and institutional affiliations.

1. Introduction

High-precision rotor position information is the basis of high-precision vector control of the permanent-magnet synchronous motor. As the key component of the angle measurement system, the performance of the angular position sensor directly affects the working stability, conversion speed and measurement accuracy of the whole angle measurement system [1–3]. Rotary inductosyn is a multipole electromagnetic induction angular position sensor, which has the characteristics of resistance to harsh environment, low installation accuracy requirements, high accuracy, fast operation speed, strong anti-electromagnetic interference ability and low cost [4–6]. The special multipole structure enables the inductosyn to achieve a precision much higher than that of other electromagnetic measuring elements, such as rotary transformer and hall position sensor [7,8]. Compared with other optical sensors, inductosyn has stronger adaptability and reliability in extreme environments [9,10].

With the continuous development of space technology, the accuracy of an angle measurement system is of increasing necessity in various fields [11–13]. Therefore, how to improve the measurement accuracy is the main problem that inductosyn currently faces. The measurement accuracy of the inductosyn is mainly determined by the zero-position accuracy [14,15]. In addition to external factors such as installation error, conductor marking deviation, coupling potential of non-effective conductor and capacitive potential are the main factors which determine the zero-position error of inductosyn [16]. Among them, the conductor marking deviation is mainly determined by the manufacturing process, and the coupling of non-effective conductors is difficult to avoid in any structure. Therefore, the most effective method to improve the accuracy of the inductosyn system is to reduce the capacitive potential [17].

The capacitive potential is generated by the coupling capacitance between the stator and the rotor, and the current forms a loop through the coupling capacitance, generating an error potential at the output end [18,19]. The phase of the error potential and the effective potential is almost the same, which is harmful to the zero-position accuracy of the inductosyn and difficult to separate. As early as the initial stage of inductosyn research, scholars paid attention to this error factor and proposed many measures to reduce capacitive coupling [20]. These measures were based on the connection of conductors in different ways, which complicated the ends of windings and inevitably affected the electromagnetic coupling of primary and secondary windings [21].

In this study, the generation principle and influencing factors of the capacitive potential of inductosyn were analyzed. The capacitance network model of inductosyn was constructed, and the measures to suppress capacitive potential were studied by combining theoretical calculation with the finite element method. Finally, the experimental measurement was carried out by capacitive potential error separation technology, which verified the accuracy of the theoretical calculation and finite element analysis.

2. Generation Mechanism of Capacitive Potential of the Inductosyn

The rotating inductosyn is spread along the circumference, and the sine and cosine windings of the stator differ by half-pole distance in space. When the position of the sinusoidal winding and the rotor conductor completely coincide, the cosine winding is located in the center of the rotor conductor, as shown in Figure 1. When the sinusoidal excitation current is applied to the rotor winding, the two-phase windings of the stator will generate mutually orthogonal induced potentials.

$$\begin{cases} U_1 = U_m \sin \omega t \\ U_{2s} = U_0 \cos \omega t \sin \theta \\ U_{2c} = U_0 \cos \omega t \cos \theta \end{cases} \tag{1}$$

Equation (1) shows the output effective potential of the inductosyn. However, due to the formation of a plate-like capacitor structure between the stator and rotor windings, a small part of the current flows through the distributed capacitance between the stator and rotor conductor, and generates an ineffective potential on the stator winding, as shown in Figure 1.

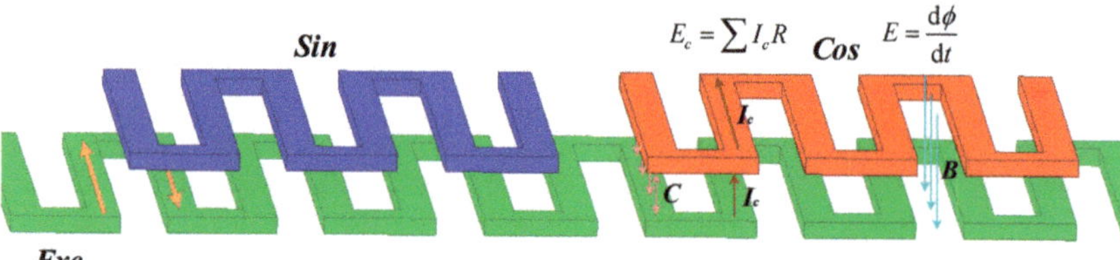

Figure 1. Generation principle of capacitive potential in the inductosyn.

Generally speaking, the amplitude of the effective potential in the inductosyn is much higher than that of the non-effective potential generated by the distributed capacitance. However, every time the rotor rotates an angle of the pole distance, the effective potential of stator winding will pass through the zero position, while the capacitive potential is not so sensitive to the angle change. Therefore, when the induced potential of the stator winding approaches zero, the proportion of capacitive ineffective potential in the output potential increases significantly. It can be seen that the capacitive potential will increase the zero-position error of the inductosyn, which may seriously affect the accuracy of the inductosyn angle measurement system.

3. Capacitance Network Model and Analytical Calculation of the Inductosyn

The inductosyn has the characteristic that $X_C \gg R \gg X_L$. Therefore, the inductance parameter can be ignored in the analysis process, and the impedance network model of the capacitor and resistance can be directly established in the inductosyn.

3.1. Model of Stator and Rotor with Common Grounding Point

When the stator and rotor have a common grounding point, the capacitive current only passes through the air gap once to form a loop. At this time, each conductor of the rotor can be equivalent to a voltage source. The equivalent model is shown in Figure 2. The path of capacitive current is relatively simple, and it is refluxed by the grounding point of stator winding.

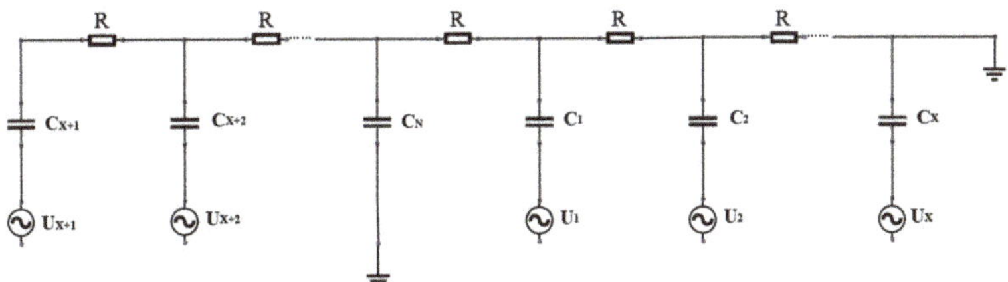

Figure 2. Network model of stator and rotor with common grounding point.

From the *i* branch on the right, the following formula can be obtained:

$$U_i = (N - X + i)U_0 (i \le X)$$

$$U_i = (i - X)U_0 (i > X) \tag{2}$$

$$R_i = iR \tag{3}$$

$$Z_1 = Z_2 = \ldots\ldots = Z_N = \frac{1}{-j\omega c} \tag{4}$$

$$E = \sum_{i=1}^{N} \frac{U_i}{Z_i} R_i = \frac{U_0 R}{Z} \sum_{i=1}^{X} (N - X + i)i + \sum_{i=X+1}^{N} (i - x)i \tag{5}$$

$$= -j\omega c U_0 RN \left(\frac{X^2}{2} - \frac{N}{2}X + \frac{N^2}{3} \right)$$

As can be seen from Equation (5), when the stator and rotor of inductosyn have a common grounding point, the generated capacitive potential grows in quadratic form, and has a large DC component. Moreover, the capacitive potential shows high order growth with the increase in the pole number.

3.2. Model of Stator and Rotor without Common Grounding Point

If one end of the stator and rotor in the inductosyn floats to the ground, the capacitive current passes through the air gap twice to form a loop. For this network, it can be divided into two impedance networks, where Q is of order *X-1* and *P* is of order *N-X-1*, as shown in Figure 3.

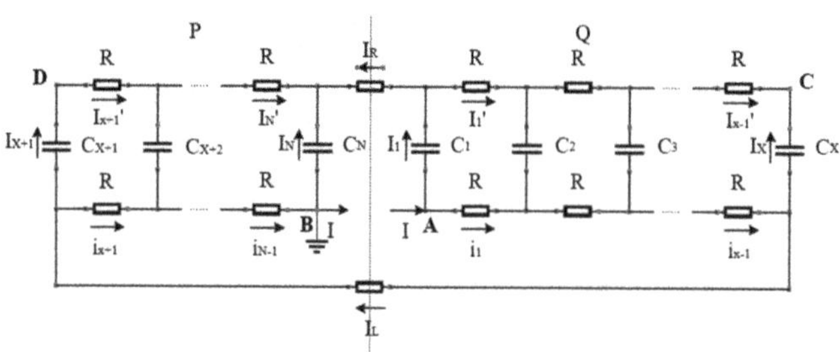

Figure 3. Network model of stator and rotor without common grounding point.

For the unit network shown in Figure 4, the following formulas can be obtained.

$$I_{k-1}X_c + I'_{k-1}R - I_kX_c - i_{k-1}R = 0 \tag{6}$$

$$I_kX_c + I'_{k+1}R - I_{k+1}X_c - i_kR = 0 \tag{7}$$

$$I_k = I'_k - I'_{k-1} = -(i_k - i_{k-1}) \tag{8}$$

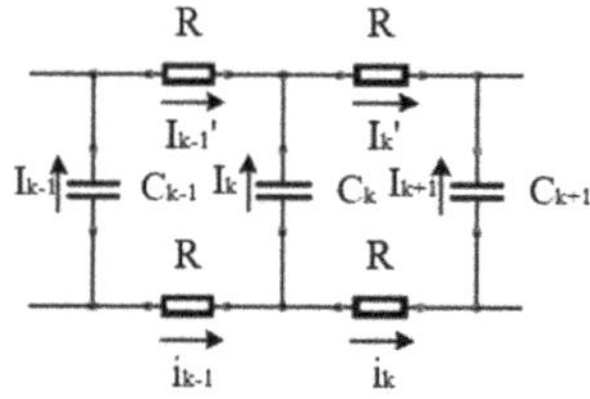

Figure 4. Network unit.

By combining Equations (6)–(8), we obtain Equation (9):

$$I_{k+1} = (2 + 2j\lambda)I_k - I_{k-1}$$

$$\lambda = R\omega c \tag{9}$$

The result of Equation (9) is as follows:

$$I_k = \frac{1}{\alpha-\beta}[(I_2 - \beta I_1)\alpha^{k-1} - (I_2 - \alpha I_1)\beta^{k-1}](1 \le k \le X)$$

$$I_k = \frac{1}{\alpha-\beta}[(I_{x+2} - \beta I_{x+1})\alpha^{k-x-1} - (I_{x+2} - \alpha I_{x+1})\beta^{k-x-1}](X+1 \le k \le N) \tag{10}$$

$$\alpha = 1 + j\lambda + \sqrt{2j\lambda - \lambda^2}$$

$$\beta = 1 + j\lambda - \sqrt{2j\lambda - \lambda^2}$$

By combining Equations (8) and (10), the following equation is obtained:

$$I'_k = \frac{1}{\alpha-\beta}[(I_2 - \beta I_1)\frac{1-\alpha^k}{1-\alpha} - (I_2 - \alpha I_1)\frac{1-\beta^k}{1-\beta}] - I_R(1 \le k \le X)$$

$$I'_k = \frac{1}{\alpha-\beta}[(I_{x+2} - \beta I_{x+1})\frac{1-\alpha^{k-x}}{1-\alpha} - (I_2 - \alpha I_1)\frac{1-\beta^{k-x}}{1-\beta}](X+1 \le k \le N) \tag{11}$$

The boundary conditions of the network in area Q are as follows:

$$I'_x = \frac{1}{\alpha - \beta}[(I_2 - \beta I_1)\frac{1 - \alpha^x}{1 - \alpha} - (I_2 - \alpha I_1)\frac{1 - \beta^x}{1 - \beta}] - I_R = 0 \tag{12}$$

$$I_1 X_c + I_1' R - I_2 X_c - i_1 R = 0 \tag{13}$$

According to the current continuity theorem, Equation (14) can be obtained:

$$I = I_L + I_R \tag{14}$$

By combining Equations (12) and (13), Equation (15) is obtained:

$$I_1 = \frac{1}{2} I \left(1 - \frac{\alpha - \beta + \alpha^{x-1} - \beta^{x-1}}{\alpha^x - \beta^x}\right) + \frac{1}{2} I_R \left(1 + \frac{\alpha - \beta - (\alpha^{x-1} - \beta^{x-1})}{\alpha^x - \beta^x}\right) \tag{15}$$

The boundary conditions of the network in area P are as follows:

$$I_N' = \frac{1}{\alpha - \beta}\left[(I_{x+2} - \beta I_{x+1})\frac{1 - \alpha^{N-x}}{1 - \alpha} - (I_2 - \alpha I_1)\frac{1 - \beta^{N-x}}{1 - \beta}\right] = -I_R \tag{16}$$

$$I_{x+1} X_c + I_{x+1}' R - I_{x+2} X_c - i_{x+1} R = 0 \tag{17}$$

By combining Equations (14), (16) and (17), we obtain Equation (18):

$$I_{x+1} = \frac{1}{2} I \left(1 - \frac{\alpha - \beta + \alpha^{x-1} - \beta^{x-1}}{\alpha^x - \beta^x}\right) - \frac{1}{2} I_R \left(1 + \frac{\alpha - \beta - (\alpha^{x-1} - \beta^{x-1})}{\alpha^x - \beta^x}\right) \tag{18}$$

In the following equations, U_{AB} is solved according to the network shown in Figure 3. By the loops $i_1, i_2, \ldots, I_L, i_{x+1}, i_{x+2}, \ldots, i_N$ and loops $I_1, I_1', I_2', \ldots, I_{x-1}', I_x, I_L, I_{x+1}, I_{x+1}', I_{x+2}', \ldots, I_{N-1}', I_N$. It can be obtained by integrating them separately:

$$\begin{aligned} U_{AB} &= (i_1 + i_2 + \ldots + i_{x-1} + I_L + i_{x+1} + i_{x+2} + \ldots + i_{N-1})R \\ &= (I_1' + I_2' + \ldots + I_{x-1}' + I_L + I_{x+1}' + I_{x+2}' + \ldots + I_{N-1}')R + (I_1 - I_x + I_{x+1} - I_N)X_c \end{aligned} \tag{19}$$

Another form of U_{AB} can be obtained from Equation (20):

$$\begin{aligned} U_{AB} &= \tfrac{1}{2}(i_1 + i_2 + \ldots + i_{x-1} + I_L + i_{x+1} + i_{x+2} + \ldots + i_{N-1} + \\ & I_1' + I_2' + \ldots + I_{x-1}' + I_L + I_{x+1}' + I_{x+2}' + \ldots + I_{N-1}')R + \tfrac{1}{2}(I_1 - I_x + I_{x+1} - I_N)X_c \end{aligned} \tag{20}$$

At the same time, it can be also obtained by solving U_{AB} along the loop I_1, I_R, I_N:

$$U_{AB} = (I_1 - I_N)X_C - I_R R \tag{21}$$

From the above equation, we can find the expression of I_R:

$$I_R = \frac{N(\alpha - \beta)(\alpha + \beta - 2)(\alpha^x - \beta^x)(\alpha^{N-x} - \beta^{N-x})}{F} I$$

$$\begin{aligned} F &= (\alpha + \beta - 2)[(N+2)(\alpha - \beta) - (\alpha^{N-x-1} - \beta^{N-x-1}) - (\alpha^{x-1} - \beta^{x-1})](\alpha^x - \beta^x)(\alpha^{N-x} - \beta^{N-x}) \\ &\quad + [\alpha - \beta + \alpha^{N-x} - \beta^{N-x} - (\alpha^{N-x-1} - \beta^{N-x-1})]2(\alpha^x - \beta^x) \\ &\quad + [\alpha^{x-1} - \beta^{x-1} - (\alpha - \beta) - (\alpha^x - \beta^x)]2(\alpha^{N-x} - \beta^{N-x}) \end{aligned} \tag{22}$$

Through Equation (22), the electric potential expression of U_{CD} at the output end can be obtained:

$$
\begin{aligned}
U_{CD} &= \sum_{j=1}^{X-1} I'_j R - I_R R + \sum_{j=X+1}^{N-1} I'_j R \\
&= \frac{1}{\alpha-\beta}\left[\frac{I_2-\beta I_1}{1-\alpha}\left[X-1-\frac{\alpha(1-\alpha^{X-1})}{1-\alpha}\right] - \frac{I_2-\alpha I_1}{1-\beta}\left[X-1-\frac{\beta(1-\beta^{X-1})}{1-\beta}\right]\right] - I_R R \\
&\quad + \frac{1}{\alpha-\beta}\left[\frac{I_{x+2}-\beta I_{x+1}}{1-\alpha}\left[N-X-1-\frac{\alpha(1-\alpha^{N-X-1})}{1-\alpha}\right] - \frac{I_{x+2}-\alpha I_{x+1}}{1-\beta}\left[N-X-1-\frac{\beta(1-\beta^{N-X-1})}{1-\beta}\right]\right]
\end{aligned}
\tag{23}
$$

The distribution law of capacitive potential of induction synchronizer under different grounding modes is shown in Figure 5. It can be seen that the relationship between capacitive potential of induction synchronizer and rotor position is sinusoidal. When there is a common ground point, the capacitive potential has a large DC component, which disappears when the rotor ends float to the ground. Therefore, in practical application, it is necessary to ensure that at least one end of the induction synchronizer is floating or that an isolation transformer is used.

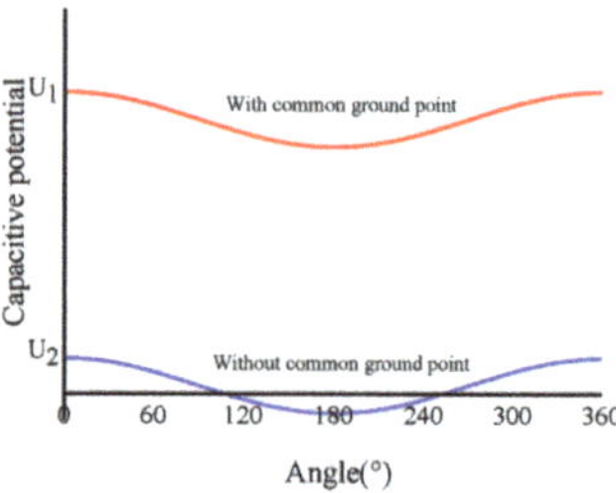

Figure 5. Capacitive potential of inductosyn.

4. Influencing Factors and Restraining Measures of Capacitive Potential

The analysis above is based on some assumptions. In practical application, the capacitive potential of inductosyn is much more complex than that in the analytical model. If the capacitive potential cannot be effectively suppressed, the accuracy of the inductosyn will be greatly reduced. According to the above analysis, we have established that the capacitive potential can be effectively suppressed by isolation transformer. When the capacitive current passes through the air gap twice, the ineffective potential generated by it will be greatly reduced. In addition, we also need other methods to suppress the capacitive potential.

4.1. Aluminum Substrate of Stator

With the continuous development of PCB (Printed Circuit Board) technology, the cost and manufacturing cycle of inductosyn can be greatly reduced. However, organic materials are used as substrates in conventional PCB processes, and these have a certain impact on the capacitive ineffective potential. The analysis in the third section is based on the absence of any metal substrate. When the aluminum substrate is used on the stator side, the capacitance between the stator conductor and the substrate is connected in series with the capacitance between the stator and rotor, which reduces the capacitive potential on the stator conductor. The effect is related to the thickness of the stator insulation layer and the resistance of the stator substrate to the ground.

The relationship between stator substrate and capacitive potential was analyzed in this paper by the joint simulation of ANSYS Maxwell and Simplorer. The conductor structure of inductosyn is too complex; therefore, we established a simplified model with two effective conductors as shown in Figure 6, which ignored the weak coupling with other surrounding conductors, to study the influence of stator substrate on capacitive potential. Stator aluminum substrate, the distance between stator winding and substrate and the resistance of stator aluminum substrate to ground were studied as variables In this simulation, the results are shown in Figure 7. It can be seen that using aluminum substrate can effectively reduce the capacitive potential, and the smaller the distance between the effective conductor plane of the stator and the aluminum substrate, the smaller the capacitive potential. In addition, on the basis of ensuring good insulation of stator winding, the smaller the resistance of stator substrate to ground, the lower the capacitive potential.

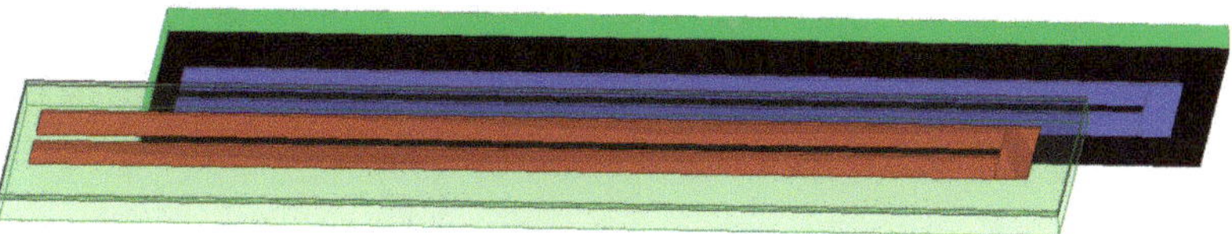

Figure 6. Simplified simulation model of inductosyn.

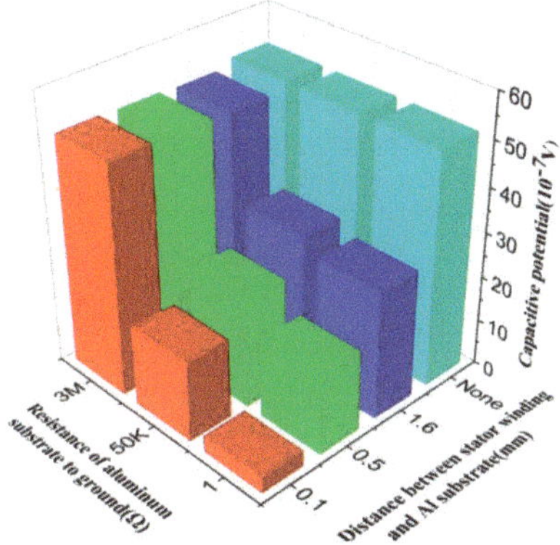

Figure 7. Influence of stator substrate on capacitive potential.

From the above analysis, it can be seen that the aluminum substrate structure needs to be adopted in the stator of inductosyn, and the insulation layer should be as thin as possible to effectively suppress the capacitive potential. This also verifies that the angle measurement accuracy of inductosyn made by traditional PCB process is very low. The organic substrate is too thick, so the capacitive potential is difficult to fully suppress. In order to solve these problems, we should use aluminum-based PCB processes or flexible PCB processes to produce inductosyn.

4.2. Aluminum Shielding Film

Aluminum shielding film, which covers the inductosyn conductor, is another method to suppress capacitive potential. The distributed capacitance between the stator and rotor of the inductosyn will change with the angle. However, it will become a constant value after using an aluminum shielding film. Moreover, part of the capacitive current will flow back through the shielding film when the aluminum shielding film is grounded, which will weaken the capacitive potential on the stator winding. The capacitor network model of inductosyn with shielding film is shown in Figure 8.

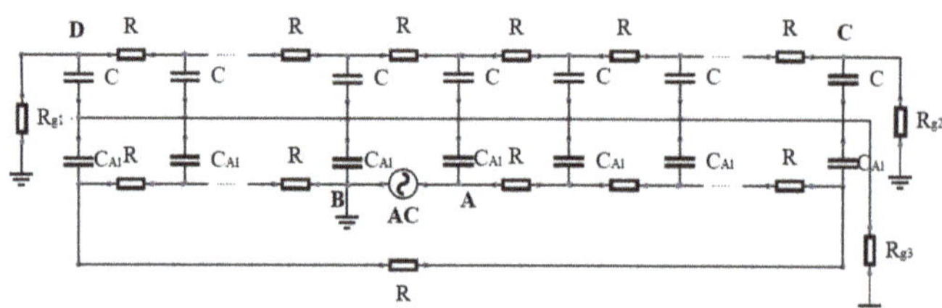

Figure 8. Capacitance network model with shielding film.

The capacitance between the shielding film and the exciting winding can be expressed as follows:

$$C_{Al} = \frac{K_1 S}{D} \tag{24}$$

Therefore, the impedance of capacitive current flowing back through the shielding film can be obtained:

$$X_c = \frac{K_2 D}{j\omega S} + R_{g3} \tag{25}$$

In order to make more capacitive current return from the aluminum shielding film, the impedance shown in Equation (25) should be as small as possible. Therefore, R_{g3} should be zero, which means the shielding film needs to be well grounded. Moreover, the distance D between the shielding film and the exciting winding should be as small as possible. Therefore, the aluminum shielding film should cover the side of the excitation winding.

5. Capacitive Potential Extraction and Separation Technology of the Inductosyn

5.1. Capacitive Potential Extraction

The induced potential and capacitive potential always exist at the same time, so it has been impossible to measure the size of the capacitive potential and related factors alone. Yongping Lu and others of Harbin Institute of technology estimated the capacitive potential by connecting the two ports of excitation to both ends of the stator and rotor wingdings, respectively, in early years [22]. However, the whole rotor is in a high potential state at this time, which is different from the distribution from 0 to power supply voltage in actual work. Therefore, the measurement results are not only larger, but also irrelevant to the relative position of stator and rotor. A special rotor that can separate the capacitive potential of the inductosyn was designed in this paper, and its principle is shown in Figure 9.

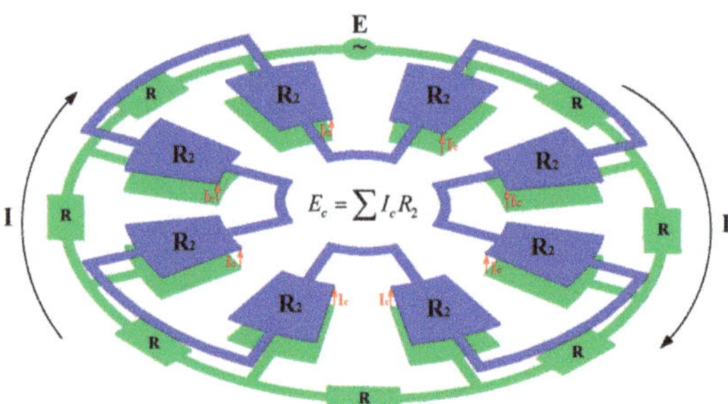

Figure 9. Extraction of capacitive potential in the inductosyn.

In the special rotor designed in this paper, the current mainly flows through the outer resistance, and the resistance value of each resistance is the same as that of each conductor of the inductosyn rotor. As long as the safe distance between the outer resistance and the inner conductor is ensured, the stator will not generate any induced potential. The potential on the stator side will be all capacitive potential at this time. Therefore, the capacitive non-effective potential of the inductosyn can be separated and accurately measured through this special rotor, which is shown in Figure 10.

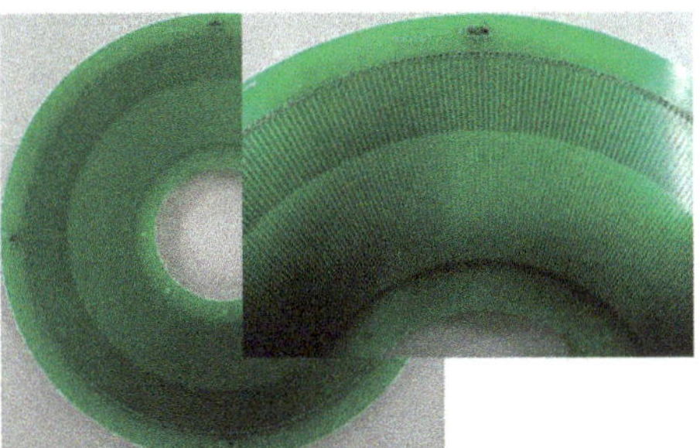

Figure 10. Special rotor for Extracting capacitive potential.

5.2. Experimental Results of Capacitive Potential SEPARATION

The test platform of inductosyn is shown in Figure 11. In the experiment, the accuracy of the precision optical turntable is 0.5 arcsec. The capacitive potential can be measured directly with the low-noise amplification module and oscilloscope. The zero-position error of the inductosyn can be measured by rotating the fine-adjustment knob with a resolution of 0.05 arcsec of the turntable. When the angle were adjusted to make the amplified signal reach the minimum value, the difference between the displayed angle and the standard position was the zero-position error. Firstly, the capacitive potential generated by different grounding methods of stator and rotor was measured. The stator adopts a 1.6 mm thick PCB board and was fixed on the aluminum substrate. The results are shown in Figure 12. It can be seen that when the inductosyn stator and rotor have a common grounding point, a large capacitive voltage drop will be generated. Therefore, in practical application, at least one side of the stator or rotor needs to be floating to the ground, so as to significantly reduce the capacitive potential.

Figure 11. Test platform of inductosyn.

In the experiment of the 360 poles of inductosyn made by PCB processes, 1.6 mm PCB stator without aluminum substrate, 1.6 mm PCB stator with aluminum substrate and 0.1 mm flexible PCB stator with aluminum substrate were selected as the experimental objects, respectively, and ensure that at least one side of the stator and rotor was floating on the ground. The relationship between stator capacitive non-effective potential and air gap is shown in Figure 13. The experimental results are in good agreement with the simulation results. The capacitive potential decreases significantly after using the aluminum substrate. The smaller the distance between the stator conductor and the aluminum substrate, the smaller the capacitive potential, and the faster the attenuation with the increase in air gap. Therefore, the stator winding of the inductosyn should be made directly on the aluminum substrate as far as possible. If the PCB process is to be adopted, the aluminum-based PCB process or flexible PCB process fixed on the aluminum substrate are the appropriate choices.

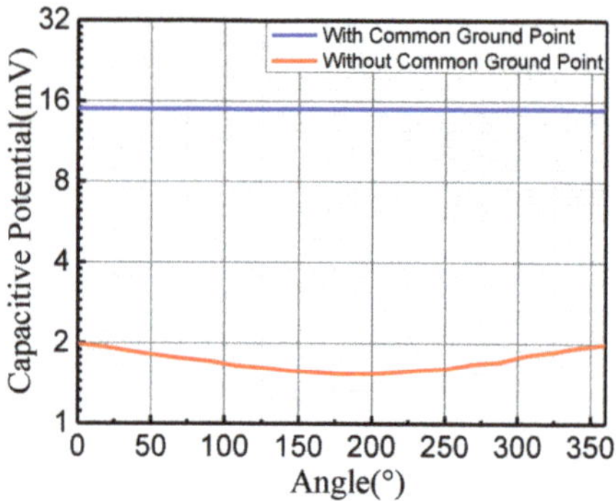

Figure 12. Influence of grounding mode.

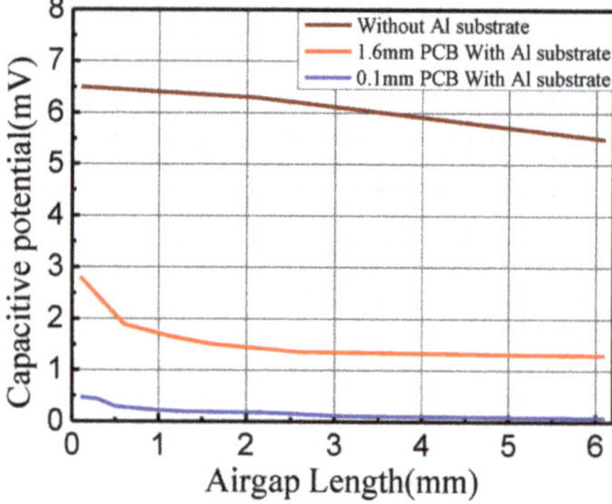

Figure 13. Influence of substrate.

In this paper, on the premise of using a flexible aluminum substrate PCB and ensuring that at least one end of stator and rotor is floating, the influence of grounding aluminum shielding film on the capacitive potential of inductosyn was measured, as shown in Figure 14. The capacitive potential of inductosyn with a shielding film was significantly reduced, and did not change with the relative position of stator and rotor.

The accuracy of the inductosyn mainly depends on the zero-position accuracy, and non-effective potential will make the zero-position error of the inductosyn jump in parity. Under the above conditions, the zero-parity errors of the inductosyn before and after the use of the shielding film were measured, as shown in Figure 15. The zero-parity error of the inductosyn before using the shielding film was very serious, and it changed greatly with the angle. After using the shielding film, the parity error decreased from about $1'$ to $10''$ and remains stable.

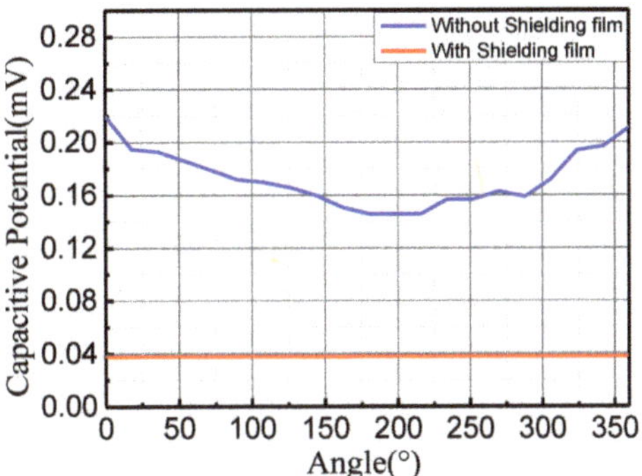

Figure 14. Influence of aluminum shielding film.

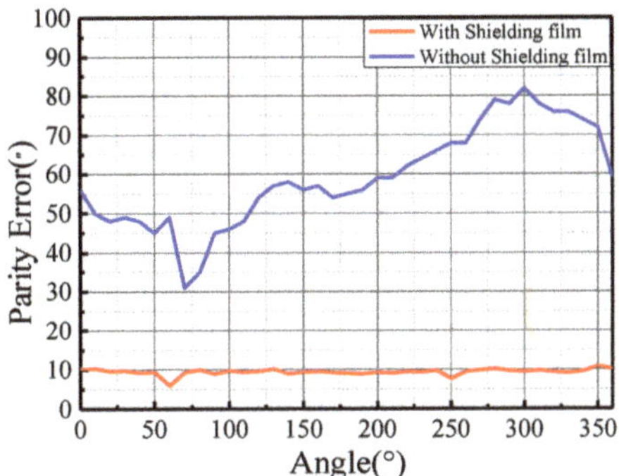

Figure 15. Parity error of inductosyn.

6. Conclusions

In this paper, the formation mechanism of capacitive potential and its harm to the accuracy of inductosyn were analyzed. Furthermore, the capacitance network model of inductosyn was established and the analytical calculation method was proposed, which reflected the distribution law of capacitive potential. This laid a foundation for the suppression of capacitive potential of inductosyn.

When the stator and rotor of inductosyn have a common grounding point, a large capacitive voltage drop is generated, which makes the inductosyn difficult to work with with high accuracy. Therefore, an isolation transformer is required to ensure that one end of the stator and rotor is floating on the ground. In addition, the capacitive potential error extraction technology was also proposed in this paper. The influence of various factors on the capacitive potential was accurately measured, which provided guidance for the error separation.

Capacitive potential was effectively suppressed by stator aluminum substrate and rotor aluminum shielding film. By using both of them at the same time and ensuring the good floating of stator and rotor, we could control the capacitive potential in a very small range. The experimental results were consistent with the theoretical calculation and simulation analysis, which verified the above argument.

Author Contributions: Conceptualization, J.S. (Jianfei Sun); investigation, C.L.; Data curation, J.S. (Jianfei Sun) and Y.Z.; writing—original draft, J.S. (Jianfei Sun); writing—review and editing, C.L.; supervision, J.S. (Jing Shang). All authors have read and agreed to the published version of the manuscript.

Funding: This research received no external funding.

Data Availability Statement: Data is contained within the article.

Conflicts of Interest: The authors declare no conflict of interest.

References

1. Dinulovic, D.; Hermann, D.; Fluegge, J.; Gatzen, H. Development of a Linear Micro-Inductosyn Sensor. *IEEE Trans. Magn.* **2006**, *42*, 2830–2832. [CrossRef]
2. Liu, C.; Xu, F. A Novel Absolute Angle-Measuring Method with Rotary Inductosyn. In Proceedings of the 2018 Eighth International Conference on Instrumentation & Measurement, Computer, Harbin, China, 19–21 July 2018., Communication and Control (IMCCC).
3. Li, T.; Lin, J.; Chen, W.; Xia, Z.; Hu, J. High Precision Fusion of Multi-channel Induction Synchronizer and Its FPGA Implementation. *Astronaut. Meas. Technol.* **2021**, *41*, 57–62.
4. Zhan, H.; Hu, Y.; Hai, L.; Zhang, G.; Zhang, H. Study on Winding Variation of Induction Synchronizer Based on Sensitivity Analysis. *Mech. Electr. Technol.* **2020**, *3*, 18–20.
5. Li, X.; Gao, X.; Zhou, C.; Wei, X. A Method of Power Frequency Noise Interference Suppression for Induction Synchronizer Circuit. *Power Electron. Technol.* **2020**, *54*, 29–32.
6. Ho, Y.; Ren, S.; Li, W. Measurement Method of Sector Scoring Error of Circular Induction Synchronizer. *Electr. Mach. Control* **2019**, *23*, 19–26.
7. Zeng, L. Linear induction synchronizer. *Trend Sci. Technol.* **2018**, *34*, 144.
8. Huang, J. Design of High Precision Electromechanical Encoder Based on Circular Induction Synchronizer. *Micromotor* **2018**, *51*, 10–14.
9. Wang, F.; Fu, J.; Zhu, Y.; Han, C. Coarse-fine Data Fusion of Absolute Circular Induction Synchronizer. *Chin. J. Sens. Technol.* **2018**, *31*, 213–217.
10. Li, H.; Zhang, R.; Han, F. Error Measurement and Compensation of Angle Measuring System of Induction Synchronizer. *J. Tsinghua Univ. Sci. Technol. Ed.* **2016**, *56*, 611–616.
11. Yuan, H.; Liu, Z.; Li, Z.; Cui, K.; Xie, Y. Dynamic Extraction and Compensation of System Error of Circular Induction synchronizer. *Opt. Precis. Eng.* **2015**, *23*, 794–802. [CrossRef]
12. Li, N. Correlation between Stability and Measurement Error of Capacitive Load and Pressure Sensor. *Sci. Technol. Innov. Rev.* **2018**, *15*, 1–3.
13. Yang, J.; Niu, Y.; Lv, A.; Chen, B.; Gao, Y.; Wang, L. *Research on Capacitive Equipment Live Detection System*; Electronics World: Gainesville, FL, USA, 2018; pp. 180–181.
14. Shi, J. Research on Angle Measurement Technology Based on Optical Fiber Angle Sensor. *Appl. Laser* **2021**, *41*, 1047–1054.
15. He, S.; Wu, Y. New Practical Fall Detection Experimental System Based on Six-axis Attitude Angle Sensor. *Inf. Comput. Theory Ed.* **2022**, *34*, 124–127.
16. Bin, H. A High Precision Angle Sensor Based on Dual Channel Rotary Transformer. *Electron. Qual.* **2022**, *2*, 23–26.
17. Liu, Z.; Zhang, J.; Fu, Z. Research on Angle Measurement Technology of High Voltage Disconnector Based on Optical fiber Angle Sensor. *High Volt. Electr. Appar.* **2021**, *57*, 9–17.
18. Ashby, D.; Sargent, T.; Cox, D.; Rosato, J.; Brynnel, J.G. The Large Binocular Telescope azimuth and elevation encoder system. *Adv. Softw. Control. Astron. II* **2008**, *7019*, 701920.
19. Alhamadi, M.; Benammar, M.; Ben-brahim, L. Precise Method for Linearizing Sine and Cosine Signals in Resolvers and Quadrature encoders Applications. In Proceedings of the 30th Annual Conference of IEEE on Industrial Electronics Society, Busan, Korea, 2–6 November 2004; pp. 1935–1940.
20. Liu, C.; Zhu, M. Error Separation Method of Inductosyn. In Proceedings of the 2013 Third International Conference on Instrumentation, Measurement, Computer, Communication and Control, Shenyang, China, 21–23 September 2013.
21. Wang, X.; Min, W.; Ge, Z.; Feng, J. Error Separation and Compensation of Inductosyn Angle Measuring System. In Proceedings of the 2010 International Conference on Measuring Technology and Mechatronics Automation, Changsha, China, 13–14 March 2010.
22. Yongping, L.; Wenyuan, C. *Inductosyn and System*; National Defense Industry Press: Beijing, China, 1985.

Article

Dual Three-Phase Permanent Magnet Synchronous Machines Vector Control Based on Triple Rotating Reference Frame

Jian-Ya Zhang [1], Qiang Zhou [1,*] and Kai Wang [2]

1. The Sixty-Third Research Institute, National University of Defense Technology, Nanjing 210007, China
2. College of Automation Engineering, Nanjing University of Aeronautics and Astronautics, Nanjing 211106, China
* Correspondence: zhouqiang63@nudt.edu.cn

Abstract: This paper presents a triple rotating coordinate transformed vector control method for dual three-phase permanent magnet (PM) machines. In the proposed scheme, the control variables are converted to three sets of $\alpha\beta$ components directly, which are 120° electric degrees different from each other. It omits the complicated six-dimensional transformed matrix and reduces the computation greatly. The relationship with vector space (VSD) control was mathematically analyzed. By ensuring the consistency of control variables in the three stationary reference frames, the suggested method can not only achieve the same fundamental control performance as VSD but compensate for the imbalance current caused by the harmonics in the back electromotive force. In addition, the proposed method belongs to multi redundancy control in theory, which is maybe a good solution for fault-tolerant operation. Finally, a prototype dual three-phase PM machine was tested. The experimental results are in good agreement with the theoretical analysis.

Keywords: mechanical antenna; PM machine; vector control; triple rotating reference frame; imbalance current

Citation: Zhang, J.-Y.; Zhou, Q.; Wang, K. Dual Three-Phase Permanent Magnet Synchronous Machines Vector Control Based on Triple Rotating Reference Frame. *Energies* **2022**, *15*, 7286. https://doi.org/10.3390/en15197286

Academic Editors: Quntao An, Bing Tian and Xinghe Fu

Received: 30 July 2022
Accepted: 23 September 2022
Published: 4 October 2022

Publisher's Note: MDPI stays neutral with regard to jurisdictional claims in published maps and institutional affiliations.

1. Introduction

The ability to penetrate seawater has long been a major problem in wireless communication [1]. Though low-frequency electromagnetic waves have great potentiality in the underwater covert communication field, the most widely studied low-frequency transmitting antenna belongs to electrically small antennas (ESA) with the features of bulky size, low radiation efficiency, high transmitting power, and energy consumption. It becomes an important bottleneck to the further development of low-frequency electromagnetic communication [2].

Mechanical Antenna (MA) was initially developed to achieve the miniaturization and low power consumption of low-frequency transmitting systems by the US Defense Advanced Research Projects Agency (DARPA) [3]. MA directly generates a strong static electric or magnetic field with the mechanical movement of special, exciting materials, eliminating the matching tuning network and its corresponding additional losses. It is expected to break through the physical scale limit of traditional low-frequency ESA [4]. Depending on different exciting materials and mechanical movement modes, MA can be broadly divided into four classes: vibrating-electret based MA (VEBMA), vibrating-magnet based MA (VMBMA), rotating-electret based MA (REBMA), and rotating-magnet based MA (RMBMA). Among them, RMBMA attracts domestic and foreign attention that benefits from the mature application of rare earth permanent magnetic materials and servo control [5–7]. Typical RMBMA architecture consists of a permanent spinning magnet and a rotating servo drive system, as shown in Figure 1. Obviously, the transmission characteristics of RMBMA are closely related to the performance of PMSM. For example, the transmission capacity will be proportional to T_{em}/J. Hence, increasing the output torque can improve MA efficiency.

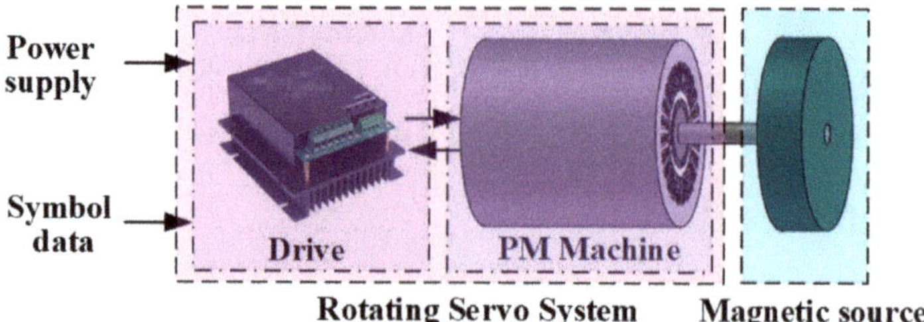

Figure 1. Typical RMBMA architecture.

Multiphase electric machines have numerous advantages over traditional three-phase machines, such as torque enhancement, high efficiency, and low torque ripple. They are obtained in wide applications; one of the most common multiphase machine structures is the dual three-phase machine [8–10]. It consists of two isolated three-phase winding sets shifted by 30° electric degrees. The special structure determines the diversity of its control methods. An effective control scheme is an essential part of dual three-phase PMSM drives, which are closely related to the performance and the whole application. However, control of dual three-phase PMSM is more challenging than other machines due to the coupling and asymmetry between two windings [11].

The vector space decomposition (VSD) method is the most widely used multiphase modeling and control approach [12,13]. However, the original VSD control suffers from imbalanced currents between two winding sets. Some improvements were suggested to solve imbalance currents [14,15]. While the solutions are overcomplicated, they have not achieved the desired effect in practice. Double d-q current control is another alternative approach for this particular winding type, which is capable of balancing currents. However, it is generally known that this alternative approach has the disadvantage of cross-coupling that induces current harmonics. Additionally, it was criticized for reducing winding utilization in fault-tolerant control [16–19].

The previous review of control strategies for dual three-phase PMSMs indicates that further improvements and better control solutions are needed for balance current, harmonic current suppression, simplicity, and high winding availability for fault-tolerant operation. Therefore, this paper aimed to develop a control scheme for dual three-phase PMSM based on a triple rotating reference frame. A simplified model to analyze the triple *dq* transformation was established. The comparison between VSD and triple *dq* transformation was analytically derived. The novelty of the proposed method was to perform the current control in a new reference frame, which solves the problems of balanced current sharing and current harmonics. It can also provide a practical means for improving the performance of dual three-phase PMSMs with redundant fault-tolerant control.

2. Mathematical Model of Dual Three-phase Machines Based on Triple Stationary Reference Frame

Due to the special structure, the dual three-phase machine can be regarded as three sets of orthogonal two-phase winding, i.e., A&Z, B&X, and C&Y, as shown in Figure 2a, even if with two isolated neutral points. Hence, the variables in the original six-dimensional system can be mapped into three virtual stationary coordinates, shifted 120° from each other. The three stationary coordinates are designated as α1-β1, α2-β2, and α3-β3.

The variables in each αβ frame can be converted to *dq* components directly by the modified Park transformation:

$$P_{3dq}(\theta_e) = \begin{bmatrix} \cos\theta_e & -\sin\theta_e \\ -\sin\theta_e & -\cos\theta_e \end{bmatrix} \tag{1}$$

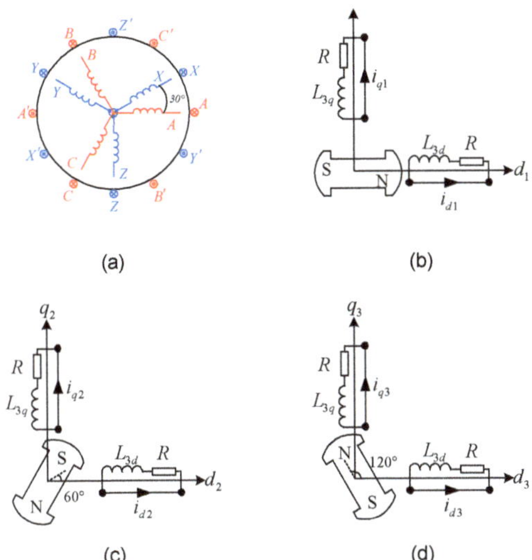

Figure 2. Decoupled three d-q coordinates, (**a**) dual three-phase windings, (**b**) d1-q1 reference frame; (**c**) d2-q2 reference frame; (**d**) d3-q3 reference frame.

Thus, the total transformation can be expressed as:

$$\begin{bmatrix} f_{d1} & f_{q1} & f_{d2} & f_{q2} & f_{d3} & f_{q3} \end{bmatrix}^{T} = T_{3dq} \begin{bmatrix} f_A & f_B & f_C & f_X & f_Y & f_Z \end{bmatrix}^{T} \tag{2}$$

$$T_{3dq} = \begin{bmatrix} P_{3dq}(\theta_1) & & \\ & P_{3dq}(\theta_2) & \\ & & P_{3dq}(\theta_3) \end{bmatrix} \tag{3}$$

where $\theta_1 = \theta_e$, $\theta_2 = \theta_e - 120°$, and $\theta_1 = \theta_e + 120°$; Symbol f is the machine variable, which can represent voltage, current, or flux.

The voltage, flux, and electrical torque equations in the original six-dimensional coordinate system can be mapped into a triple stationary reference frame by both sides multiplied T_{3dq}. By simplifying and omitting higher harmonic inductance components, the voltage equations in the new reference frame can be rewritten as follows:

$$\begin{cases} u_{d1} = Ri_{d1} + \dot{\psi}_{d1} - \omega_e \psi_{q1} \\ u_{q1} = Ri_{q1} + \dot{\psi}_{q1} + \omega_e \psi_{d1} \\ u_{d2} = Ri_{d2} + \dot{\psi}_{d2} - \omega_e \psi_{q2} \\ u_{q2} = Ri_{q2} + \dot{\psi}_{q2} + \omega_e \psi_{d2} \\ u_{d3} = Ri_{d3} + \dot{\psi}_{d3} - \omega_e \psi_{q3} \\ u_{q3} = Ri_{q3} + \dot{\psi}_{q3} + \omega_e \psi_{d3} \end{cases} \tag{4}$$

Additionally, the flux equations:

$$\begin{cases} \psi_{d1} = (L_{aad} + L_{aal})i_{d1} + L_{aad}(i_{d2} + i_{d3}) + \psi_{fd} \\ \psi_{q1} = (L_{aaq} + L_{aal})i_{q1} + L_{aaq}(i_{q2} + i_{q3}) \\ \psi_{d2} = (L_{aad} + L_{aal})i_{d2} + L_{aad}(i_{d1} + i_{d3}) + \psi_{fd} \\ \psi_{q2} = (L_{aaq} + L_{aal})i_{q2} + L_{aaq}(i_{q1} + i_{q3}) \\ \psi_{d3} = (L_{aad} + L_{aal})i_{d3} + L_{aad}(i_{d1} + i_{d2}) + \psi_{fd} \\ \psi_{q3} = (L_{aaq} + L_{aal})i_{q3} + L_{aaq}(i_{q1} + i_{q2}) \end{cases} \tag{5}$$

Based on the electrical machinery theory with Virtual Displacement, the electromagnetic torque T_e can be represented as the derivative of magnetic co-energy W_c with respect to mechanical position θ_m:

$$T_e = \frac{\partial W_c}{\partial \theta_m} = p_n \left(\frac{1}{2} I_s^T \frac{\partial L_s}{\partial \theta_e} I_s + I_s^T \frac{\partial \psi_s}{\partial \theta_e} \right) \tag{6}$$

The electromagnetic torque in the new reference frame is obtained by substituting (3)–(5) into (6):

$$T_e = p_n \left[\begin{array}{c} (L_{aad} - L_{aaq})(i_{d1} + i_{d2} + i_{d3})(i_{q1} + i_{q2} + i_{q3}) \\ + \psi_{fd}(i_{q1} + i_{q2} + i_{q3}) \end{array} \right] \tag{7}$$

From the flux Equation (5), it can be seen that there are complex cross couplings among the three d-q reference frames. It is therefore desirable to thoroughly understand the variables in the new model by relating them with variables in the conventional double d-q model, which is able to interpret physically. This can easily be performed by comparing the transformation matrices between the two models.

The Clark–Park transformation of the conventional double d-q model is given as:

$$\begin{cases} \begin{bmatrix} f_{D1} & f_{Q1} \end{bmatrix}^T = T_{DQ1} \begin{bmatrix} f_A & f_B & f_C \end{bmatrix}^T \\ \begin{bmatrix} f_{D2} & f_{Q2} \end{bmatrix}^T = T_{DQ2} \begin{bmatrix} f_X & f_Y & f_Z \end{bmatrix}^T \end{cases} \tag{8}$$

With

$$T_{DQ1} = \frac{2}{3} \begin{bmatrix} \cos\theta_e & -\sin\theta_e \\ \cos(\theta_e - 120°) & -\sin(\theta_e - 120°) \\ \cos(\theta_e + 120°) & -\sin(\theta_e + 120°) \end{bmatrix}^T \tag{9}$$

$$T_{DQ2} = \frac{2}{3} \begin{bmatrix} \cos(\theta_e - 30°) & -\sin(\theta_e - 30°) \\ \cos(\theta_e - 150°) & -\sin(\theta_e - 150°) \\ \cos(\theta_e + 90°) & -\sin(\theta_e + 90°) \end{bmatrix}^T \tag{10}$$

Then, the variables in the two models can be associated with the original six-dimensional space:

$$\begin{bmatrix} f_A & f_B & f_C & f_X & f_Y & f_Z \end{bmatrix}^T = T_{dq3}^{-1} \begin{bmatrix} f_{d1} & f_{q1} & f_{d2} & f_{q2} & f_{d3} & f_{q3} \end{bmatrix}^T \tag{11}$$

$$\begin{cases} \begin{bmatrix} f_A & f_B & f_C \end{bmatrix}^T = T_{DQ1}^{-1} \begin{bmatrix} f_{D1} & f_{Q1} \end{bmatrix}^T \\ \begin{bmatrix} f_X & f_Y & f_Z \end{bmatrix}^T = T_{DQ1}^{-2} \begin{bmatrix} f_{D2} & f_{Q2} \end{bmatrix}^T \end{cases} \tag{12}$$

By substituting (3), (8), and (9) into (10) and (11), the relationship can be obtained as

$$\begin{cases} f_{D1} = f_{d1} = f_{d2} = f_{d3} \\ f_{Q1} = f_{q1} = f_{q2} = f_{q3} \\ f_{D2} = f_{d1} = f_{d2} = f_{d3} \\ f_{Q2} = f_{q1} = f_{q2} = f_{q3} \end{cases} \tag{13}$$

It indicates that the variables in three d-q reference frames are always balanced, even if there is no careful regulation for current control or distribution. In addition, the same conclusion can be drawn by observing Figure 2a. The three rectangular coordinate systems consist of phase windings from two different sets. Each axis is only related to one phase winding, α-axes being coincident with phases of A, B, and C, and β-axes being opposite to phases of X, Y, and Z. Variables in the $\alpha\beta$ reference frame can be represented by their equivalence in the d-q reference frame by Park transformation, such as the q-axis components are consistent with amplitudes of $\alpha\beta$-axis components. Hence, the components in the three d-q reference frames should be symmetric.

By comparing (12) with (5), the flux can be rewritten as

$$\begin{cases} \psi_{d1} = L_d i_{d1} + \psi_{fd} \\ \psi_{q1} = L_q i_{q1} \\ \psi_{d2} = L_d i_{d2} + \psi_{fd} \\ \psi_{q2} = L_q i_{q2} \\ \psi_{d3} = L_d i_{d3} + \psi_{fd} \\ \psi_{q3} = L_q i_{q3} \end{cases} \tag{14}$$

where $L_d = 3L_{aad} + L_{aal}$ and $L_q = 3L_{aaq} + L_{aal}$. It can be observed that the three d-q reference frames are totally decoupled with respect to each other, which yields a very simple form for the machine equations, as shown in Figure 2b–d.

Therefore, the voltage equations are rewritten as:

$$\begin{cases} u_{d1} = R i_{d1} + L_d \frac{di_{d1}}{dt} - \omega_e L_q i_{q1} \\ u_{q1} = R i_{q1} + L_q \frac{di_{q1}}{dt} + \omega_e L_d i_{d1} + \omega_e \psi_{fd} \\ u_{d2} = R i_{d2} + L_d \frac{di_{d2}}{dt} - \omega_e L_q i_{q2} \\ u_{q2} = R i_{q2} + L_q \frac{di_{q2}}{dt} + \omega_e L_d i_{d2} + \omega_e \psi_{fd} \\ u_{d3} = R i_{d3} + L_d \frac{di_{d3}}{dt} - \omega_e L_q i_{q3} \\ u_{q3} = R i_{q3} + L_q \frac{di_{q3}}{dt} + \omega_e L_d i_{d3} + \omega_e \psi_{fd} \end{cases} \tag{15}$$

The electromagnetic torque can be rewritten as:

$$T_e = p_n \begin{cases} i_{q1}\left[(L_d - L_q)i_{d1} + \psi_{fd}\right] \\ +i_{q2}\left[(L_d - L_q)i_{d2} + \psi_{fd}\right] \\ +i_{q3}\left[(L_d - L_q)i_{d3} + \psi_{fd}\right] \end{cases} \tag{16}$$

It can be seen that the total output torque is the sum of torque generated by the three sets of two orthogonal windings. Thus, the dual three-phase machine can be regarded as three separate sets of orthogonal two-phase winding. It is assumed that the machine is surfaced mounted, $L_{aad} \approx L_{aaq}$, T_e can be simplified as

$$T_e = p_n \psi_{fd}(i_{q1} + i_{q2} + i_{q3}) = 3p_n \psi_{fd} I_T \tag{17}$$

where I_T is designated as torque current to be referred to torque capacity.

3. Interpretation of Phase Currents Using VSD and Triple Stationary Transformation Modeling Approaches

In the early research of multiphase machines, VSD is widely regarded as the most effective motor vector control approach [20]. It is, therefore, desirable to provide a better theoretical interpretation of the proposed method by relating it with the VSD control model. The VSD current transformed matrix is expressed as:

$$\begin{bmatrix} i_\alpha \\ i_\beta \\ i_{z1} \\ i_{z2} \end{bmatrix} = \frac{1}{3} \begin{bmatrix} 1 & -\frac{1}{2} & -\frac{1}{2} & \frac{\sqrt{3}}{2} & -\frac{\sqrt{3}}{2} & 0 \\ 0 & \frac{\sqrt{3}}{2} & -\frac{\sqrt{3}}{2} & \frac{1}{2} & \frac{1}{2} & -1 \\ 1 & -\frac{1}{2} & -\frac{1}{2} & -\frac{\sqrt{3}}{2} & \frac{\sqrt{3}}{2} & 0 \\ 0 & -\frac{\sqrt{3}}{2} & \frac{\sqrt{3}}{2} & \frac{1}{2} & \frac{1}{2} & -1 \end{bmatrix} \begin{bmatrix} i_A \\ i_B \\ i_C \\ i_X \\ i_Y \\ i_Z \end{bmatrix} \tag{18}$$

It is obvious that each phase current component involves five phase currents, which presents great complexity compared with the triple stationary transformation.

By substituting (17) into (2), the VSD variables can be mapped to a triple stationary reference frame:

$$\begin{cases} i_\alpha = \frac{1}{3}(i_{\alpha 1} - \frac{1}{2}i_{\alpha 2} - \frac{1}{2}i_{\alpha 3} - \frac{\sqrt{3}}{2}i_{\beta 2} - \frac{1}{2}i_{\beta 3}) \\ i_\beta = \frac{1}{3}(\frac{\sqrt{3}}{2}i_{\alpha 2} - \frac{\sqrt{3}}{2}i_{\alpha 3} + i_{\beta 1} - \frac{1}{2}i_{\beta 2} - \frac{1}{2}i_{\beta 3}) \\ i_{z1} = \frac{1}{3}(i_{\alpha 1} - \frac{1}{2}i_{\alpha 2} - \frac{1}{2}i_{\alpha 3} + \frac{\sqrt{3}}{2}i_{\beta 2} - \frac{\sqrt{3}}{2}i_{\beta 3}) \\ i_{z2} = \frac{1}{3}(-\frac{\sqrt{3}}{2}i_{\alpha 2} + \frac{\sqrt{3}}{2}i_{\alpha 3} + i_{\beta 1} - \frac{1}{2}i_{\beta 2} - \frac{1}{2}i_{\beta 3}) \end{cases} \tag{19}$$

In order to comprehend the relation intuitively and thoroughly, these variables are converted to dq reference frame:

$$\begin{cases} i_d = \frac{1}{3}(i_{d1} + i_{d2} + i_{d3}) \\ i_q = \frac{1}{3}(i_{q1} + i_{q2} + i_{q3}) \\ i_{z1} = \left(\frac{1}{3}i_{d1} - \frac{1}{6}i_{d2} - \frac{1}{6}i_{d3} - \frac{\sqrt{3}}{6}i_{q2} + \frac{\sqrt{3}}{6}i_{q3}\right)\cos\theta \\ \qquad + \left(-\frac{1}{3}i_{q1} + \frac{1}{6}i_{q2} + \frac{1}{6}i_{q3} - \frac{\sqrt{3}}{6}i_{d2} + \frac{\sqrt{3}}{6}i_{d3}\right)\sin\theta \\ i_{z2} = \left(\frac{1}{3}i_{q1} - \frac{1}{6}i_{q2} - \frac{1}{6}i_{q3} + \frac{\sqrt{3}}{6}i_{d2} - \frac{\sqrt{3}}{6}i_{d3}\right)\cos\theta \\ \qquad + \left(\frac{1}{3}i_{d1} - \frac{1}{6}i_{d2} - \frac{1}{6}i_{d3} - \frac{\sqrt{3}}{6}i_{q2} + \frac{\sqrt{3}}{6}i_{q3}\right)\sin\theta \end{cases} \tag{20}$$

The *dq* current components in VSD are proportional to the sum of the $d_1, d_2, d_3,$ and $q_1, q_2,$ and q_3 components in the triple stationary reference frame. On the other hand, $z1z2$ current harmonic components are associated not only with the triple stationery transformed variables but also with the rotor position. It can be shown that if the sum of d_1, d_2, d_3 and q_1, q_2, q_3 components are three times that in VSD, then the triple stationary control could have the same fundamental control performance as the VSD control without the stability issues. For restraining the harmonic components, i_{z1} and i_{z2} need to be zero, which is equivalent as:

$$\begin{cases} \frac{1}{3}i_{d1} - \frac{1}{6}i_{d2} - \frac{1}{6}i_{d3} - \frac{\sqrt{3}}{6}i_{q2} + \frac{\sqrt{3}}{6}i_{q3} = 0 \\ \frac{1}{3}i_{q1} - \frac{1}{6}i_{q2} - \frac{1}{6}i_{q3} + \frac{\sqrt{3}}{6}i_{d2} - \frac{\sqrt{3}}{6}i_{d3} = 0 \end{cases} \tag{21}$$

Considering the sum current of the neutral point is zero, we have

$$\begin{cases} i_A + i_B + i_C = 0 \\ i_X + i_Y + i_Z = 0 \end{cases} \tag{22}$$

Converted to the d-q axis by Park transformation, the neutral currents can be rewritten as:

$$\begin{aligned} &(i_{d1} - \tfrac{1}{2}i_{d2} - \tfrac{1}{2}i_{d3} + \tfrac{\sqrt{3}}{2}i_{q2} - \tfrac{\sqrt{3}}{2}i_{q3})\cos\theta \\ &-(i_{q1} - \tfrac{1}{2}i_{q2} - \tfrac{1}{2}i_{q3} - \tfrac{\sqrt{3}}{2}i_{d2} + \tfrac{\sqrt{3}}{2}i_{d3})\sin\theta = 0 \end{aligned} \tag{23}$$

Thus, the relationship between currents in *the d1q1, d2q2,* and *d3q3* axes can be written as:

$$\begin{cases} i_{d1} - \frac{1}{2}i_{d2} - \frac{1}{2}i_{d3} + \frac{\sqrt{3}}{2}i_{q2} - \frac{\sqrt{3}}{2}i_{q3} = 0 \\ i_{q1} - \frac{1}{2}i_{q2} - \frac{1}{2}i_{q3} - \frac{\sqrt{3}}{2}i_{d2} + \frac{\sqrt{3}}{2}i_{d3} = 0 \end{cases} \tag{24}$$

By solving Equations (11) and (14), the following can be obtained:

$$\begin{cases} i_{d1} = i_{d2} = i_{d3} \\ i_{q1} = i_{q2} = i_{q3} \end{cases} \tag{25}$$

Therefore, it can be concluded that as long as the components in the three stationary reference frames are balanced, the control performance is the same as the VSD, and the harmonics can be removed. The suggested control scheme guarantees balanced current sharing between the winding sets.

4. Control Scheme of Triple Stationary Reference Frame

The suggested control scheme is illustrated in Figure 3. These three α-β reference frames are independent of each other. It converts the feedback currents to three sets of *dq* components via Park transformation. As mentioned, the *d1*, *d2*, *d3* and *q1*, *q2*, *q3* components are dc variables. The cross-coupling between *d*- and *q*-axis can be eliminated by the decoupling voltage terms feed-forward compensation for the output of the current controllers:

$$\begin{cases} \Delta u_{d1} = -\omega_e L_q i_{q1} \\ \Delta u_{q1} = \omega_e\left(L_d i_{d1} + \psi_{fd}\right) \\ \Delta u_{d2} = -\omega_e L_q i_{q2} \\ \Delta u_{q2} = \omega_e\left(L_d i_{d2} + \psi_{fd}\right) \\ \Delta u_{d3} = -\omega_e L_q i_{q3} \\ \Delta u_{q3} = \omega_e\left(L_d i_{d3} + \psi_{fd}\right) \end{cases} \tag{26}$$

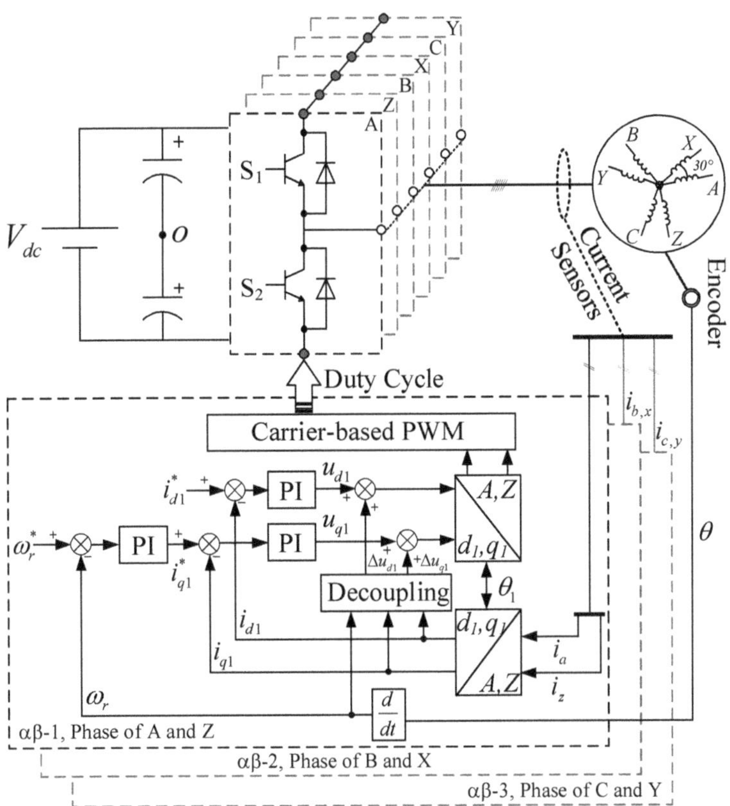

Figure 3. Control scheme of triple stationary frame.

Hence, the *dq*-currents can be effectively regulated without static errors by simple PI controllers. For surface-mounted PM machines, the reference currents of *d1,2,3*-axes set to zero are equivalent to maximum torque per ampere (MTPA), and the output of the speed regulator is as the common reference currents of *q1,2,3*-axes. It is obvious that the current PI controllers output phase voltages directly. The conventional stationary coordinate transformations between current PI controllers and phase voltages are avoided, which reduces the computation complexity. For synthesizing the required reference voltage vectors in the three d-q reference frames by a VSI, a carried-based PWM strategy was adopted since six different voltage vector components (u_{d1}, u_{q1}, u_{d2}, u_{q2}, u_{d3}, u_{q3}) are modulated in three decoupled parts simultaneously. Hence, each $\alpha\beta$ reference frame can be seen as a control unit, which contains a speed regulator, current controllers, PWM, and two-phase inverter, respectively. It realizes the triple-redundancy control of dual three-phase machines.

Compared with conventional double-redundancy control, the proposed scheme has many advantageous properties. More redundancies mean better post-fault control performance. Hence, the most obvious one is the ease of implementation of high-performance fault-tolerant control. Furthermore, both neutrals can be preserved under open-phase faults. The current constraints can especially be released by changing the neutral connection. It provides the possibility of improving the post-fault control performance, which conventional double d-q control has lacked.

5. Experimental Results

5.1. Experimental Setup

In order to verify the effectiveness of the proposed fault-tolerant control, an experiment platform was developed, which consists of two three-phase VSIs with a common dc link, a dual three-phase PM machine, and a dynamometer acting as the system load. Table 1 presents the parameters of the dual three-phase PM machine. Carrier-based PWM strategies are employed to generate PWM duties for each set. The VSIs operate at a switching frequency of 10 kHz. The setup includes an encoder for the rotor angle and speed feedback. The hardware platform based on dSPACE-1007 is shown in Figure 4.

Table 1. Parameters of dual three-phase PM machine.

Parameter	Value
Resistance (Ω)	0.48
Leakage inductance (mH)	0.262
d-axis self-inductance (mH)	0.28
q-axis self-inductance (mH)	0.28
PM flux linkage (Wb)	0.07
Pole pairs	5
DC link voltage (V)	50

5.2. Dynamic Performance Observation

The dynamic preference was tested by step response of given speed. During this test, the reference speed was changed from 0 to 250 r/min, rising to 500 r/min, then falling to 0. At last, the prototype runs the continuous speed transient state (500 r/min, 1 Hz) without load. Figure 5 shows the experimental results. Obviously, the proposed method achieves high-speed tracking accuracy, and even the distinction cannot be seen in Figure 5.

In order to showcase the rise time of speed step response, the given speed was increased from 0 to 500 r/min directly, without load. The speed response curve and q-axis currents are shown in Figure 6. The rise time was 0.028 s, and the currents changed rapidly and consistently. This control strategy presents good dynamic performance.

5.3. Compensation of Imbalance Currents

Previous studies show that asymmetric phase resistance and back EMF harmonics may induce currents in the z1z2 sub-plane, which results in imbalanced currents between two winding sets and additional copper loss [21]. However, the conventional VSD is incompetent to eliminate currents in the z1z2 sub-plane. The prototype possesses nonsinusoidal back EMFs, which are suited to verify the balance control.

During this test, the machine runs with a rated speed (500 r/min) and rated load (3.5 N.m). The conventional VSD experimental results are shown in Figure 7a. The currents between two winding sets present asymmetry. One peak current is 3.4A, while the other is about 4.1A. The phase currents are not purely sinusoidal, as shown in Figure 7b; there is a fifth harmonic in the spectrum. It is evident that the conventional VSD is not capable of regulating currents in the z1z2 sub-plane effectively.

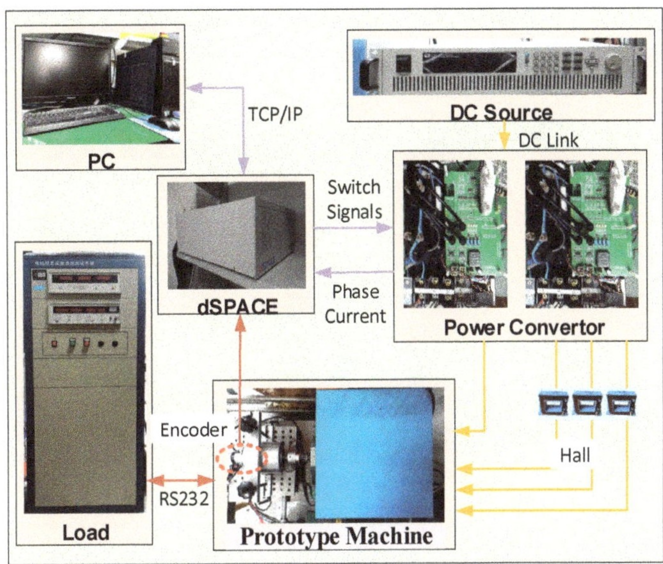

Figure 4. Experiment setup for dual three-phase PMSM drive testing.

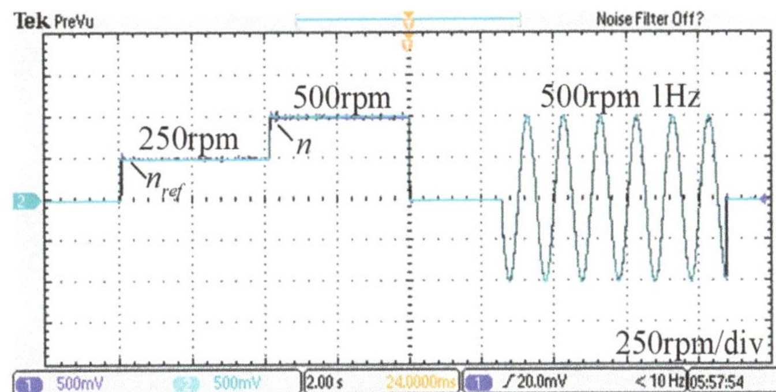

Figure 5. Continuous speed transient state: Channel 1: actual speed, Channel 2: reference speed (250 rpm/div), Horizontal: Time (2 s/div).

As a comparison, Figure 8a,b also shows the phase currents where the prototype was operated with the suggested control method in this paper. The phase peak currents of A and Z are both 3.7A. There are no current harmonics in the spectrum. It was concluded that the imbalance currents could be corrected with the triple stationary frame control scheme.

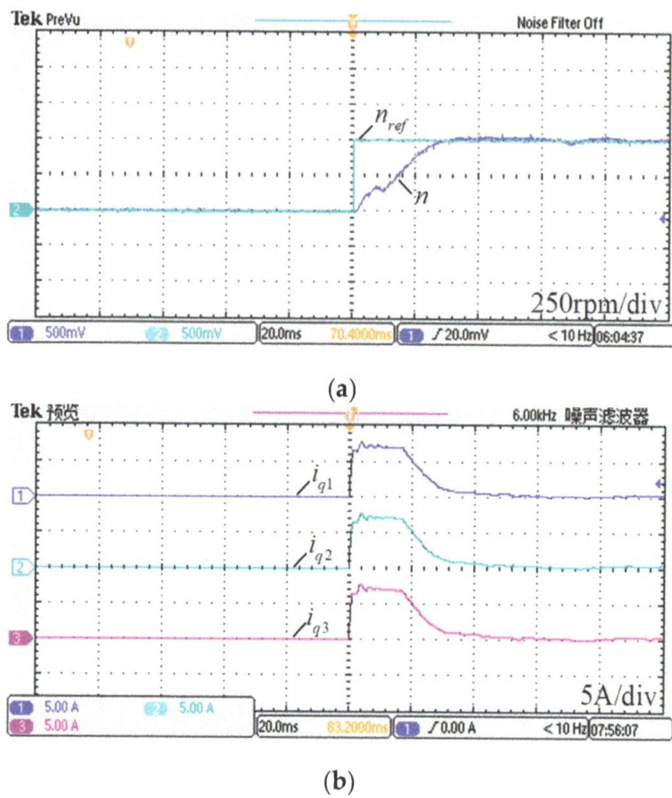

(**a**)

(**b**)

Figure 6. Step response of given speed, (**a**) speed: Channel 1: actual speed, Channel 2: reference speed (250 rpm/div), Horizontal: Time (0.02 s/div); (**b**) q-axis currents: Channel 1: q1, Channel 2: q2, Channel 3: q3, (5A/div), Horizontal: Time (0.02 s/div). (PS: "预览" means preview, "噪声滤波器" means Noise Filter).

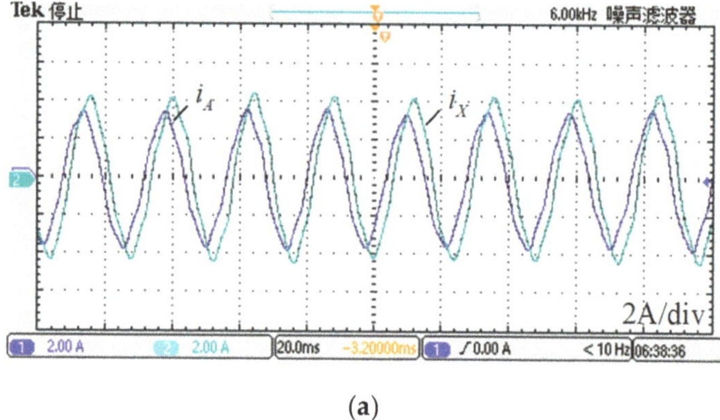

(**a**)

Figure 7. *Cont.*

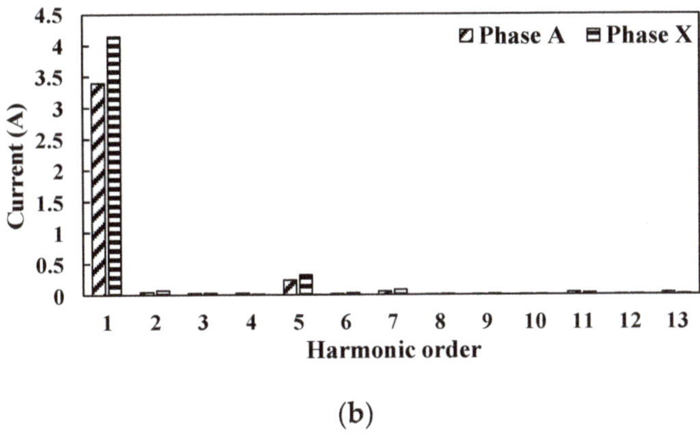

(b)

Figure 7. Experimental results with conventional VSD, (**a**) currents: Channel 1: phase of A, Channel 2: phase of X (2A/div), Horizontal: Time (0.02 s/div); (**b**) corresponding harmonics analysis. (PS: "停止" means stop, "噪声滤波器" means Noise Filter).

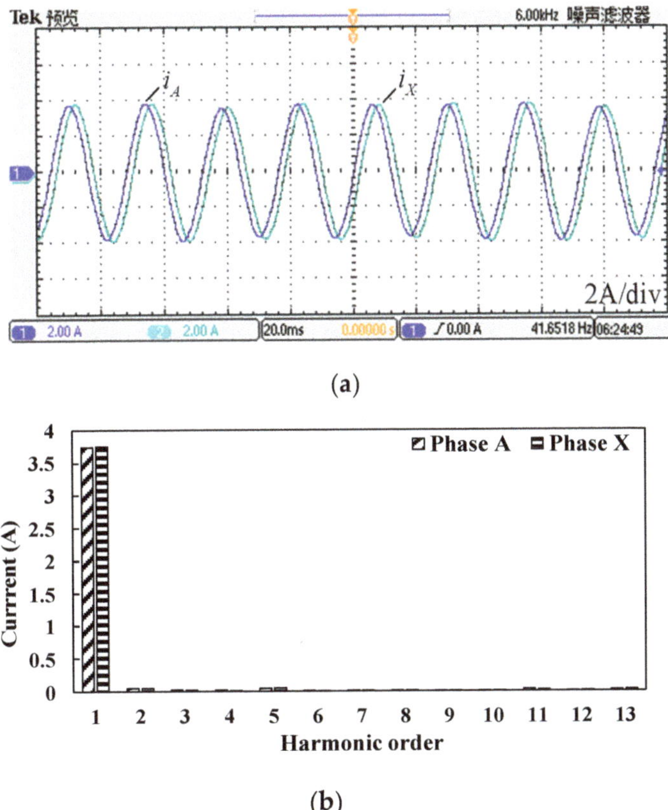

(a)

(b)

Figure 8. Experimental results with triple stationary control, (**a**) currents: Channel 1: phase of A, Channel 2: phase of X (2A/div), Horizontal: Time (0.02 s/div); (**b**) corresponding harmonics analysis. (PS: "预览" means preview, "噪声滤波器" means Noise Filter).

6. Conclusions

This paper proposes a vector control method for dual three-phase PM machines based on a triple stationary reference frame. The novel coordinate transformation and mathematical model were discussed. The fast dynamic response and imbalance of current inhibition were demonstrated by experiments. Compared with previous studies, the proposed method adopted three Park transformations to obtain the feedback of currents directly, omitting the complicated six dimensions transformed matrix. The vector control algorithm was simplified without sacrificing performance. The imbalanced currents caused by harmonic EMFs were suppressed effectively. Additionally, this suggested scheme theoretically achieves triple-redundancy control for dual three-phase machines. Although there is a lack of experimental verification, it is still a good solution for fault tolerance. Therefore, the proposed control scheme is suitable for high-speed machine applications, especially for the RMBMA system.

Author Contributions: Writing—original draft, J.-Y.Z.; Writing—review & editing, Q.Z. and K.W. All authors have read and agreed to the published version of the manuscript.

Funding: This work was supported in part by the National Natural Science Foundation of China (Project 61971431).

Conflicts of Interest: The authors declare no conflict of interest.

Nomenclature

u	Voltage	L_{aaq}	Q-axis self-inductance
i	Current	T_e	Torque
ψ	Flux	p_n	Pole pairs
R	Resistance	θ_e	Rotor position
W_c	magnetic co energy	θ_m	Mechanical position
L_s	inductance matrix	ω_e	Electric angular frequency
L_{aal}	Leakage inductance	I_s	Current matrix
L_{aad}	D-axis self-inductance	ψ_s	Flux matrix

References

1. Barani, N.; Sarabandi, K. Mechanical Antennas: Emerging Solution for Very-Low Frequency (VLF) Communication. In Proceedings of the 2018 IEEE International Symposium on Antennas and Propagation & USNC/URSI National Radio Science Meeting, Boston, MA, USA, 8–13 July 2018; pp. 95–96. [CrossRef]
2. Bickford, J.A.; Duwel, A.E.; Weinberg, M.S.; McNabb, R.S.; Freeman, D.K.; Ward, P.A. Performance of Electrically Small Conventional and Mechanical Antennas. *IEEE Trans. Antennas Propag.* **2019**, *67*, 2209–2223. [CrossRef]
3. Gong, S.; Liu, Y.; Liu, Y. A Rotating-magnet based mechanical antenna (RMBMA) for ELF-ULF wireless communication. *Prog. Electromagn. Res. M* **2018**, *72*, 125–133. [CrossRef]
4. Prasad, M.N.S.; Huang, Y.K.; Wang, E. Going beyond Chu Harrington limit: ULF radiation with a spinning magnet array. In Proceedings of the 2017 XXXIInd General Assembly and Scientific Symposium of the International Union of Radio Science (URSI GASS), Montreal, QC, Canada, 19–26 August 2017.
5. Selvin, S.; Prasad, M.S.; Huang, Y.; Wang, E. Spinning magnet antenna for VLF transmitting. In Proceedings of the 2017 IEEE International Symposium on Antennas and Propagation & USNC/URSI National Radio Science Meeting, San Diego, CA, USA, 9–14 July 2017.
6. Prasad, M.S.; Selvin, S.; Tok, R.U.; Huang, Y.; Wang, Y. Directly modulated spinning magnet arrays for ULF communications. In Proceedings of the 2018 IEEE Radio and Wireless Symposium (RWS), Anaheim, CA, USA, 15–18 January 2018; pp. 171–173.
7. Fawole, O.C.; Tabib-Azar, M. An electromechanically modulated permanent magnet antenna for wireless communication in Harsh electromagnetic environments. *IEEE Trans. Antennas Propag.* **2017**, *65*, 6927–6936. [CrossRef]
8. Lyra, R.O.C.; Lipo, T.A. Torque density improvement in a six-phase induction motor with third harmonic current in-jection. *IEEE Trans. Ind. Appl.* **2022**, *38*, 1351–1360. [CrossRef]
9. Hu, Y.S.; Zhu, Z.Q.; Odavic, M. Torque capability enhancement of dual three-phase pmsm drive with fifth and seventh current harmonics injection. *IEEE Trans. Ind. Appl.* **2017**, *53*, 4526–4535. [CrossRef]
10. Wang, K.; Zhang, J.Y.; Gu, Z.Y.; Sun, H.Y.; Zhu, Z.Q. Torque Improvement of Dual Three-Phase Permanent Magnet Machine Using Zero Sequence Components. *IEEE Trans. Magn.* **2017**, *53*, 8109004. [CrossRef]
11. Levi, E. Multiphase Electric Machines for Variable-Speed Applications. *IEEE Trans. Ind. Electron.* **2008**, *55*, 1893–1909. [CrossRef]

12. Eldeeb, H.M.; Abdel-Khalik, A.S.; Hackl, C.M. Dynamic Modeling of Dual Three-Phase IPMSM Drives with Different Neutral Configurations. *IEEE Trans. Ind. Electron.* **2019**, *66*, 141–151. [CrossRef]

13. Karttunen, J.; Kallio, S.; Peltoniemi, P.; Silventoinen, P.; Pyrhonen, O. Decoupled Vector Control Scheme for Dual Three-Phase Permanent Magnet Synchronous Machines. *IEEE Trans. Ind. Electron.* **2014**, *61*, 2185–2196. [CrossRef]

14. Che, H.S.; Levi, E.; Jones, M.; Hew, W.-P.; Rahim, N.A. Current Control Methods for an Asymmetrical Six-Phase Induction Motor Drive. *IEEE Trans. Power Electron.* **2014**, *29*, 407–417. [CrossRef]

15. Bojoi, I.R.; Tenconi, A.; Griva, G.; Profumo, F. Vector control of dual-three phase induction motor drives using two current sensors. *IEEE Trans. Ind. Appl.* **2006**, *42*, 1284–1292. [CrossRef]

16. Singh, G.; Nam, K.; Lim, S. A Simple Indirect Field-Oriented Control Scheme for Multiphase Induction Machine. *IEEE Trans. Ind. Electron.* **2005**, *52*, 1177–1184. [CrossRef]

17. Duran, M.; Kouro, S.; Wu, B.; Levi, E.; Barrero, F.; Alepuz, S. Six-phase PMSG wind energy conversion system based on medium-voltage multilevel converter. In Proceedings of the 2011 14th European Conference on Power Electronics and Applications, Birmingham, UK, 30 August–1 September 2011; pp. 1–10.

18. He, Y.; Wang, Y.; Wu, J.; Feng, Y.; Liu, J. A simple current sharing scheme for dual three-phase permanent-magnet synchronous motor drives. In Proceedings of the 2010 Twenty-Fifth Annual IEEE Applied Power Electronics Conference and Exposition (APEC), Palm Springs, CA, USA, 21–25 February 2010; pp. 1093–1096. [CrossRef]

19. Wang, X.; Wang, Z.; Gu, M.; Xiao, D.; He, J.; Emadi, A. Diagnosis-Free Self-Healing Scheme for Open-Circuit Faults in Dual Three-Phase PMSM Drives. *IEEE Trans. Power Electron.* **2020**, *35*, 12053–12071. [CrossRef]

20. Barrero, F.; Duran, M.J. Recent advances in the design, modeling, and control of multiphase machines—Part I. *IEEE Trans. Ind. Electron.* **2016**, *63*, 449–458. [CrossRef]

21. Hu, Y.; Zhu, Z.-Q.; Liu, K. Current Control for Dual Three-Phase Permanent Magnet Synchronous Motors Accounting for Current Unbalance and Harmonics. *IEEE J. Emerg. Sel. Top. Power Electron.* **2014**, *2*, 272–284. [CrossRef]

Article

Influence of the Shielding Winding on the Bearing Voltage in a Permanent Magnet Synchronous Machine

Sebastian Berhausen [1,*], Tomasz Jarek [2] and Petr Orság [3]

[1] Faculty of Electrical Engineering, Silesian University of Technology, 44-100 Gliwice, Poland
[2] Łukasiewicz Research Network—Institute of Electrical Drives and Machines KOMEL, 40-203 Katowice, Poland
[3] Faculty of Electrical Engineering and Computer Science, VSB-Technical University of Ostrava, 17. listopadu 15, 708 00 Ostrava, Czech Republic
*   Correspondence: sebastian.berhausen@polsl.pl

Abstract: This article presents selected methods of limiting the bearing voltages of synchronous machines with permanent magnets supplied from power electronic converters. The authors analyzed methods based on the use of various shielding windings placed in slot wedges and mounted in the stator end-winding region. The values of the parasitic capacitances of the machine, on which the levels of bearing voltages depend, were determined using the finite element method. Three-dimensional simulation models were used for the calculations. The analysis of the influence of the shielding windings on the bearing voltage waveforms was conducted on the basis of circuit models with two- and three-level converters. The obtained calculation results indicate a high potential in limiting bearing voltages.

Keywords: bearing voltages; PMSM; parasitic capacitances; FEM 3D; shielding winding

Citation: Berhausen, S.; Jarek, T.; Orság, P. Influence of the Shielding Winding on the Bearing Voltage in a Permanent Magnet Synchronous Machine. *Energies* **2022**, *15*, 8001. https://doi.org/10.3390/en15218001

Academic Editors: Quntao An, Bing Tian and Xinghe Fu

Received: 5 September 2022
Accepted: 25 October 2022
Published: 27 October 2022

Publisher's Note: MDPI stays neutral with regard to jurisdictional claims in published maps and institutional affiliations.

1. Introduction

Electric machines have been among the largest consumers of electricity for many years. They are used in many industries as key elements of many drive systems. It is also difficult to imagine everyday life without the possibility of using basic appliances with electric motors in households. The market of electric vehicles [1] has been developing rapidly for several years now—not only passenger cars but also public transport vehicles (e.g., buses [2]) and inland waterway boats [3]. This development was possible owing to, among other things, the development of highly efficient electric motors, particularly synchronous motors with permanent magnets, which are characterized by a high power-to-weight ratio, a high efficiency and good dynamic parameters [4]. Modern drive systems often operate at high rotational speeds, which means that the stator windings must be supplied with high-frequency voltage. In modern drive systems, in order to supply permanent magnet motors, power electronic converters are used, which enable the implementation of complex control strategies. The most frequently used converter systems are those based on PWM modulation, in which the regulation of output parameters (voltage, current and frequency) is performed by modulating the pulse width and changing the switching frequency of power electronic transistors. This control enables the smooth regulation of the frequency and amplitude of the first harmonic of the voltage supplying the electric machine. Undoubtedly, the advantages of using converters have led to their widespread use in electric drives. Unfortunately, these converters introduce higher harmonics into the voltage supplying the motor, which, in turn, triggers a number of unfavorable phenomena. The use of converters contributes to, among other things, an increase in the generated noise and an increase in the amplitudes of forces causing vibrations of the motor [5], leading to a higher risk of bearing damage. Another negative result caused by converter usage is the appearance of bearing voltages and currents [6]. There is a small layer of oil film between the raceways and rolling elements of the bearings. The bearing voltage can break the oil film, causing the current flow in the circuit consisting of the bearings, the frame and the machine shaft. Due to the small contact area between raceways and rolling elements, the current flow can reach densities that can damage the raceway surfaces. This current causes local pitting on the raceway surfaces and on the rolling elements of the bearings. An example of damage to the rolling element of a bearing resulting from the flow of the bearing current is shown in Figure 1. The bearing voltage is the main source of bearing currents. On the contact surfaces, that is, where rolling elements touch the raceways, an effect similar to electrical arc welding occurs. The current flowing through the bearing often causes local heating of the bearing, even to the melting point. In places where the metal has been melted, traces of discoloration or pitting of various shapes and sizes are formed [7]. The further flow of bearing currents in the presence of precipitated metal filings results in a deepening of the bearing damage process. The intensity of the damage largely depends on the intensity of the current, the time of its flow, the rotational speed, the mechanical load and the type of lubricant used [8].

Figure 1. Example of damage to a rolling element of a bearing due to the flow of the bearing current.

The available publications provide comprehensive information on bearing failures by categorizing the causes of their occurrence and present photographs of the consequences of bearing current flow [7,9–11]. Bearing current density J_b is defined as the ratio of the bearing current amplitude $I_{b(peak)}$ to the Hertzian contact area A_{Hz} (1):

$$J_b = \frac{I_{b(peak)}}{A_{Hz}}.$$

(1)

The available literature provides different values of the permissible bearing current densities, but in general, it can be assumed that values below $J_b < 0.1$ A/mm^2 do not threaten the bearing service life, while the bearing current with a density of $J_b > 1$ A/mm^2 significantly reduces the failure-free time of its operation [12]. For this reason, sliding bearings, which are used in high-power machines, are less sensitive to the phenomenon of bearing currents due to the much larger contact area between the bearing elements [12].

It is worth emphasizing here that bearing currents may appear not only when motors are supplied from power electronic converters but also when the supply voltage is sinusoidal [13]. The mechanisms for the generation of bearing currents are different for each case. When the machine is supplied with sinusoidal voltage, bearing currents are characterized by a low frequency ($f < 1$ kHz) and occur in machines with asymmetry in the electromagnetic circuit [14]. This asymmetry may be caused by, among other things, the asymmetry of the air gap (static or dynamic), damage to the magnets placed in the rotor, the asymmetrical connection of windings, the short-circuit of the stator sheets, the anisotropy of the magnetic circuit or the segmentation of the stator core, which occurs in large-sized electrical machines. Another reason for the formation of magnetic circuit asymmetry may be the asymmetrical arrangement of ventilation ducts or stator core mounting welds on the magnetic circuit. In such cases, a circular flux appears in the stator yoke, surrounding the machine shaft [15]. It induces a potential difference between the ends of the shaft, under the influence of which current flows in a closed circuit consisting of the frame, bearings and shaft [16]. For a symmetrical electromagnetic circuit of the machine, the distribution of the magnetic field across the cross-section is the same within each polar pitch, so circular flux does not occur.

In the case of machines powered by power electronic converters, the main role is played by the interaction of the fast-changing common-mode voltage occurring at the winding star point with the system of parasitic capacitances of the machine. In both cases, the bearing currents can cause severe damage to the raceways and rolling elements. The damage intensity depends mainly on the bearing current, its duration, its bearing load and its rotational speed. Therefore, methods of limiting bearing voltages in electrical machines are still being developed. Considering that the vast majority of permanent magnet synchronous machines work with power electronic converters, this article focuses on methods of limiting bearing voltages and currents under such power supply. More detailed information related to the discussed issues is described in the second chapter of this article.

The aim of the article is to present the idea of limiting bearing voltages in electric machines, particularly in synchronous machines with permanent magnets placed inside the rotor. The research presented in the article is a continuation of the publication [17], in which the authors conducted a detailed analysis of the impact of machine design solutions on parasitic capacitances. It was limited to the simulation of two-dimensional models, and the subject of the research was the comparison of, among other things, the shape of the stator slot opening, the number of shielding wires used in the slot opening space and the demonstration of their influence on parasitic capacitances and bearing voltages. The conclusions were used to conduct research aimed at determining the internal capacitances of the machine on the basis of a three-dimensional model of an electric machine with an IPM rotor.

2. Bearing Voltages and Currents in Electric Machines Powered by Power Electronic Converters

The vast majority of currently used power electronic converters work with PWM (Pulse Width Modulation), for which the SVM (Space Vector Modulation) method is most frequently used. It consists in forming the output voltage of the converter using vector relationships. In the case of a two-level power electronic converter, the diagram of which is shown in Figure 2a, each of the operating states of the system is represented by voltage vectors: six active vectors and two zero vectors. In the SVM algorithm, the vector of the set converter output voltage is generated for each transistor switching period as a combination of zero vectors and two adjacent active vectors. The transistor switching combinations corresponding to each control vector are shown in Figure 2b.

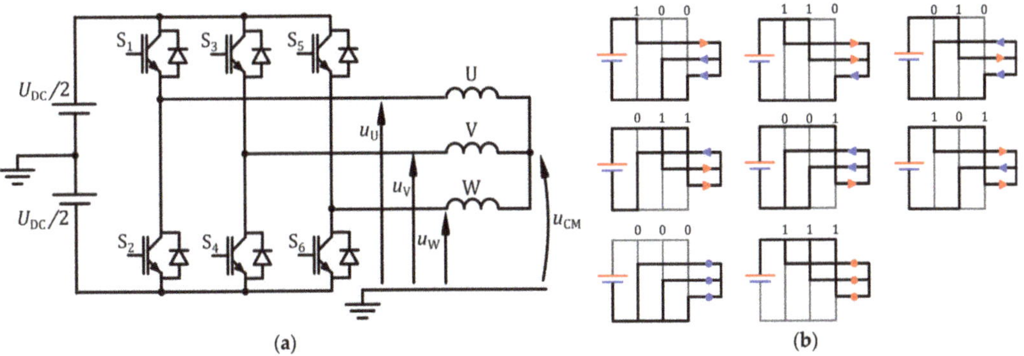

Figure 2. Diagram of a two-level DC-AC converter (**a**) and transistor switching combinations (**b**).

The sequence and switching times of the individual transistors are controlled by a vector modulator in the control system. The examples of the output voltages of a two-level converter, u_U, u_V and u_W, are shown in Figure 3. The voltages on the individual phases of the winding assume values that depend on the voltage in the DC-link circuit U_{DC}: $U_{DC}/2$ or $-U_{DC}/2$. The result of the voltages at the output of the converter formed in such a way is the presence of a non-zero voltage at the neutral point of the winding called the common-mode voltage u_{CM}, the value of which is equal to the average value of the voltages in the individual phases according to formula (2) [18]:

$$u_{CM} = \frac{u_U + u_V + u_W}{3}.$$

(2)

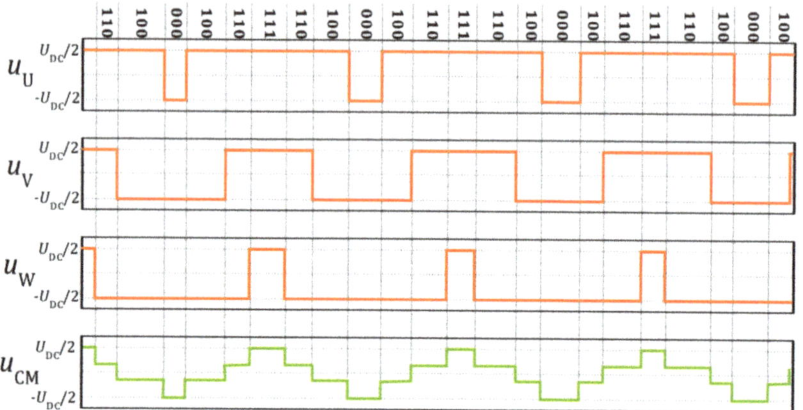

Figure 3. Examples of waveforms of phase voltages at the output terminals of a two-level DC/AC converter and the common-mode voltage u_{CM}.

The turn-on times of transistors depend on the amplitude of the voltage vector and the sector in which it is located. The construction of the voltage vector V_{ref} can be described by Equation (3) [19]:

$$\vec{V}_{ref} = \frac{T_0}{T_S}\vec{V}_0 + \frac{T_1}{T_S}\vec{V}_1 + \frac{T_2}{T_S}\vec{V}_2,$$
$$T_S = T_0 + T_1 + T_2,$$

(3)

where:

V_0—zero vector;

V_1, V_2—active vectors;

T_s—switching period of transistors ($1/f_s$);

T_0—duration of the zero vector;

T_1, T_2—durations of active vectors V_1 and V_2.

The common-mode voltage when the machine is fed from a two-level converter has the shape of a stepped curve, with a frequency equal to the switching frequency of the transistors and a steepness resulting from the switching rate. The u_{CM} voltage assumes values of $-U_{DC}/2$, $-U_{DC}/6$, $U_{DC}/6$ and $U_{DC}/2$, with amplitudes falling on zero vectors, as shown in Figure 3.

As shown above, powering the motor from a power electronic converter contributes to the occurrence of a non-zero common-mode voltage u_{CM} in the stator winding. The presence of this voltage can contribute to capacitive or discharge-bearing current (EDM) [15]. Capacitive current refers to the case where the bearing current $i_b = (du_b/dt) \cdot C_b$ flows through the oil film of the bearing. Due to the small capacitance of the bearing C_b, capacitive currents are usually harmless to the bearing. The second type of bearing current (EDM) occurs when the oil film is punctured, and it appears as a point short-circuit arc that destructively affects the bearing raceway surfaces. The main elements through which the bearing current flows are the metal structural parts of the machine. Since the resistances of the shaft, bearing and frame are very small, the oil film has a decisive influence on the value of the bearing current. Its thickness decreases with increasing temperature. If its electrical strength is exceeded, then a short-term current pulse flows through the bearing, which is the main cause of pitting on the bearing raceways and rolling elements. Both types of currents occur when there is a stepped, rapidly changing common-mode voltage in the machine winding that affects the parasitic capacitances of the machine.

2.1. Parasitic Capacitances of the Machine and Their Equivalent Diagram

As described above, the power supply of a machine from a power electronic converter involves the occurrence of common-mode voltage u_{CM}, whose rapid changes du_{CM}/dt reaching up to several kV/μs stimulate the capacitive couplings present in an electric machine and play an important role in the mechanism of generating bearing currents. These capacitances occur between three characteristic elements (frame, winding and rotor) isolated from each other. In order to analyze the equivalent circuit that takes into account the parasitic capacitances of the machine, it is necessary to distinguish its following design elements:

- Stator iron core, placed in the grounded frame of the machine;
- Stator winding, isolated from the inner surface of the stator slots and supplied from an external voltage source;
- Rotor installed inside the frame by bearings, isolated from the structure due to the presence of an oil film.

Bearing in mind the insulation between the stator winding and the ferromagnetic circuit, as well as the isolation of the rotor from the other structural components of the machine (due to the presence of an oil film in the bearings), one can formulate its equivalent circuit diagram that consists of parasitic capacitances [20]. In a simplified model of an electric machine, capacitances can be distinguished: C_{ws}—the capacitance between the stator winding and the grounded stator, C_{wr}—the capacitance between the stator winding and the rotor, C_{sr}—the capacitance between the rotor and the grounded stator, C_{b1} and C_{b2}—the capacitances of the bearings on the drive and non-drive sides [21]. It is worth noting that the capacitance C_{wr} occurs both along the active length of the stator iron core and in the end-windings region. Thus, it is possible to distinguish two components of C_{wr} capacitance connected in parallel: $C_{wr(ew)}$ and $C_{wr(slot)}$, which correspond to the capacitance between the windings located in the stator slots and the end-winding capacitance, respectively [22]. The location of the above-mentioned parasitic capacitances is illustrated in a cross-sectional view of the electric machine in Figure 4, whereas the equivalent diagram with lumped parameters is shown in Figure 5. It is worth noting here that, in the case of supplying the stator windings with a sinusoidal voltage at the frequency of f = 50 Hz, these capacitances do not play any role in generating bearing voltages due to the large value of reactance $X_c = 1/(2\pi f C)$.

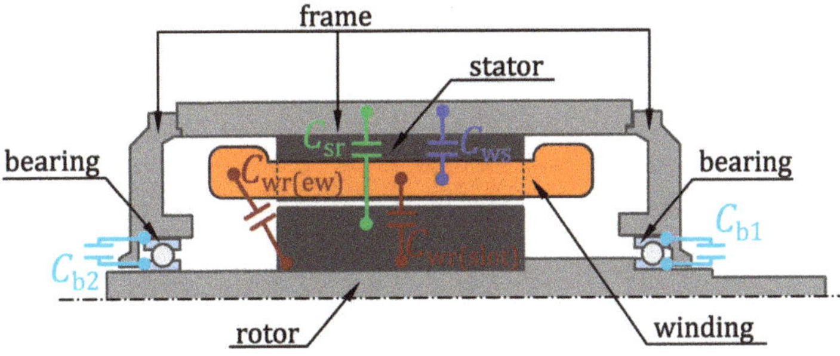

Figure 4. Distribution of parasitic capacitances in an electric machine.

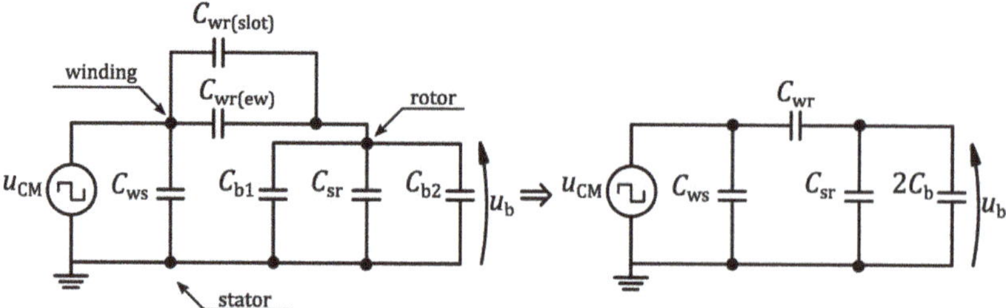

Figure 5. Equivalent diagram of parasitic capacitance with a specification of common-mode voltage u_{CM} and bearing voltage u_b.

The value of the C_{ws} capacitance depends on, among other things, the geometrical dimensions of the machine and the configuration of the windings, as well as on the dielectric parameters of the stator slot insulation used. The C_{wr} value depends on the shape and dimensions of the electromagnetic circuit. The article omits the bearing resistances, limiting it to the equivalent diagram that consists only of parasitic capacitances.

The equivalent circuit diagram shown in Figure 5 explains that the source of the voltage u_b that occurs between the raceways of the bearing is the common-mode voltage u_{CM}. The ratio of these voltages (BVR—Bearing Voltage Ratio) is given by the equation [23,24]:

$$\mathrm{BVR} = \frac{u_b}{u_{CM}} = \frac{C_{wr}}{C_{b1} + C_{b2} + C_{sr} + C_{wr}} = \frac{C_{wr(ew)} + C_{wr(slot)}}{C_{b1} + C_{b2} + C_{sr} + C_{wr(ew)} + C_{wr(slot)}}. \tag{4}$$

The above equation shows the relationship between the bearing voltage and the common-mode voltage, ignoring the stator-to-ground resistance. The higher the BVR coefficient, the greater the observed values of the bearing voltages, which are the main sources of currents flowing through the bearings. Various methods are used to limit the bearing voltages. They are described in the next chapter of the article.

3. Methods of Limiting Bearing Voltages

The dynamic development of electric machines and drives has contributed to the development of various methods of counteracting and reducing the effects of bearing current flow. Generally, two groups of methods can be distinguished:

- Methods based on limiting the source of bearing voltages and currents;
- Methods based on the introduction of high impedance into the potential flow path of bearing currents.

The first group of methods includes methods based on modifications to the power supply systems of the electric machine (the use of multi-level converters, the modification of control algorithms [25], the use of additional passive filters [26] or active filters [24,27]) as well as modifications to the machine structure (the use of shielding windings, the use of an oblique opening of the stator slots [28], etc.). The second group of methods is restricted to the use of insulated bearings—fully ceramic or hybrid [29]. The use of appropriate countermeasures depends primarily on the type of supply and the applicability of individual methods. Both groups of methods can be combined and applied in drive systems in order to provide protection against the destructive phenomenon of bearing voltages and currents.

As was shown in the previous chapter, the value of the bearing voltage depends on the parasitic capacitances, particularly on the capacitance between the stator winding and the rotor. Different solutions aimed at limiting the C_{wr} capacitance are presented in existing publications on the subject. One of them is to increase the distance between the stator windings located in the slots and the rotor surface [30]. Solutions based on the introduction of a grounded shield into the slot opening space [31,32] and around the stator end-winding [33–35] are highly effective. These shields are implemented in the form of, among other things, tapes of conductive material or copper wires.

The concept suggested by the authors is based on the use of special wedges closing the stator slots, in which there are copper wires along the entire length. These wires can be connected, for example, in series, thus creating a special shielding winding, one terminal of which is connected to the grounded frame. Such a solution is characterized, among other things, by the following advantages: the wires are evenly distributed in the wedges along the entire slot opening length of the stator core and are also prevented from accidentally touching the stator winding. The concept and an exemplary connection of the wires in the slot wedges are shown in Figures 6 and 7.

Figure 6. The concept of using the shielding winding located in the stator slot wedges.

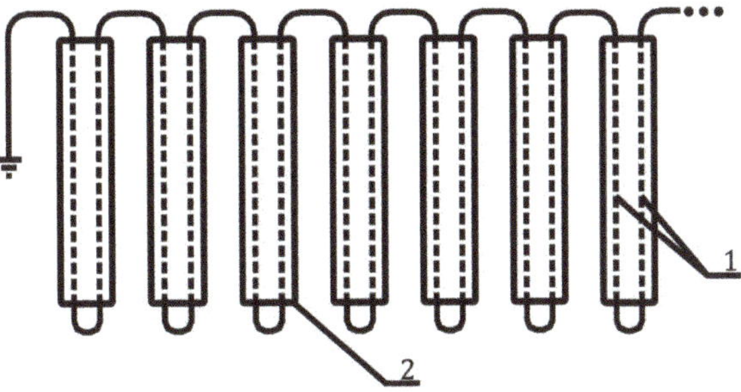

Figure 7. An example of a connection of shielding wires (1) placed in slot wedges (2).

4. Analysis of the Influence of Shielding Windings on the Parasitic Capacitances of a Permanent Magnet Synchronous Machine

In order to determine the parasitic capacitances of an electric machine, a three-dimensional simulation model of a synchronous machine with permanent magnets placed inside the rotor (IPM) was developed. The choice of such a machine model is due to the numerous advantages of such design. These include, first of all, the presence of a reluctance torque, which interacts with the torque produced by the permanent magnets [36]. The reason for the presence of the reluctance torque is the magnetic asymmetry of the rotor, since the reluctances for the armature interaction fluxes in the direct and quadrature axes are not equal. Compared to machines with SPM rotors, IPM rotors have a higher torque overload and also do not require additional protection against the effects of centrifugal force on the magnets.

Traditional methods for the analytical determination of parasitic capacitances are being replaced by modern numerical methods more and more often. Nowadays, among the increasingly used numerical methods intended, among other things, for calculating the parameters of electrical machines is the finite element method. With this method, it is possible to develop models of electrical machines that take into account multivariate design changes without the need to build costly prototypes [37]. A certain limitation in determining parameters in this way is the relatively long calculation time, especially in cases where three-dimensional models of electrical machines are investigated.

In order to analyze the effect of shielding windings on parasitic capacitances and bearing voltages in the machine, it was decided to conduct a series of analyses based on simulations of three-dimensional FEM models. First, a reference model without shielding windings was developed. Using such a model, the values of the parasitic capacitances of the machine were determined, which was the starting point for the analysis of the effect of different shielding winding designs on the parasitic capacitances and BVR coefficients. The purpose of such a comparative analysis is to try to answer the question "which of the analyzed variants has the greatest effectiveness in reducing dangerous bearing voltages".

4.1. Calculation of Parasitic Capacitances for the Reference Machine Model

The developed reference model lacks solutions that would limit the value of capacitance C_{wr}, and it is treated as a starting model for further analysis. The model was developed based on the design data of the actual machine shown in Table 1. The basic dimensions of the machine are shown in Figure 8. The stator winding was modeled as a single geometric solid, which significantly simplified the model and made it possible to reduce the number of mesh elements. It is practically impossible to model all the wires of the stator winding with their mutual alignment in a three-dimensional model. The construction of the model and calculations to determine the parasitic capacitances of the machine were performed in the Ansys Maxwell environment using an electrostatic solver. The calculations presented in the paper were carried out using a workstation equipped with a 32-core processor (Threadripper 3970×) supporting a 256 GB quad-channel DDR4 memory controller.

Table 1. List of parameters of the considered electric machine.

Parameter	Value/Type
Frame size	132 mm
Rated power	7 kVA
Rated voltage	400 V
Number of phases	3
Type of winding	Single-layer, arranged in stator slots
Number of poles	4
Air gap width	1.5 mm
Number of stator slots	36

In order to ensure reliable results of the simulation model calculations, the finite element mesh was parameterized in a way that ensures a high density of finite elements in the key spaces of the model, particularly in the space between the stator and the rotor. The generated mesh is shown in Figure 9. The number of mesh elements was approximately 8.9 million tets.

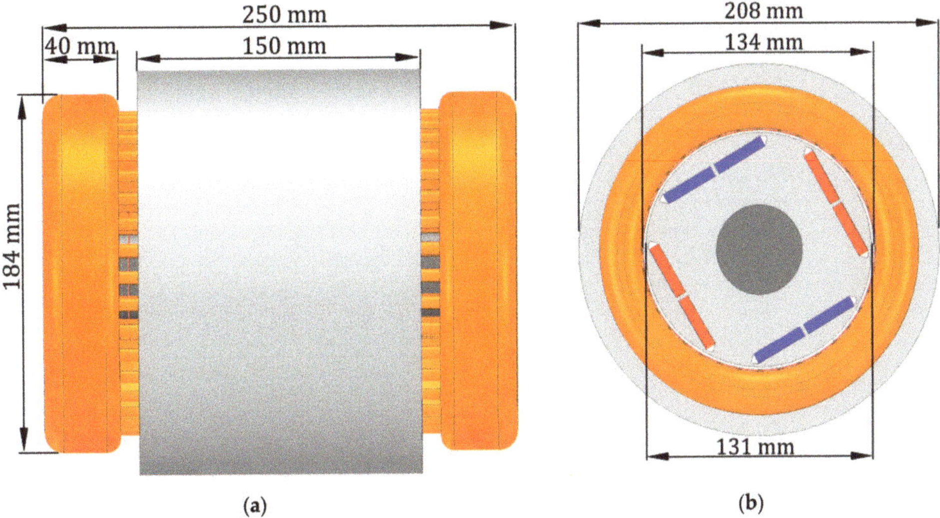

Figure 8. Reference model of a PMSM machine with an IPM rotor, with the most important geometrical dimensions marked: (**a**) side view, (**b**) front view.

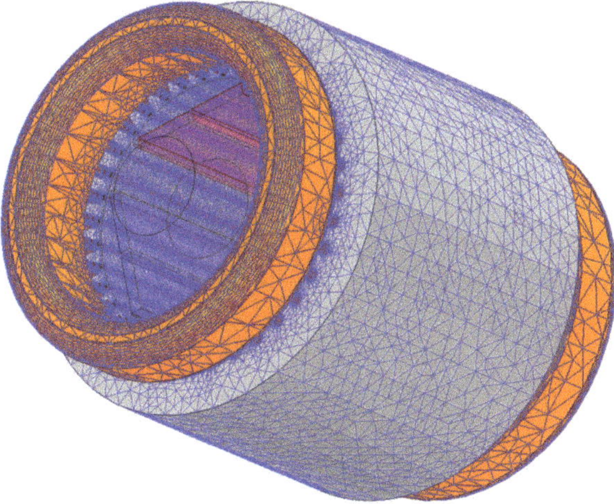

Figure 9. Finite element mesh of the reference model.

The presented model is a reference model that was used to determine the parasitic capacitances of the machine: $C_{wr(ref)}$ = 33.84 pF, $C_{ws(ref)}$ = 4921 pF, $C_{sr(ref)}$ = 324 pF. Such values of capacitances determine the contribution of the bearing voltage in the common-mode voltage defined by the parameter BVR = 6.07%. These values define the starting point for the analysis of the effectiveness of the solutions discussed in the subsequent chapters of the article. The obtained values may differ from the capacitances of the real model, particularly due to the simplification of the winding shape. However, in further simulations, this shape remains unchanged, so the obtained changes in the capacitances values will reflect the actual yield and effectiveness of the analyzed solutions. Since the article provides, among other things, an analysis of the influence of the shielding winding located in the end-windings region, it becomes reasonable to modify the reference model, by means of which it will be possible to calculate the capacitance between the end-windings and the rotor $C_{wr(ew)}$. The next subsection of the article is dedicated to this issue.

Modified Reference Model

In order to determine the contribution of the $C_{wr(slot\_ref)}$ and $C_{wr(ew\_ref)}$ components to the total capacitance $C_{wr(ref)}$ between the stator winding and the rotor of the machine under consideration, the simulation model was modified by depriving it of end-windings, as shown in Figure 10. Using the modified model created in this way made it possible to determine the capacitance $C_{wr(slot\_ref)}$ of the reference model. The difference in values for the full and modified reference model determines the value of the capacitance that exists between the stator end-windings and the rotor $C_{wr(ew\_ref)}$:

$$C_{wr(ew\_ref)} = C_{wr(ref)} - C_{wr(slot\_ref)}. \tag{5}$$

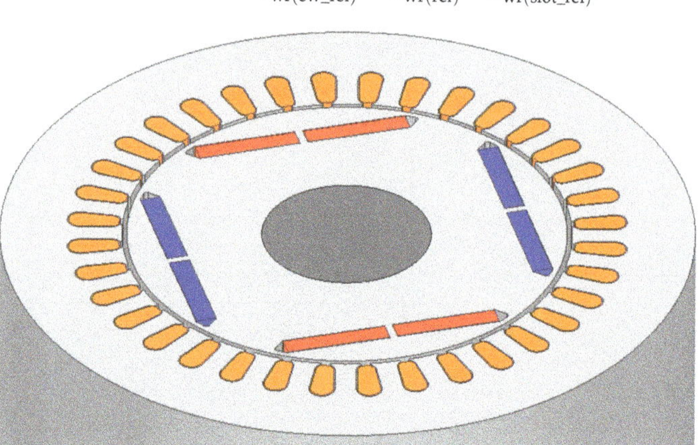

Figure 10. Reference model without the stator end-windings.

All model dimensions, except for the end-windings, remained unchanged. The values of the components of the capacitance ($C_{wr(ref)}$, $C_{wr(slot\_ref)}$, $C_{wr(ew\_ref)}$) determined for the reference model are presented in Table 2. The presented values show that the contribution of capacitance in the slot part $C_{wr(slot\_ref)}$ in relation to the total capacitance $C_{wr(ref)}$ amounts to approximately 69%. In view of the above, it is reasonable to shield this machine space, which fits well with the concept of using slot wedges with a shielding winding proposed by the authors, the effectiveness of which is the subject of the studies described in an upcoming part of the article. The ratio of the capacitance components $C_{wr(ew)}$ and $C_{wr(slot)}$ depends on the geometrical dimensions of the machine's electromagnetic circuit, particularly on the length of the iron core and the length of the end-winding overhang [38]. Thus, the ratio will be different for high-speed and low-speed machines.

Table 2. Summary of the values of the components of $C_{wr(ref)}$ capacitance: $C_{wr(slot\_ref)}$ and $C_{wr(ew\_ref)}$.

$C_{wr(ref)}$	$C_{wr(slot\_ref)}$	$C_{wr(ew\_ref)}$
33.84 pF	23.48 pF	10.36 pF

Next, the modified reference model was supplemented with shielding winding wires placed in slot wedges. Similar to the case above, the model lacked the end-windings. Simulation studies allowed us to determine the effect of using such a shielding winding on the value of capacitance $C_{wr(slot)}$. In this case, the value of capacitance $C_{wr(slot)} = 3.17$ pF was obtained. It means a reduction of this component by 86.5% with respect to the model without shielding windings, which shows the high effectiveness of the suggested solution in reducing the capacitance of $C_{wr(slot)}$.

4.2. Analysis of the Influence of the Shielding Winding Placed in the Stator Slots on the Parasitic Capacitances of the Machine

According to the previously described concept, the shielding winding was added to the reference model from Section 4.1, located in the wedges closing the stator slots. This winding consists of 72 copper wires, with 2 wires placed in each wedge, connected in series, as shown in Figure 11.

Figure 11. A 3D model of the machine with a visualized shielding winding for $l_{ew} = 5$ mm.

The analysis of the effect of the shielding winding was carried out in two variants. The first variant involved studying the effect of the diameter of the wire forming the shielding winding on the parasitic capacitances of the machine. The range of these changes was from 0.45 to 0.6 mm. Figure 12 shows the characteristics of the changes in these capacitances. The second variant of calculations concerned the analysis of the effect of the length of end-windings overhang in range from 2.5 to 15 mm. The results of these calculations are shown in Figure 13. For each of the analyzed cases, the number of mesh elements slightly differed and ranged from 22 to 28 million tets. Such a large number of elements resulted from, among other thing, the small diameters of the shielding wires used.

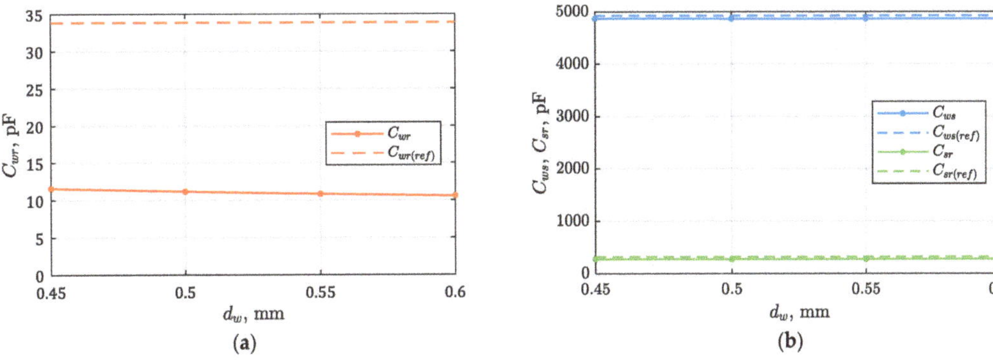

Figure 12. The results of the calculations of parasitic capacitances: C_{wr} (**a**) and C_{ws}, C_{sr} (**b**) for different diameters of the shielding wire and the length of the end-winding overhang $l_{ew} = 2.5$ mm compared to the capacitance values of the reference model.

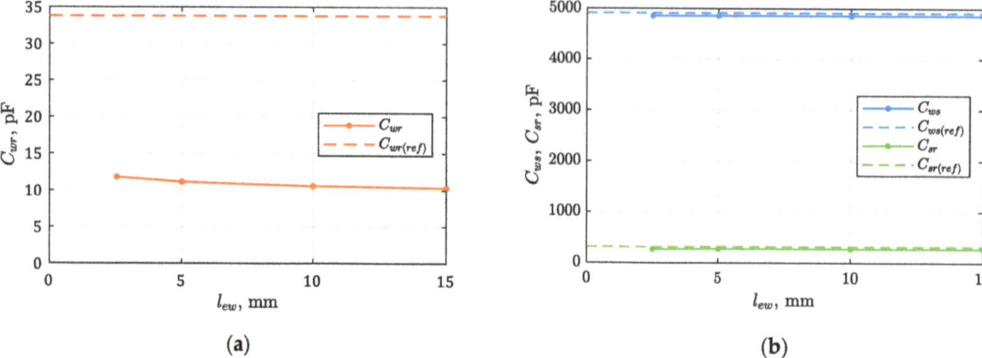

Figure 13. The results of the calculations of parasitic capacitances: C_{wr} (**a**) and C_{ws}, C_{sr} (**b**) for different overhang lengths of the shielding winding and the wire diameter $d_w = 0.5$ mm compared to the capacitance values of the reference model.

On the basis of the performed calculations, it can be concluded that the use of the shielding winding located in the stator slot wedges reduces the value of the capacitance C_{wr} compared to the same capacitance determined for the reference model. The reduction in this capacitance in relation to the reference model ranges from 65.0 to 69.3%, depending on the analyzed calculation variants. The values of the remaining capacitances C_{ws} and C_{sr} are practically insensitive to changes in the parameters l_{ew} and d_w. The reduction in the C_{wr} capacitance along with the increase in the diameter of the shielding wires result from the increasing filling of the slot opening space by the grounded shielding winding. The connections of the shielding winding are located under the stator end-windings; hence, with the increase in the l_{ew} parameter, a non-significant reduction in the capacitance C_{wr} can be observed.

4.3. Analysis of the Effect of a Shielding Winding Placed in the End-Winding Region on the Parasitic Capacitances of the Machine

The solution based on the shielding winding placed in the slot wedges, due to its location, noticeably reduces the value of the capacitance component $C_{wr(slot)}$ and only slightly reduces the value of the component $C_{wr(ew)}$. As previously shown, the $C_{wr(ew)}$ capacitance is approximately 30% of the total C_{wr} capacitance. In view of the above, it is reasonable to study the impact of the effect of an additional winding, placed at the end-windings. For this purpose, the reference model was supplemented with an additional helix winding made of copper wires located on both sides of the stator iron core, as shown in Figure 14. The diameter of the helix shielding winding is $d_w = 0.5$ mm. This helix was placed at a distance of $h_1 = 4.0$ mm from the stator iron core, and its pitch is $y_{helix} = 2$ mm. The simulations were carried out for a different number of helix turns: $n_{helix} = 3 \div 6$, and the results of the obtained parasitic capacitances of the machine are shown in Figure 15a,b.

Due to the location of the helix shielding winding (directly under the end-winding, at a distance of $h_1 = 4.0$ mm from the stator iron core), its contribution to the reduction in the capacitance component $C_{wr(slot)}$ is negligible. The change in capacitance C_{wr} caused by this winding refers to a reduction in the capacitance component of $C_{wr(ew)}$ by 5.62 pF (54%) for $n_{helix} = 3$ and by 7.70 pF (74%) for $n_{helix} = 6$.

Figure 14. A 3D model with a shielding winding in the form of a helix placed under the stator end-winding.

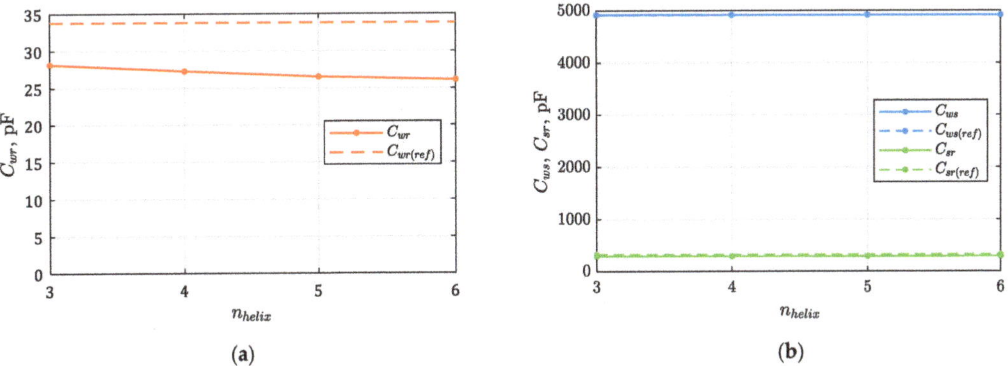

(a)

(b)

Figure 15. Results of parasitic capacitances calculations: C_{wr} (**a**) and C_{ws}, C_{sr} (**b**) depending on the number of turns of the helix shielding winding n_{helix} compared to the capacitance values of the reference model.

4.4. Determination of Parasitic Capacitances for the Final Machine Model

The last studied variant is the solution that uses both shielding windings described in the previous sections. In the case of the winding placed in the slot wedges, it was decided to have a minimum overhang length of the shielding winding $l_{ew} = 2.5$ mm (in order to provide space for the placement of the winding in the form of a helix), and the value of the diameter of the shielding wire $d_w = 0.5$ mm. In the case of the shielding winding, it was limited to a helix with the number of turns $n_{helix} = 4$. The model developed in this way is shown in Figure 16. Due to the presence of a large number of objects and small geometric dimensions, the model is characterized by a large number of mesh elements—more than 33 million tets. Since the number of finite elements is so high, the calculation time is very long; for the presented variant, the calculations lasted almost 8.5 days. The values of the obtained parasitic capacitances are shown in Table 3.

Table 3. Parasitic capacitances of the model with two shielding windings compared to the capacitance values of the reference model.

C_{wr}	$C_{wr(ref)}$	C_{ws}	$C_{ws(ref)}$	C_{sr}	$C_{sr(ref)}$
8.05 pF	33.84 pF	4860 pF	4921 pF	277.3 pF	324.0 pF

Figure 16. A 3D model with shielding windings: in the stator slot wedges and in the form of a helix under the stator end-winding.

4.5. Summary

In summary, the obtained results were limited to four variants:

- Ref.—reference model without any solutions reducing the C_{wr} capacitance;
- Var. A—model with shielding windings in stator slots, $d_w = 0.5$ mm, $l_{ew} = 2.5$ mm;
- Var. B—model with helix shielding winding, $n_{helix} = 4$;
- Var. C—model with shielding winding in stator slots, $d_w = 0.5$ mm, $l_{ew} = 2.5$ mm, and with helix shielding winding, $n_{helix} = 4$ (Var. A + Var. B).

The values of capacitance C_{wr} and BVR coefficients for the mentioned design variants are compared graphically in Figure 17a,b. They show that the effective solution is the shielding winding located in the stator slots. It makes it possible to significantly reduce the capacitance C_{wr} and, consequently, the BVR coefficient. Its effectiveness can be further improved by adding a helix winding placed from the inside of the stator end-windings. If only the helix winding is used, the effect is unsatisfactory. It is also worth noting that the overall impact on the reduction in C_{wr} for variants A and B is greater than that for variant C. To determine the value of the BVR coefficient, the bearing capacitances $C_{b1} = C_{b2} = 100$ pF were used.

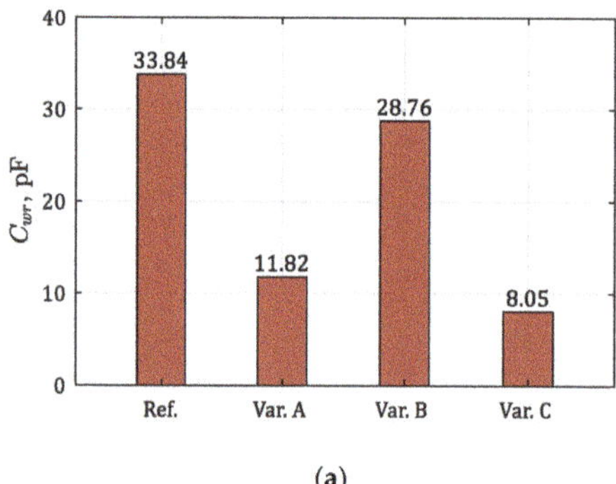

(a)

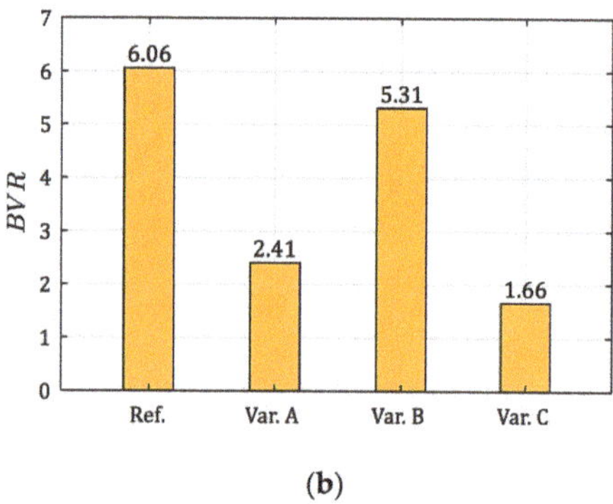

(b)

Figure 17. Comparison of capacitances C_{wr} (**a**) and BVR coefficients (**b**) for selected solution variants.

5. Analysis of the Effect of Machine Shielding Windings on Bearing Voltage Levels

As shown in Section 2.1, the bearing voltage u_b depends on the value of the common-mode voltage u_{CM} and the values of the machine's parasitic capacitances C_{wr}, C_{sr} and C_{ws}.

The calculations presented in following subsections of the paper are aimed at answering the question: what is the effect of the machine shielding windings suggested in the article on the bearing voltage levels when fed from a PWM converter? In order to answer this question, circuit models of two-level and three-level DC/AC converters were developed. The converters, along with the machine winding system and parasitic capacitances, were developed in the Matlab/Simulink environment. The following assumptions were made:

- The switching on of IGBT transistors occurs without delay;
- The switched-off transistor does not conduct current;
- The diode present in the structure of the transistor in the conducting state is replaced by an ideal diode;
- Voltages in the DC-link circuits have the same values.

5.1. Results of the Calculations of Bearing Voltages When the Machine Is Powered by a Two-Level Converter

The basic and most widely used converter for powering permanent magnet synchronous machines is a two-level converter. It consists of six transistors connected in three branches. In the developed model shown in Figure 18, the output of the converter is connected to the three-phase stator winding of the machine represented by three RL series circuits. A voltage divider consisting of capacitances C_{wr}, C_{ws}, C_{sr} and C_b is connected to the Artificial Neutral Point, which was used to determine the bearing voltage u_b. The value of the voltage in the DC-link circuit U_{DC} is 565 V. The bearing voltage waveforms obtained by simulation for the selected variants are shown in Figure 19.

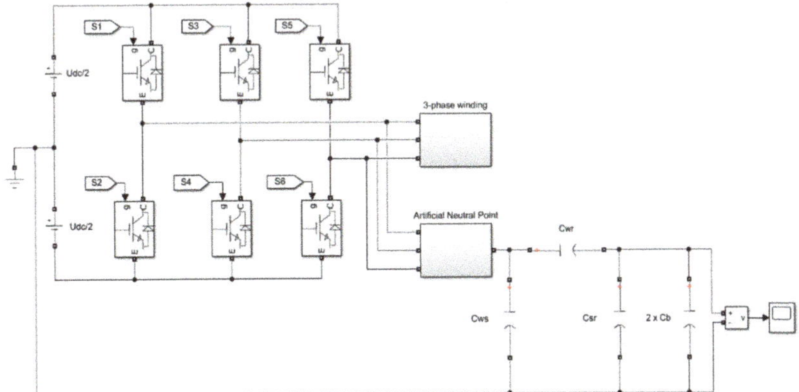

Figure 18. Simulation model with the two-level DC/AC converter.

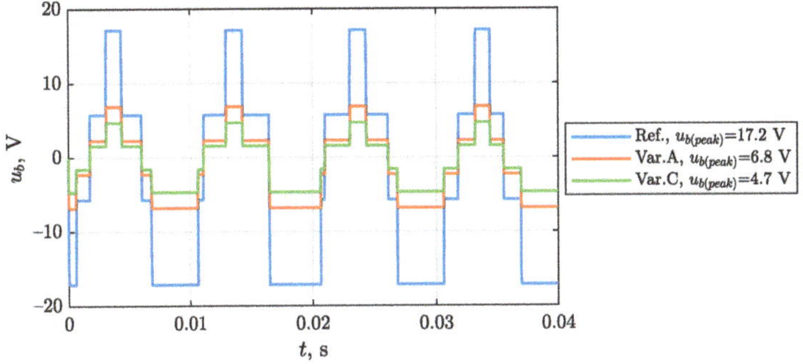

Figure 19. Bearing voltage waveforms for the two-level converter model for selected solution variants.

The bearing voltage waveforms shown in Figure 19 indicate that, for the reference variant, the bearing voltage amplitude is 17.2 V. The application of the proposed solutions based on the shielding windings contributes to lowering the bearing voltage amplitudes to 6.8 V for variant A and 4.7 V for variant C, respectively.

5.2. Results of the Calculations of Bearing Voltages When the Machine Is Powered by a Three-Level Converter

In order to complement the studies presented in the previous section, calculations of the bearing voltage were carried out when the machine was powered by a three-level converter containing neutral-point clamped (NPC) diodes [39]. This type of converter is the most common group of multilevel converters. The converter consists of 12 IGBT transistors and 6 leveling diodes. One way to limit the common-mode voltage is to eliminate the zero vectors from the control strategy, which are the source of the maximum amplitude of the common-mode voltage u_{CM} [40]. The model of such a three-level converter is shown in Figure 20. The results of model calculations for selected variants of machine design solutions are shown in Figure 21. From the presented results, a reduction in the amplitude of the bearing voltage u_b by 1/3 compared to a 2-level converter can be observed. In this case, the amplitude of the bearing voltage for the reference model is 11.4 V. The use of shielding windings in the machine contributes to reducing the amplitude of the bearing voltage to 4.5 V for variant A and to 3.1 V for variant C.

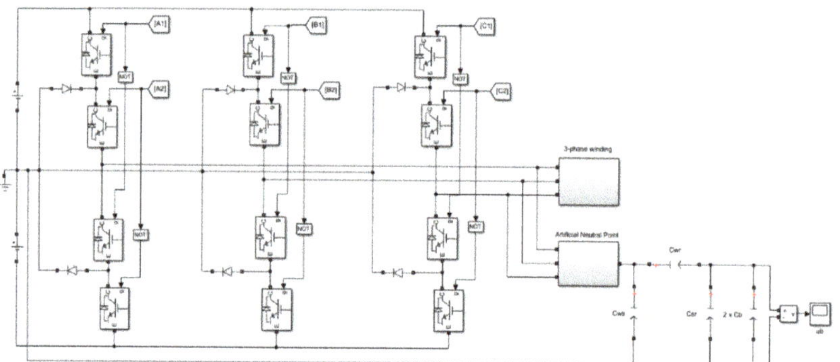

Figure 20. Simulation model with the three-level DC/AC converter.

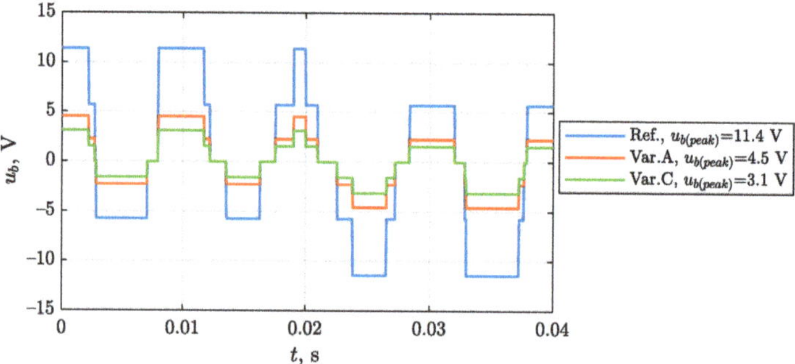

Figure 21. Bearing voltage waveforms for the three-level converter model for selected solution variants.

6. Conclusions

The source of the bearing voltage is primarily the asymmetrical nature of the voltage that occurs when the machine is powered from a power electronic converter. In order to reduce the negative effects associated with it, the authors suggest the use of two additional shielding windings in the machine. The main advantage of the suggested solutions is that there is no need to redesign the magnetic core of the machine.

The only fundamental change is to equip the machine with stator slot wedges in which the wires are placed, which, from the technological point of view, is a relatively easy task to perform. This solution is also supported by technical reasons—the grounded shielding winding is protected from short-circuiting to the main winding located in the stator slots.

From the presented calculations of bearing voltages, it can be seen that the use of a shielding winding placed in the stator slot wedges causes a reduction in the value of the capacitance C_{wr}, resulting in a decrease in the amplitude of the bearing voltage from 17.2 V to 6.8 V.

Simulation research has shown that the overhang length of the shielding winding does not significantly reduce the resultant capacitance C_{wr}. Therefore, for technological reasons, it is reasonable to use the shortest possible connections between wedges. It reduces the problems of ensuring the electrical insulation of these connections and also has a beneficial effect on their stiffness.

If the above solution turns out to be insufficient from the point of view of bearing voltage levels, equipping the machine with a helix-shaped shielding winding located in stator end-winding region may be considered. This solution, together with the shielding winding in wedges, increases the effectiveness of the bearing voltage limitation. Compared to the reference model, the bearing voltage amplitudes were reduced from 17.2 V to 4.7 V (when the machine is powered from a conventional two-level converter).

The calculations obtained using converter circuit models showed that the method of limiting bearing voltages can additionally be successfully combined with other available methods. For example, when the levels of bearing voltages cause their accelerated wear, it is possible to additionally consider the use of a three-level converter to power the machine. For this supply type, the bearing voltage amplitudes of the machine under consideration were reduced to about 3.1 V.

Due to the nature of the 3D FEM models used, capacitance calculations are time-consuming. The most time-consuming variant turned out to be the model with two shielding windings—for this model, calculations took more than 8 days. For the other cases, the calculations were correspondingly shorter.

The presented results of the simulations justify the desirability of using shielding windings in machines powered by converters, particularly where the high operational reliability of drive systems is required.

Author Contributions: Conceptualization, S.B. and T.J.; methodology, S.B. and T.J.; simulations, S.B. and T.J.; writing—review and editing, S.B., T.J. and P.O.; visualization, S.B., T.J. and P.O.; validation, S.B., T.J. and P.O. All authors have read and agreed to the published version of the manuscript.

Funding: The research is co-financed under the Program of the Ministry of Science and Higher Education "Implementation Doctorate" (Poland).

Institutional Review Board Statement: Not applicable.

Informed Consent Statement: Not applicable.

Data Availability Statement: Not applicable.

Conflicts of Interest: The authors declare no conflict of interest.

References

1. Sendek-Matysiak, E.; Łosiewicz, Z. Analysis of the Development of the Electromobility Market in Poland in the Context of the Implemented Subsidies. *Energies* **2021**, *14*, 222. [CrossRef]
2. Xu, Y.; Ai, M.; Xu, Z.; Liu, W.; Wang, Y. Analysis of Interior Permanent Magnet Synchronous Motor Used for Electric Vehicles Based on Power Matching and Driving Conditions. In Proceedings of the 2020 IEEE International Conference on Applied Superconductivity and Electromagnetic Devices (ASEMD), Tianjin, China, 16 October 2020; pp. 1–2.
3. Dupczak, B.S.; Perin, A.J.; Heldwein, M.L.; Martins, C.A.; Cros, J. PMSM Specification and Design for an Electrical Boat Propulsion System. In Proceedings of the Power Electronics South America 2012, São Paulo, Brazil, 11–13 September 2012. [CrossRef]

4. Wang, H.; Leng, J. Summary on Development of Permanent Magnet Synchronous Motor. In Proceedings of the 2018 Chinese Control And Decision Conference (CCDC), Shenyang, China, 9–11 June 2018; pp. 689–693.

5. Delgado-Arredondo, P.A.; Romero-Troncoso, R.J.; Duque-Perez, O.; Morinigo-Sotelo, D.; Osornio-Rios, R.A. Vibration, Acoustic Noise Generation and Power Quality in Inverter-Fed Induction Motors. In Proceedings of the 2019 IEEE 12th International Symposium on Diagnostics for Electrical Machines, Power Electronics and Drives (SDEMPED), Toulouse, France, 27–30 August 2019; pp. 412–418.

6. Moreno, Y.; Almandoz, G.; Egea, A.; Arribas, B.; Urdangarin, A. Analysis of Permanent Magnet Motors in High Frequency—A Review. *Appl. Sci.* **2021**, *11*, 6334. [CrossRef]

7. Schneider, V.; Behrendt, C.; Höltje, P.; Cornel, D.; Becker-Dombrowsky, F.M.; Puchtler, S.; Gutiérrez Guzmán, F.; Ponick, B.; Jacobs, G.; Kirchner, E. Electrical Bearing Damage, A Problem in the Nano- and Macro-Range. *Lubricants* **2022**, *10*, 194. [CrossRef]

8. Gonda, A.; Capan, R.; Bechev, D.; Sauer, B. The Influence of Lubricant Conductivity on Bearing Currents in the Case of Rolling Bearing Greases. *Lubricants* **2019**, *7*, 108. [CrossRef]

9. SKF. *SKF Bearing Maintenance Handbook*; SKF: Göteborg, Sweden, 2011; ISBN 978-91-978966-4-1.

10. *Timken Bearing Damage Analysis with Lubrication Reference Guide*; The Timken Company: North Canton, OH, USA, 2015.

11. *Bearing Failure: Causes and Cures*; Barden Precision Bearings: Danbury, CT, USA, 1998.

12. Gemeinder, Y.; Weicker, M. *Application Guide Bearing Currents*; Institut für Elektrische Energiewandlung, Technische Universität Darmstadt: Darmstadt, Germany, 2021.

13. Berhausen, S.; Jarek, T. Method of Limiting Shaft Voltages in AC Electric Machines. *Energies* **2021**, *14*, 3326. [CrossRef]

14. Costello, M.J. Shaft Voltages and Rotating Machinery. *IEEE Trans. Ind. Appl.* **1993**, *29*, 419–426. [CrossRef]

15. Plazenet, T.; Boileau, T.; Caironi, C.; Nahid-Mobarakeh, B. A Comprehensive Study on Shaft Voltages and Bearing Currents in Rotating Machines. *IEEE Trans. Ind. Appl.* **2018**, *54*, 3749–3759. [CrossRef]

16. Plazenet, T.; Boileau, T.; Caironi, C.; Nahid-Mobarakeh, B. An Overview of Shaft Voltages and Bearing Currents in Rotating Machines. In Proceedings of the 2016 IEEE Industry Applications Society Annual Meeting, Portland, OR, USA, 2–6 October 2016; pp. 1–8.

17. Berhausen, S.; Jarek, T. Analysis of Impact of Design Solutions of an Electric Machine with Permanent Magnets for Bearing Voltages with Inverter Power Supply. *Energies* **2022**, *15*, 4475. [CrossRef]

18. Guzinski, J. Common-Mode Voltage and Bearing Currents in PWM Inverters: Causes, Effects and Prevention. In *Power Electronics for Renewable Energy Systems, Transportation and Industrial Applications*; Abu-Rub, H., Malinowski, M., Al-Haddad, K., Eds.; John Wiley & Sons, Ltd.: Chichester, UK, 2014; pp. 664–694. ISBN 978-1-118-75552-5.

19. *PMSM FOC Motor Control Software Using XMC*; Infineon: Munich, Germany, 2019.

20. Muetze, A.; Binder, A. Calculation of Motor Capacitances for Prediction of the Voltage Across the Bearings in Machines of Inverter-Based Drive Systems. *IEEE Trans. Ind. Appl.* **2007**, *43*, 665–672. [CrossRef]

21. Stockbrügger, J.O.; Ponick, B. Analytical Determination of the Slot and the End-Winding Portion of the Winding-to-Rotor Capacitance for the Prediction of Shaft Voltage in Electrical Machines. *Energies* **2020**, *14*, 174. [CrossRef]

22. Sengamalai, U.; Thamizh Thentral, T.M.; Ramasamy, P.; Bajaj, M.; Hussain Bukhari, S.S.; Elattar, E.E.; Althobaiti, A.; Kamel, S. Mitigation of Circulating Bearing Current in Induction Motor Drive Using Modified ANN Based MRAS for Traction Application. *Mathematics* **2022**, *10*, 1220. [CrossRef]

23. Ahola, J.; Muetze, A.; Niemela, M.; Romanenko, A. Normalization-Based Approach to Electric Motor BVR Related Capaci-tances Computation. *IEEE Trans. Ind. Appl.* **2019**, *55*, 2770–2780. [CrossRef]

24. Turzyński, M.; Musznicki, P. A Review of Reduction Methods of Impact of Common-Mode Voltage on Electric Drives. *Energies* **2021**, *14*, 4003. [CrossRef]

25. Akagi, H.; Tamura, S. A Passive EMI Filter for Eliminating Both Bearing Current and Ground Leakage Current From an Inverter-Driven Motor. *IEEE Trans. Power Electron.* **2006**, *21*, 1459–1469. [CrossRef]

26. Yang, L.; Yang, Y.; Wen, J.; Jia, L.; Yang, E.; Liu, R. Suppression of Rotor-Grounding Bearing Currents Based on Matching Stator and Rotor Grounding Impedances. *Energies* **2022**, *15*, 1638. [CrossRef]

27. Anagha, E.R.; Nisha, P.V.; Sindhu, T.K. Design of an Active EMI Filter for Bearing Current Elimination in VFD. In Proceedings of the 2018 IEEE International Symposium on Electromagnetic Compatibility and 2018 IEEE Asia-Pacific Symposium on Electromagnetic Compatibility (EMC/APEMC), Singapore, 14–18 May 2018; pp. 131–134.

28. Yea, M.; Han, K.J. Modified Slot Opening for Reducing Shaft-to-Frame Voltage of AC Motors. *Energies* **2020**, *13*, 760. [CrossRef]

29. *Current-Insulating Bearings*; Schaeffler Technologies AG & Co. KG: Schweinfurt, Germany, 2020.

30. Vostrov, K.; Pyrhonen, J.; Niemela, M.; Ahola, J.; Lindh, P. Mitigating Noncirculating Bearing Currents by a Cor-rect Stator Magnetic Circuit and Winding Design. *IEEE Trans. Ind. Electron.* **2021**, *68*, 3805–3812. [CrossRef]

31. Ferreira, F.J.T.E.; Cistelecan, M.V.; de Almeida, A.T. Slot-Embedded Partial Electrostatic Shield for High-Frequency Bearing Current Mitigation in Inverter-Fed Induction Motors. In Proceedings of the IEEE XIX International Conference on Electrical Machines—ICEM 2010, Rome, Italy, 6–8 September 2010; pp. 1–6.

32. Alves Êvo, M.T.; Paula, H. Electrostatic Shielding for Bearings Discharge Currents Attenuation: Analysis of Its Effectiveness, Losses and Impact on the Motor Performance—A Study for Design Guidelines. *IET Electr. Power Appl.* **2020**, *14*, 1050–1059. [CrossRef]

33. Gerber, S.; Wangi, R.-J. Reduction of Inverter-Induced Shaft Voltages Using Electrostatic Shielding. In Proceedings of the 2019 Southern African Universities Power Engineering Conference/Robotics and Mechatronics/Pattern Recognition Association of South Africa (SAUPEC/RobMech/PRASA), Bloemfontein, South Africa, 28–30 January 2019; pp. 310–315.

34. Magdun, O.; Gemeinder, Y.; Binder, A. Prevention of Harmful EDM Currents in Inverter-Fed AC Machines by Use of Electrostatic Shields in the Stator Winding Overhang. In Proceedings of the IECON 2010—36th Annual Conference on IEEE Industrial Electronics Society, Glendale, AZ, USA, 7–10 November 2010; pp. 962–967.

35. Vostrov, K.; Pyrhonen, J.; Ahola, J. Shielding the End Windings to Reduce Bearing Currents. In Proceedings of the 2020 International Conference on Electrical Machines (ICEM), Gothenburg, Sweden, 23–26 August 23 2020; pp. 1431–1437.

36. Du, G.; Li, N.; Zhou, Q.; Gao, W.; Wang, L.; Pu, T. Multi-Physics Comparison of Surface-Mounted and Interior Permanent Magnet Synchronous Motor for High-Speed Applications. *Machines* **2022**, *10*, 700. [CrossRef]

37. Berhausen, S.; Paszek, S. Determination of the Leakage Reactance of End Windings of a High-Power Synchro-nous Generator Stator Winding Using the Finite Element Method. *Energies* **2021**, *14*, 7091. [CrossRef]

38. Vostrov, K.; Pyrhonen, J.; Ahola, J. The Role of End-Winding in Building Up Parasitic Capacitances in Induction Motors. In Proceedings of the 2019 IEEE International Electric Machines & Drives Conference (IEMDC), San Diego, CA, USA, 12–15 May 2019; pp. 154–159.

39. Zhang, M.; Cui, Y.; Wang, Q.; Tao, J.; Wang, X.; Zhao, H.; Li, G. A Study on Neutral-Point Potential in Three-Level NPC Converters. *Energies* **2019**, *12*, 3367. [CrossRef]

40. Zare, F.; Adabi, J.; Ghosh, A. Different Approaches to Reduce Shaft Voltages in AC Generators. In Proceedings of the 2009 13th European Conference on Power Electronics and Applications, Barcelona, Spain, 8–10 September 2010.

Article

Virtual Modeling and Experimental Validation of the Line-Start Permanent Magnet Motor in the Presence of Harmonics

Jonathan Muñoz Tabora [1,*], Bendict Katukula Tshoombe [1], Wellington da Silva Fonseca [1], Maria Emília de Lima Tostes [1], Edson Ortiz de Matos [1], Ubiratan Holanda Bezerra [1] and Marcelo de Oliveira e Silva [2]

[1] Amazon Energy Efficiency Center (CEAMAZON), Federal University of Pará, Belém 66075-110, Brazil
[2] Postgraduate Program in Mechanical Engineering, Federal University of Pará, Belém 66075-110, Brazil
* Correspondence: jonathan.tabora@itec.ufpa.br

Abstract: The world is experiencing an accelerated energy transition that is driven by the climate goals to be met and that has driven the growth of different potential sectors such as electric mobility powered by electric motors, which continue to be the largest load globally. However, new needs in relation to power density, weight, and efficiency have led manufacturers to experiment with new technologies, such as rare earth elements (REEs). The permanent magnet motor is a candidate to be the substitute for the conventional induction motor considering the new editions of the IEC 60034-30-1, for which study and evaluation continue to be focused on identifying the weaknesses and benefits of its application on a large scale in industry and electric mobility. This work presents a FEM model to assess the line-start permanent magnet motor (LSPMM), aiming to simulate the behavior of the LSPMM under supply conditions with distorted voltages (harmonic content) and evaluate its thermal and magnetic performance. The model created in the FEM software is then validated by bench tests in order to constitute an alternative analysis tool that can be used for studies in previous project phases and even to implement predictive maintenance schemes in industries.

Keywords: line-start permanent magnet motor; voltage harmonics; FEMM; temperature model; computational simulations

Citation: Tabora, J.M.; Tshoombe, B.K.; Fonseca, W.d.S.; Tostes, M.E.d.L.; Matos, E.O.d.; Bezerra, U.H.; Silva, M.d.O.e. Virtual Modeling and Experimental Validation of the Line-Start Permanent Magnet Motor in the Presence of Harmonics. *Energies* **2022**, *15*, 8603. https://doi.org/10.3390/en15228603

Academic Editors: Bing Tian, Xinghe Fu and Quntao An

Received: 22 September 2022
Accepted: 7 November 2022
Published: 17 November 2022

Publisher's Note: MDPI stays neutral with regard to jurisdictional claims in published maps and institutional affiliations.

1. Introduction

The electric motor market has become more competitive in recent years, as the growth of new areas such as electric vehicles requires more efficient motors with higher power density, which has led manufacturers to introduce new technologies that have lower energy consumption for the same power output. This technological evolution in motor drive systems has also allowed many countries to implement policies and incentives for the substitution of old/non-efficient motors with higher efficiency motors, which brings a benefit both to end-users through savings in consumption and to the country's energy matrix through reductions in energy consumption and demand. Within current technologies, permanent magnets have gained a greater field of use among manufacturers despite being a recent and immature technology. Its use extends not only to the residential, commercial, and industrial sectors but also to electric vehicles and boats given the great benefits it brings in terms of economy, weight, power density, etc.

Different works analyzing the construction, rotor topologies, operating principles, benefits, and challenges of the line-start permanent magnet motor (LSPMM) have been presented in [1–6], highlighting the main trends in the development of these technologies. Furthermore, since electric motors are subjected to different disturbances in electrical systems, different authors have presented experimental results on the LSPMM performance in the presence of disturbances such as harmonics, subharmonics, voltage unbalance, and voltage variation. Upon comparison with its predecessors, as shown in [6–13], in ideal conditions, the LSPMM presents lower consumption, temperatures, and quieter operations when compared to the squirrel cage induction motor (SCIM). However, these conditions change in the presence of voltage harmonics and voltage unbalance, with lower performances and higher temperatures, which may be attributed to the harmonic distortion of the magnets as well as to greater harmonic losses. Another fact related to the high-efficiency motors and the LSPMM is the smaller leakage and magnetization impedance as a product of its constructive improvements in relation to the old/non-efficient motors but with less ability to filter the harmonics present in the power supply compared to previous technologies.

Concerning temperature, thermography has been one of the most used methods in the industry for the identification of faults in electric motors as well as predictive maintenance. The works presented in [7,14–18] show theoretical and practical studies of electric motor monitoring techniques, including statistical and computational intelligence analysis, with useful results for the identification of faults of different natures, such as electrical, mechanical, thermal, and environmental.

In the literature, many works have focused on evaluating motors' performances and losses using techniques based on numerical-computation-based models. The works presented in [19,20] use the same LSPMM technology used in this work to model and validate the LSPMM model through experimental tests. The studies use the parameters of the LSPMM equivalent circuit calculated empirically as well as numerical methods that are later validated by means of experimental measurements without, however, considering harmonics in the motor supply.

The demand for higher efficiencies in electric motors is increasing considering the new categories such as electric vehicles and that, in this sense, both manufacturers and researchers are constantly searching and updating for better operational projects from the design phase. In this line, different works analyzing the design and optimization of electric motors were presented in [21–24]. The work in [22] presented a complete evaluation of permanent magnet synchronous hub motors (PMSHMs) from optimization to experimental validation. With this goal, a multi-objective optimization strategy was proposed in combination with a multi-objective genetic optimization algorithm (nondominated sorting genetic algorithm, NSGA III) to improve the operation of the PMSHMs in relation to torque ripple and higher maximum torque, all at a lower computational cost, starting from a sequential subspace optimization strategy; the results are validated using the FEM software that is also used in this work. The study was finally validated from measurements on the prototype built from the optimized model, constituting it a tool for engine optimization at low computational cost. Other works using numerical method models and software such as the finite element method (FEM) of the LSPMM were presented in [7,25–29].

However, despite the accomplished evaluation carried out by the authors in the aforementioned works, there are few works that analyze the modeling and validation of the LSPMM in the presence of voltage harmonics and that, given their different impact on motors according to the order and percentage, should be considered in the optimization aiming to increase the manufacturing tolerances in the face of these disturbances.

Considering that permanent magnet electric motors are emerging as a good candidate to replace induction electric motors, their operation must be evaluated in real electrical power conditions as well as from the computational point of view, which aims to find improvement opportunities through experimental validation. Given this scenario, the present work evaluates a 0.75 kW line-start permanent magnet motor in the presence of 2nd- and 5th-order voltage harmonics through numerical-computation-based models and then validates it through experimental tests in the CEAMAZON laboratory. To this end, the following goals have been defined:

- Model the LSPMM in the FEM software based on its physical dimensions;
- Apply the harmonic waveforms to the electric motor through the FEM software;
- Validate the results obtained from measurement campaigns in CEAMAZON with an LSPMM of 0.75 kW.

The selected harmonics for this study (2nd and 5th) are called negative sequence harmonics, known to result in greater negative impacts on IMs when compared to zero and positive sequence harmonics [8] and due to which their consequences should be evaluated by manufacturers and end-users aiming to propose solutions from the project phase or even as preventive and predictive maintenance schemes in industries, constituting it as an alternative analysis tool.

2. Induction Motors

2.1. Minimum Energy Performance Standards

The next edition of the IEC 60034-30-1 [30] will define the IE5 efficiency class although there are already commercial proposals from manufacturers for this efficiency class; globally, the maximum established efficiency class is the IE3 class, which has not yet been achieved by all countries, as presented in Figure 1. Europe has taken a step forward by defining the month of July 2023 as the date for the introduction of the IE4 class for motors between 75 kW and 200 kW [31,32].

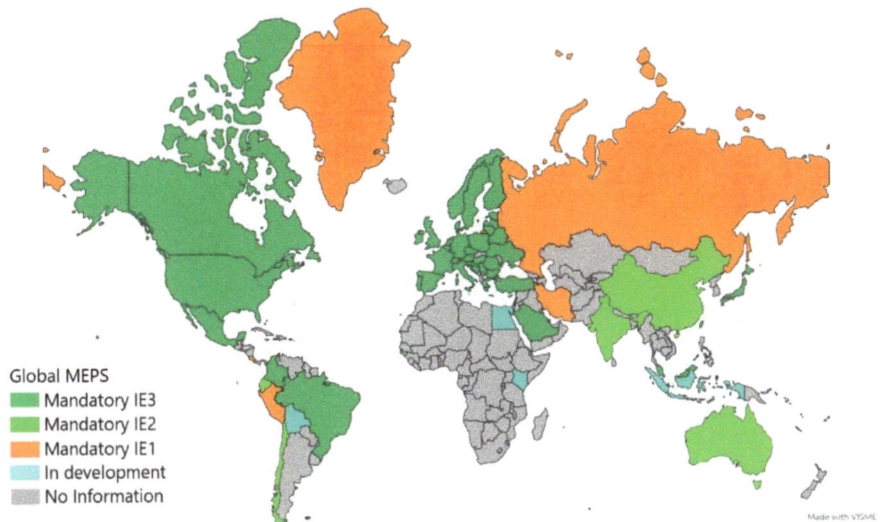

Figure 1. Countries with MEPS for electric motors [33,34].

The adoption of MEPS in the countries translates into great advantages at energy, economic, and ecological levels due to the savings with the substitution of old/inefficient motors for more efficient ones mainly in those cases in which the lack of knowledge leads users to be guided by the initial cost of the motor at the time of purchase and not by the nominal efficiency, which is more related to operating costs that represent more than 95% of the total costs of the motor throughout its useful life as compared to 2% that represents the initial cost [35].

Considering the new editions of IEC 60034-30-1, higher efficiencies will be required for the IE5 class, for which new technologies such as permanent magnet motors will take on greater importance as substitutes for conventional induction motors given their technical and economic benefits. The presence of permanent magnet motors is increasingly present in electric mobility and is growing in the industry as well. The magnetic field of permanent magnets contributes to the reduction of the magnetization current, which is necessary to create the magnetic fields in the machine core and whose value can reach 50% of the nominal motor current in low-power motors operating with no load [36]. The motor's input current is the sum of two components: the magnetizing current and the current due to the load. The reduction in the magnetizing current due to the magnets contributes to these motors, presenting lower nominal currents and lower temperatures as well as higher efficiencies.

Constructively, the main difference between a SCIM and an LSPMM is the presence of permanent magnets in the rotor, which can be on the surface or internal to the rotor. Figure 2 presents the individual magnetic fields generated in the rotor, stator, and in a combined way.

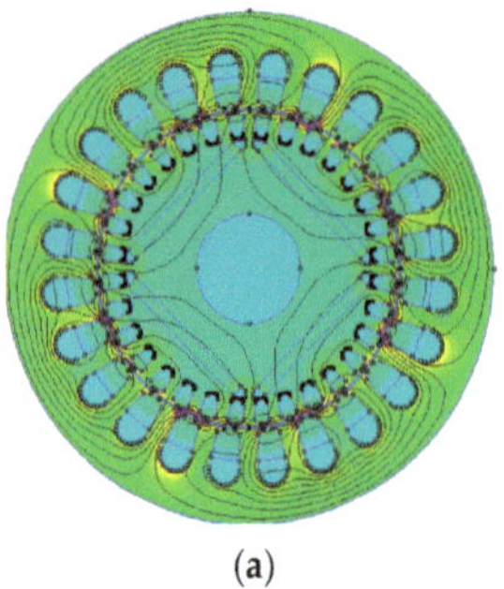

(a)

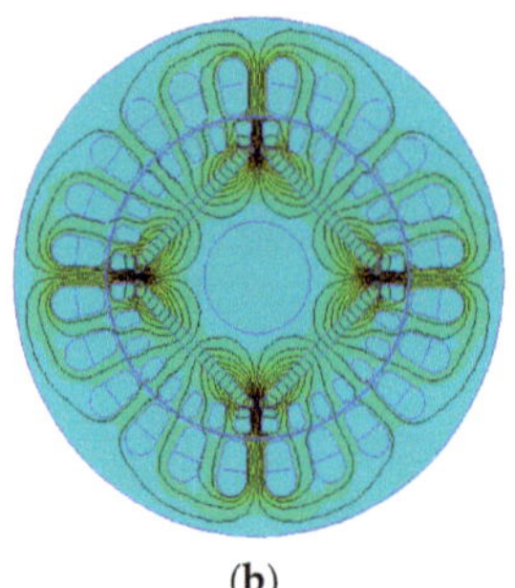

(b)

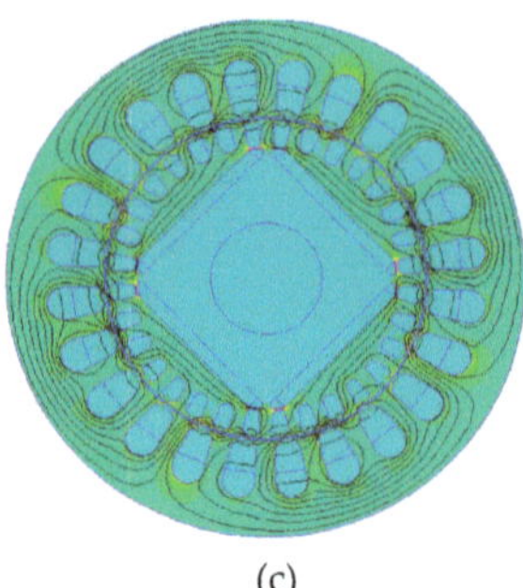

(c)

Figure 2. Line-start permanent magnet motor magnetic fields [7]: (**a**) stator winding fields; (**b**) permanent magnets fields; (**c**) stator and rotor magnetic fields.

2.2. Finite Elements Formulation for Time-Harmonic Magnetic Problems

The finite element method (FEM) is a numerical method used to solve boundary value problems in engineering. For convenience, the FEM procedure permits the domain to be discretized into a finite number of parts (or elements) and emphasizes that the characteristics of the continuous domain may be estimated by assembling similar properties of discretized elements per node [8].

In order to solve for the unknow vector within the global matrix of the domain, boundary conditions need to be imposed on the solution domain. The two most important boundary conditions in finite element analysis are the Dirichlet and the periodical boundary conditions. The governing field and motion equations are listed below.

The field equation is derived from Maxwell's equations as follows in (1)–(3):

$$\nabla \times H = J_0 + J_e \tag{1}$$

$$\nabla \times E = -\left(\frac{\partial B}{\partial t}\right), \tag{2}$$

$$\nabla \cdot B = 0. \tag{3}$$

with the constitutive relationship in (4):

$$J = \sigma E. \tag{4}$$

and the magnetic flux density (B) in (5),

$$B = \mu H + \mu_0 M_r, \tag{5}$$

where M_r is the remanent magnetization of the permanent magnets, J_0 is the magnetization current density, E is the electric field intensity, H is the magnetic field intensity, B is the magnetic field density and σ is the specific electric conductivity.

Combining Equations (2) and (4) and the magnetic vector potential A yields the following current density vector equation (J_e):

$$B = \nabla \times A, \tag{6}$$

$$J_e = -\sigma \frac{\partial A}{\partial t} + \sigma \frac{V_{tz}}{l}, \tag{7}$$

where V_{tz} is electric scalar potential, and l is the conductor length.

The current density in Equation (7) includes two parts: one is the induced quantity produced by electromagnetic induction in squirrel cage bars, and the other one is caused by the effect of charge buildup at the conductor end. Observe that the moving frame is considered as the reference frame.

Permanent magnets, which are represented by (5), can yield the following field equation:

$$\nabla \times H = \nabla \times (v \nabla \times A) - \nabla \times (v\mu_0 M_r), \tag{8}$$

where $v = \frac{1}{\mu}$ is the magnetic reluctivity.

Combining (1), (7), and (8) yields the governing equation for permanent magnet:

$$\nabla \times (v\nabla \times A) = J_0 + J_e + \nabla \times (v\mu_0 M_r). \tag{9}$$

Equation (9) is the fundamental governing equation for modeling various field problems. In the analysis of electric machines, it is common to evaluate the field solution in two dimensions, considering that the current density J and the magnetic vector potential A have only z-directed invariant components. Below is the differential form of Equation (9).

$$\frac{\partial}{\partial x}\left(v\frac{\partial A}{\partial x}\right) + \frac{\partial}{\partial y}\left(v\frac{\partial A}{\partial y}\right) = \sigma\frac{\partial A}{\partial y} - \sigma(v \times B) - \sigma\frac{V_{ts}}{l} - \nabla \times (v\mu_0 M_r). \tag{10}$$

The stator phase equation is presented in (11):

$$V_s = R_s I_s + L_e\frac{dI_s}{dt} + \frac{l}{ms}\left(\iint_{\Omega+} \frac{\partial A}{\partial t}dxdy - \iint_{\Omega-} \frac{\partial A}{\partial t}dxdy, \tag{11}$$

V_s is the applied stator phase voltage, I_s, the stator phase current, R_s is the total equivalent resistance per phase, L_e is the total equivalent inductance of end winding, l is the length of stator windings in z-direction and usually uses the same value as the axial length of stator iron core, m is the number of stator winding branches in parallel connection, and s is the equivalent cross-section area of one turn of stator windings. The equation for the squirrel cage bars is presented in (12):

$$V_{bk} = R_{bk}I_{bk} + \frac{l_{bk}}{s_{bk}} \iint_{s_{bk}} \frac{\partial A}{\partial t}dxdy, \tag{12}$$

where l_{bk} is the length of the k_{th} bar in z-direction, and s_{bk} is the bar cross-section area.

For the mechanical motion, the equations presented in (13) and (14) were used:

$$J_r\frac{d\omega_m}{dt} = T_{em} - T_f - D_f\omega_m \tag{13}$$

$$\frac{d\theta_m}{dt} = \omega_m \tag{14}$$

where J_r = the rotor moment of inertia, θ_m = the rotor position, ω_m = the mechanical motor speed, T_{em} = the electromagnetic torque, T_f = the applied load torque, and D_f = the coefficient of friction.

2.3. Finite Elements Formulation for Thermal Analysis

The heat flow static problems addressed by FEM are heat-conduction problems. These problems are represented by a temperature gradient, G, and heat flux density, F [37]. The heat flux density must obey Gauss' law, which says that the heat flux out of any closed volume is equal to the heat generation within the volume; this law is represented in differential form as

$$\nabla \cdot F = q, \tag{15}$$

where q represents volume heat generation.

Temperature gradient and heat flux density are also related to one another via the constitutive relationship:

$$F = kG, \tag{16}$$

where k is the coefficient of thermal conductivity.

FEM allows for the variation of conductivity as an arbitrary function of temperature. Usually, the goal is to find the temperature, T, rather than the heat flux density or temperature gradient. Temperature is related to the temperature gradient by

$$G = -\nabla T. \tag{17}$$

Substituting (17) into Gauss' law and applying the constitutive relationship yields the fifth-order partial differential Equation (18):

$$-\nabla \cdot (k\nabla T) = q. \tag{18}$$

Thus, to perform the motor thermal analysis, FEMM solves the problem around the Biot–Fourier equation [37], and Equation (19) is used to obtain the temperature and gradient values of the model:

$$pc_p\frac{dT}{dt} - \nabla \cdot (k\nabla T) = q \tag{19}$$

where pc_p is the volumetric heat capacity, with p being the mass density and c_p the specific heat capacity. FEMM employs Euler´s implicit discretizing scheme to solve Equation (19), mentioned above.

Having calculated the temperature values at all finite elements of the specific domain by numerically solving Equation (19), the software updates the values of electric conductivity using the equation in (20) [38].

$$\sigma(T) = \frac{1}{p_o(1 + \alpha(T - T_0)} \tag{20}$$

where p_o is the electrical resistivity at 0 °C and α the coefficient of thermal conductivity; T_0 is the initial room temperature, and T is the temperature at each time step of the simulation.

3. Methodology

Aiming to validate the model as well as to identify the operational characteristics of the LSPMM, the methodology was separated into two parts: an experimental one and a computational one, as presented in Figure 3. Initially, experimental measurement campaigns were carried out with the LSPMM under ideal conditions until reaching thermal equilibrium according to IEC 60034-2-1 [39], and then, in the presence of voltage harmonics, it increased by 2% every 10 min and then up to 25%. Under these conditions, the motor input parameters as well as thermographic images were recorded for subsequent processing and results.

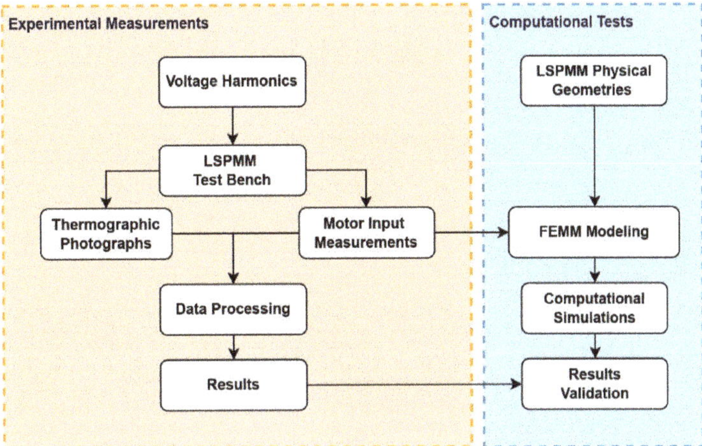

Figure 3. Methodology Flowchart.

At the same time, the motor dimensions were modeled in the Finite Elements Method Magnetics (FEMM) software [40] and the materials defined from the motor manufacturer information. After modeling and inserting the voltage harmonics, computational simulations were carried out to verify the performance of this technology in the presence of harmonics, and then, the experimental results were validated.

The experimental measurements were carried out using the test bench shown in Figure 4. The test bench is composed of a programmable source (1) in which the harmonics under study were generated. The LSPMM input parameters were measured using a power quality analyzer (2), while the load on the output was controlled by an eddy current brake (3), and finally, the motor data (4) are presented in Table 1. The thermographic measurements were made using a model T620 thermographic camera from the FLIR^{TM} manufacturer with an emissivity of 0.94, which corresponds to the ink of the LSPMM. The images were captured at the two angles presented in Figure 5, every two minutes from thermal equilibrium and then processed in the camera software.

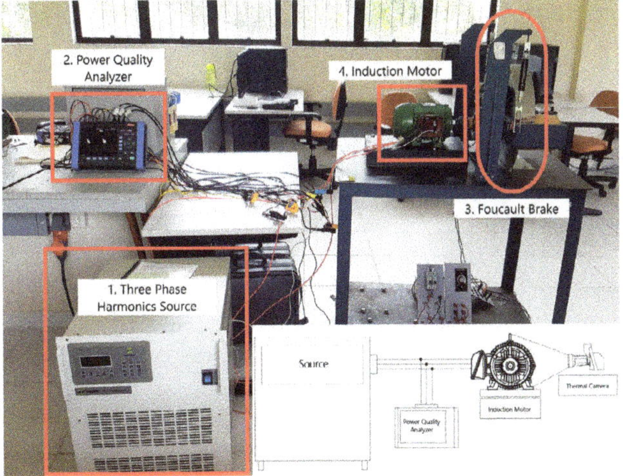

Figure 4. General test setup [8].

Table 1. Line-start Permanent Magnet Motor Parameters.

Induction Motor Class	IE4 Class
Technology	LSPMM
Power	0.75 kW
Voltage	220 V/380 V
Speed (rpm)	1800
Torque (Nm)	3.96
Current (A)	3.08/1.78
Efficiency (%)	87.4
Power Factor	0.73

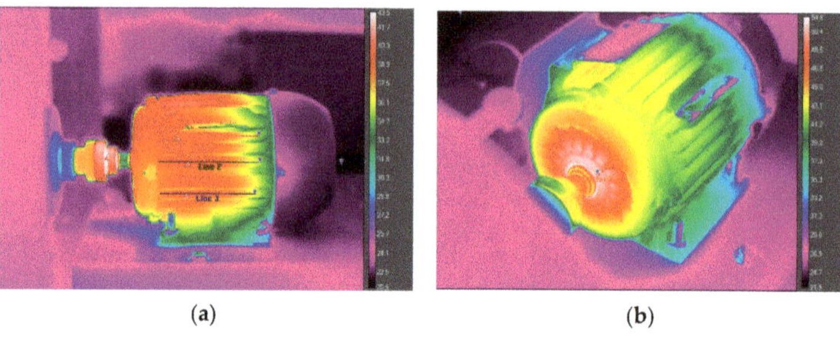

(a)   (b)

Figure 5. Thermographic images of the line-start permanent magnet motor in presence of voltage harmonics [8]: (**a**) lateral view; (**b**) front view.

The simulation was performed by considering the strong coupling between the thermal and magnetic effects. The Multiphysics coupling was achieved through FEM, and it was built in the Lua console component, where the calculations were performed interactively over increasing harmonic distortion percentages.

Each iteration is composed of a magnetic simulation and sequentially a thermal simulation; the result of each simulation was used as input to the next, thus achieving strong coupling. At the start of every iteration, the magnetic simulation was realized; Equation (10) was numerically solved to calculate the magnetic vector potential A at all elements that form the analysis region. The resistive losses at the copper strands were also calculated. The next step was the thermal simulation. Equation (19) was numerically solved at all finite elements using the resistive losses calculated in the previous magnetic simulation as heat sources. Electric conductivities were updated to the new temperatures according to (20) and were reinserted into the magnetic simulation of the next time step. This process was repeated iteratively until the simulated timespan was completed, and it is presented in Figure 6.

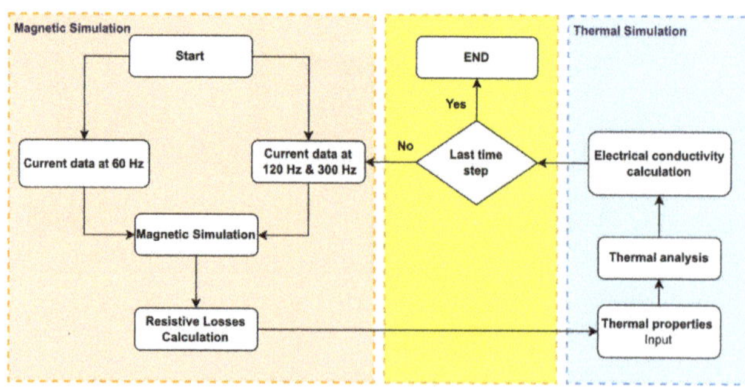

Figure 6. Flowchart for numerical simulations with strong thermo-magnetic coupling.

4. Results

4.1. Experimental Results

The second- and fifth-order voltage harmonics are classified as negative sequence harmonics and are characterized by generating a magnetic field opposite to the resulting magnetic field in the motor due to the fundamental frequency. This opposite torque produces vibrations and caused a decrease in speed as well as an increase in consumption and motor temperature. In Figures 7 and 8, the input line currents (a) and the active, reactive, and total power (b) are presented for the LSPMM in the presence of 2nd- and 5th-order voltage harmonics.

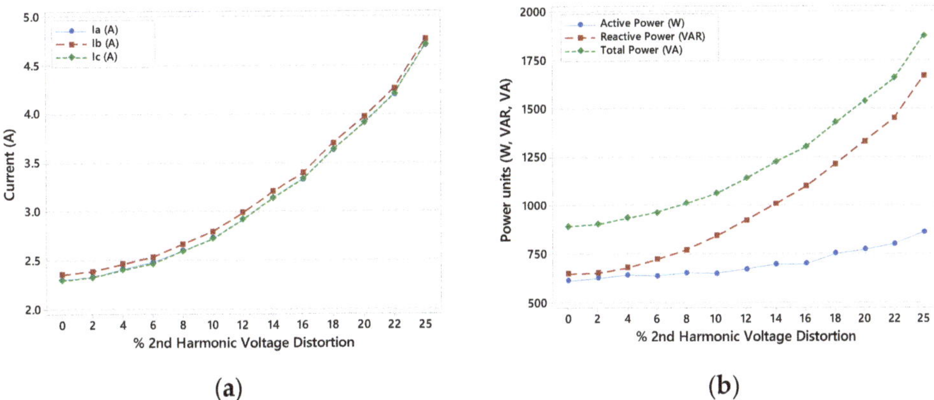

(a)   **(b)**

Figure 7. LSPMM in the presence of second voltage harmonic: (**a**) input line currents; (**b**) active, reactive, and total Power.

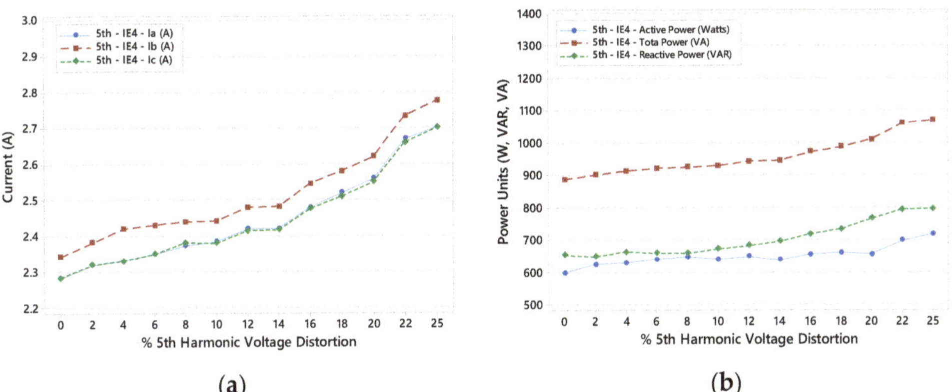

(a)   **(b)**

Figure 8. LSPMM in the presence of fifth voltage harmonic: (**a**) input line currents; (**b**) active, reactive and total Power.

It can be seen how the total current increases as a result of the negative sequence order voltage harmonics being increased every 10 min, and the same happens for the active, reactive, and total power. However, it is important to clarify that these increases are not due to the rise in the output load of the motor, which maintains a constant load, but due to harmonics since only the fundamental frequency can provide real power at the output [41]. Additionally, this is also verified with the increase in the reactive power of the motor due to the increase in distortion harmonic percentage. In the figures below, it can be observed how the 2nd-order voltage harmonic results in higher increases in the reactive power, mainly due to magnetic losses, as will be presented in Figure 8.

As it was observed in Figures 7 and 8, harmonics result in increases in the LSPMM input power, and this increase translates into losses due to overloads that can reduce the motor's useful life as well as reduce efficiency and increase consumption, which translate into higher operating costs for users. Figures 9 and 10 present these losses, as observed in the LSPMM thermography images, for the thermal equilibrium condition without the presence of harmonics (thermograms (a) and (c)) and the same angles in the presence of 25% of 2nd- and 5th-order voltage harmonic distortions. The deviation produced by the negative sequence voltage harmonic disturbance is considerable and not recommended since it can degrade the insulation in the motor windings and produce internal short circuits.

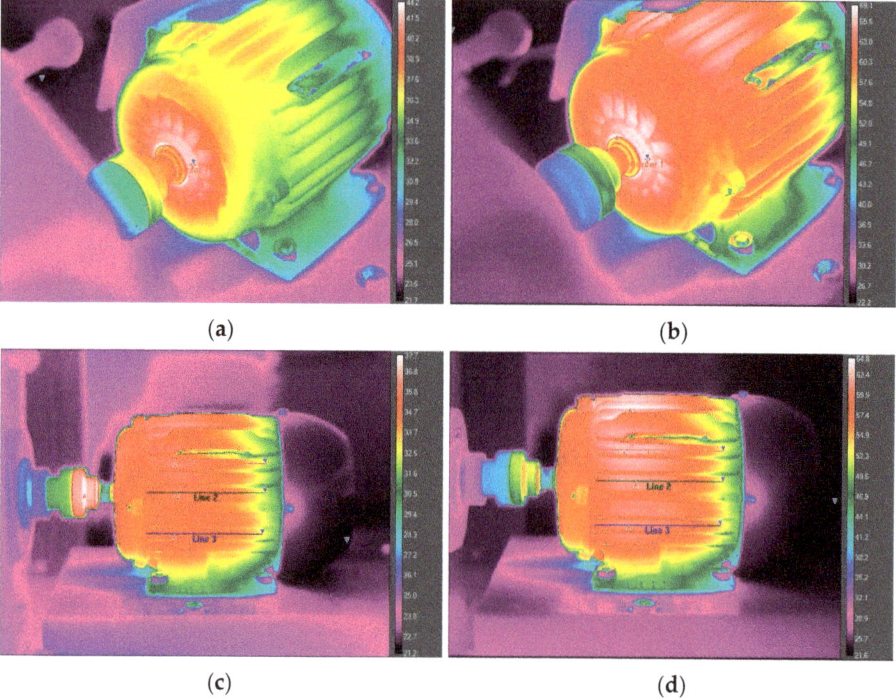

Figure 9. Thermographic images of the line-start permanent magnet motor in presence of 2nd-order voltage harmonics in frontal and lateral view: (**a**) thermal equilibrium frontal view; (**b**) 25% of 2nd-order voltage harmonic in frontal view; (**c**) thermal equilibrium lateral view; (**d**) 25% of 2nd-order voltage harmonic in lateral view.

In the measurements analysis, it was observed that for this motor, in the presence of only the 2nd- and 5th-order voltage harmonics, new current harmonics appeared, as shown in Figure 11, mainly due to the motor being in the saturation zone. The additional harmonics observed in the analysis for the 2nd-order voltage harmonic are (Figure 11a) 5th-order harmonics of negative sequence as well as 4th- and 7th-order harmonics of positive sequence. These harmonics result in higher oscillations and harmonic losses in the motor, which increase consumption and temperature, two important parameters for the end-user concerning with economic and useful life terms.

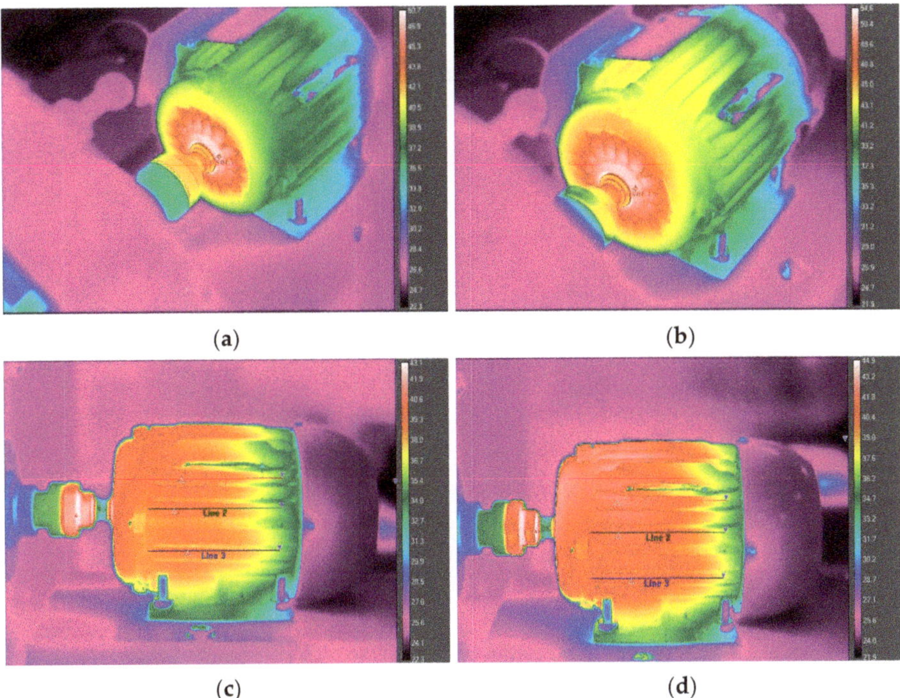

Figure 10. Thermographic images of the line-start permanent magnet motor in presence of 5th-order voltage harmonics in frontal and lateral view: (**a**) thermal equilibrium frontal view; (**b**) 25% of 5th-order voltage harmonic in frontal view; (**c**) thermal equilibrium lateral view; (**d**) 25% of 5th-order voltage harmonic in lateral view.

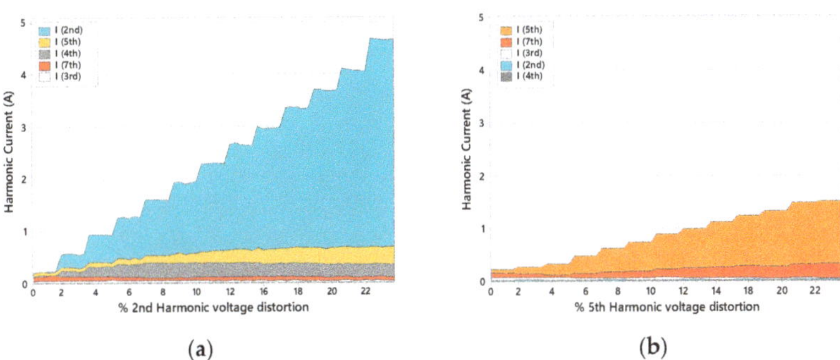

Figure 11. Harmonic currents present in LSPMM in the presence of (**a**) 2nd-order voltage harmonic and (**b**) 5th-order voltage harmonic [13].

For the 5th-order sequence harmonic, a 7th-order harmonic of positive sequence appears, as presented in Figure 11b. This harmonic produces a positive magnetic field that, in combination with the 5th-order magnetic field, results in higher oscillations, vibrations, and noise in the LSPMM for higher harmonic percentages. This fact was considered during the LSPMM modeling in the FEMM software, aiming at a more exact representation of it, as will be presented in the next sections.

4.2. Finite Elements Analysis Results

For the LSPMM modelling and simulation, initially, the problem was defined in the software FEMM with information such as frequency and depth since the software presents a 2D visualization; however, the problem solution is based on the 3D dimension. After defining the problem, the geometry was inserted, created from the physical dimensions of the motor (rotor diameter, number of slots, air gap distance, etc.). For this study, the complete motor was considered to better visualize the paths of flux lines in each harmonic distortion condition. The materials were inserted based on the materials obtained from manufacturer information for the engine analyzed (Figure 12a).

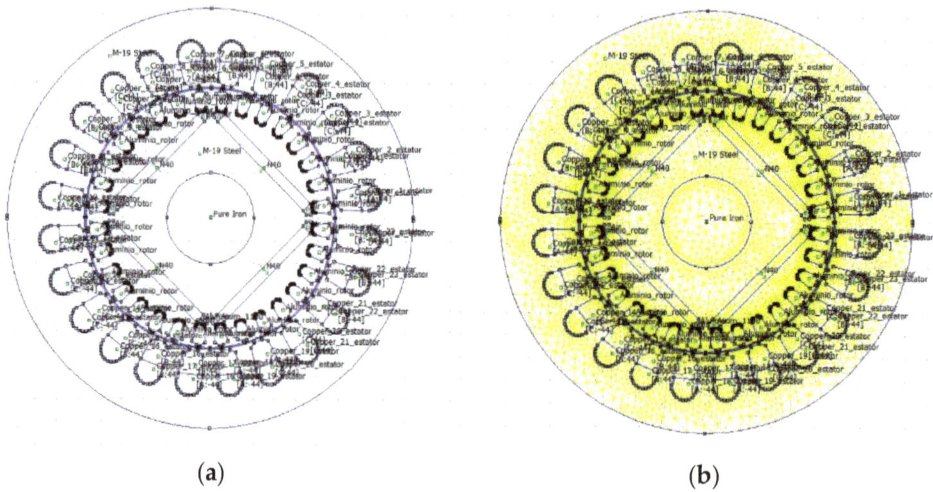

(a)   (b)

Figure 12. Line-start permanent magnet motor simulation on FEMM: (**a**) LSPMM geometry and materials and (**b**) LSPMM mesh.

The boundary conditions, which are useful to help direct the motor response in the simulation (magnetic flux response, current response, etc.), were also defined during the motor modelling step, and then the mesh was created and refined until a constant response was obtained (Figure 12b); finally, the simulation was performed, from which the results presented in this section were obtained.

The presence of voltage harmonics results in additional harmonic currents, which induce harmonic voltages in the rotor bars that produce harmonic currents circulating in the rotor squirrel cage. These additional harmonic components produce additional magnetic fields that result in opposite torques, which are translated into a reduction in speed for asynchronous electric motors.

Figure 13 presents the magnetic field lines in the LSPMM in nominal conditions (Figure 13a), 2nd-order voltage harmonic (Figure 13b), and 5th-order voltage harmonic (Figure 13c). It can be seen how the harmonics result in a larger number of flux lines, observed through the magnetic field density, but it is also important to note how the trajectory of the flux lines changes with the presence of harmonics, with special focus on the 2nd-order harmonic flux lines that present longer trajectories, which in turn translates into higher reluctance and consequently higher magnetic losses.

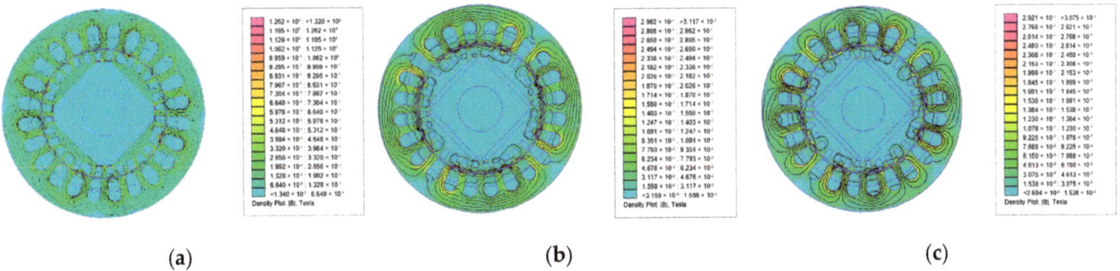

(a)   **(b)**   **(c)**

Figure 13. Density flux plot for (**a**) nominal conditions; (**b**) 2nd-order voltage harmonics and (**c**) 5th-order voltage harmonics.

However, the LSPMM speed does not vary for any of the analyzed harmonics, which does not mean that the harmonics do not impact other variables such as consumption, temperature, and power factor.

The objective of using the finite element analysis is to conduct a thorough diagnosis of the motor electromagnetic behavior using a method that is non-invasive and has no interference with the operation of the motor being analyzed. Using this method, it was possible to extract the resistive losses in the stator armature as well as the torque via the weighted stress tensor method [42]. These simulations were performed considering that the motor is operating at no load and with only the fault brakes, which in this case are neglected. To perform the thermal analysis, FEMM solves the problem around the transient temperature model using (19). From that, it is possible to obtain the temperature and gradient values of the temperature model.

In order to account for temperature exchange with the air, we applied the convection boundary condition. Here, we define the thermal transfer coefficient for the inner parts as well as the exterior part of the motor from the parameters used in reference [43] and adjusted based on the motor output power given that the one considered in this study presented a lower power (0.75 kW). For the exterior part of the frame, the coefficient was 30 W/m^2 °C. As for the inner parts of the motor, the air gap, and cooling air vents, the coefficient is higher, at 65 W/m^2 °C in this case, and the gaps between the stator/rotor armature and coil ends were not considered, as their value is negligible. Figure 14 illustrates the areas where these boundary conditions were applied.

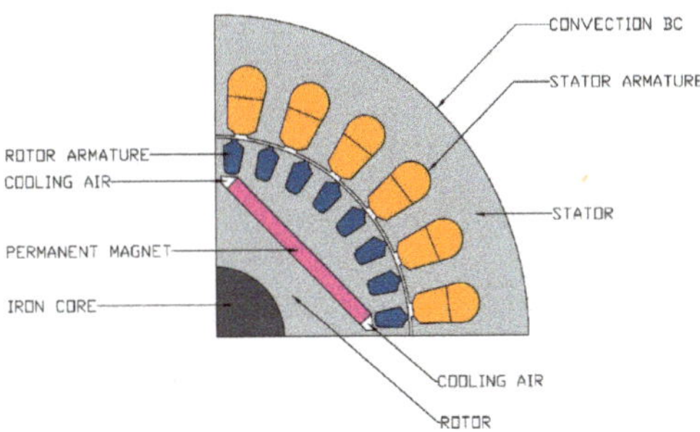

Figure 14. Quarter section of motor illustrating the areas with convection boundary conditions. The temperature distribution in the motor at full load for each voltage harmonic is presented in Figure 15. For the second-order voltage harmonic (Figure 15a), it is observed how the rotor presents the higher temperatures, which, despite the synchronism (zero slip), can be justified by the second-order harmonic component configured at the motor input as well as the new harmonics that appear as a result of the saturation in the ferromagnetic core. Regarding the fifth-order harmonic (Figure 15b), a similar pattern is observed, however, with lower temperatures, which shows that this harmonic is less detrimental to the LSPMM.

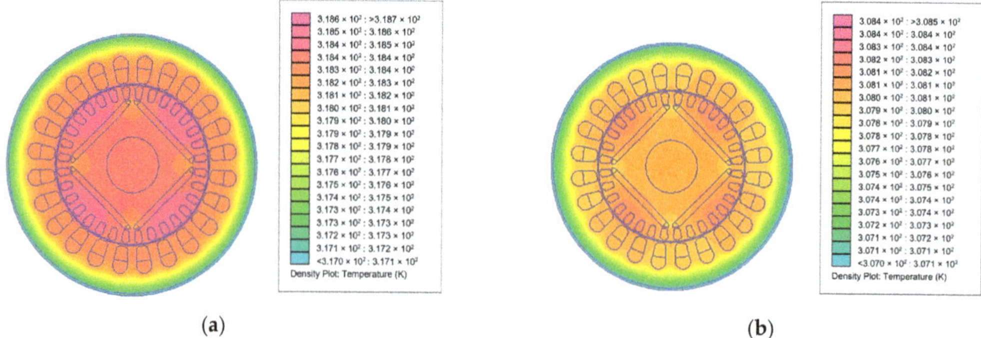

(a)   (b)

Figure 15. Temperature distribution (in Kelvin) in the motor from the FEMM thermal simulation for: (**a**) second-order voltage harmonic and (**b**) fifth-order voltage harmonic.

Analogous to the experiment results, the harmonic analysis was carried out for the simulation. Figure 16 shows the LSPMM lateral temperature as simulated in FEMM, with 25% of the voltage harmonic, and its comparison with the experimentally measured temperature in the lateral view of Figure 5a. It should, however, be noted that there is a large discrepancy due to the number of elements in the mesh and also due to not having considered the influence of the permanent magnets-magnetization being affected.

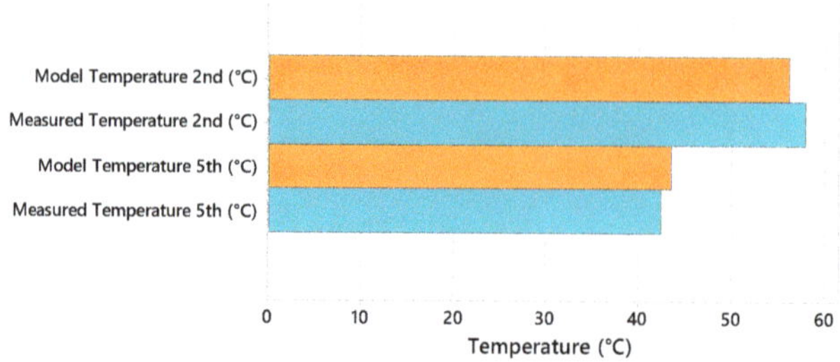

Figure 16. Comparison between the model and measured temperature for 25% voltage harmonic distortion of 2nd- and 5th-order harmonics.

Based on the results obtained from the computational simulation as well as from the experimental measurements, errors of 6.3% and 2.8% for the 2nd- and 5th-order harmonics, respectively, were found, which validates the model proposed on the line-start permanent magnet motor in this work.

5. Conclusions

The present study aimed to analyze the line-start permanent magnet motor experimentally and through computational simulations using FEMM in the presence of second- and fifth-order voltage harmonics. With this objective, measurement campaigns were carried out on a 0.75 kW permanent magnet motor and simulations based on the physical dimensions and construction characteristics of the machine.

From the experimental results, it was possible to observe the considerable impact that the 2nd- and 5th-order harmonics of the negative sequence have on the LSPMM. Concerning currents, increases were observed mainly due to the reactive power increases as a product of the high harmonic frequencies in the leakage reactance and the magnetization branch reactance, while the active power presents small increments, which will result in lower power factors for each harmonic and higher temperatures. This impact can be justified in the construction of more efficient motors, in which the magnetization and leakage reactance present lower values, which will decrease losses but also reduce the filtering of external harmonics in the motor. The computational simulation, on the other hand, showed that the created model reliably represents the performance of the motor in non-ideal conditions, such as in the presence of the negative sequence voltage harmonic, showing its considerable effect on the internal thermal distribution of the motor.

The simulations in the FEMM software also allowed a better evaluation of the magnetic losses in the presence of voltage harmonics. From the magnetic flux paths changes in the presence of harmonics that produce additional temperatures in the motor, observed in the experimental measures and validated from the model built based on the motor geometry, and that can be certainly useful to determine the main effects of these disturbances on components such as motor bearings or insulation degradation, aiming to program predictive maintenance with adequate frequencies to the disturbances present in the motor.

The creation and validation of the LSPMM model allow evaluating the state of the motors from the experimental data using numerical methods that will certainly give a clearer scenario of the thermal-magnetic behavior of the motor, which can be easily extended to a predictive maintenance product with wide application in the industry and in electric mobility in general.

Author Contributions: Conceptualization, J.M.T. and W.d.S.F.; methodology, J.M.T. and W.d.S.F.; software, W.d.S.F. and B.K.T.; validation, J.M.T., W.d.S.F. and B.K.T.; formal analysis, J.M.T. and W.d.S.F.; investigation, J.M.T. and W.d.S.F.; resources, W.d.S.F.; data curation, J.M.T. and B.K.T.; writing—original draft preparation, J.M.T., W.d.S.F. and B.K.T.; writing—review and editing, U.H.B., M.E.d.L.T., E.O.d.M. and W.d.S.F.; supervision, U.H.B., M.E.d.L.T., E.O.d.M., M.d.O.e.S. and W.d.S.F.; funding acquisition, W.d.S.F. and M.d.O.e.S. All authors have read and agreed to the published version of the manuscript.

Funding: This research was funded by the National Academic Cooperation Program in the Amazon (PROCAD), Process No. 88887.200548/2018-00.

Institutional Review Board Statement: Not applicable.

Informed Consent Statement: Not applicable.

Data Availability Statement: Not applicable.

Acknowledgments: This work was supported by the Brazilian National Council of Scientific and Technological Research (CNPq) as well as the Paulo Freire + Program.

Conflicts of Interest: The authors declare no conflict of interest.

References

1. Sarac, V.J.; Stefanov, G. Various Rotor Topologies of Line-Start Synchronous Motor for Efficiency Improvement. *Power Electron. Drives* **2019**, *5*, 83–95. [CrossRef]
2. Zeng, X.; Quan, L.; Zhu, X.; Xu, L.; Liu, F. Investigation of an Asymmetrical Rotor Hybrid Permanent Magnet Motor for Approaching Maximum Output Torque. *IEEE Trans. Appl. Supercond.* **2019**, *29*, 1–4. [CrossRef]
3. Study on Line-Start Permanent Magnet Assistance Synchronous Reluctance Motor for Improving Efficiency and Power Factor. Available online: https://www.mdpi.com/1996-1073/13/2/384 (accessed on 16 March 2022).
4. Zöhra, B.; Akar, M. Design Trends for Line Start Permanent Magnet Synchronous Motors. In Proceedings of the 2019 3rd International Symposium on Multidisciplinary Studies and Innovative Technologies (ISMSIT), Ankara, Turkey, 11–13 October 2019; pp. 1–7.
5. Lyskawinski, W. Comparative Analysis of Energy Performance of Squirrel Cage Induction Motor, Line-Start Synchronous Reluctance and Permanent Magnet Motors Employing the Same Stator Design. *Arch. Electr. Eng.* **2020**, *69*, 967–981. [CrossRef]
6. Ferreira, F.J.T.E.; Leprettre, B.; de Almeida, A.T. Comparison of Protection Requirements in IE2-, IE3-, and IE4-Class Motors. *IEEE Trans. Ind. Appl.* **2016**, *52*, 3603–3610. [CrossRef]
7. Tshoombe, B.K.; Muñoz Tabora, J.; da Silva Fonseca, W.; Emília Lima Tostes, M.; de Matos, E.O. Voltage Harmonic Impacts on Line Start Permanent Magnet Motor. In Proceedings of the 2021 14th IEEE International Conference on Industry Applications (INDUSCON), São Paulo, Brazil, 16–18 August 2021; pp. 962–968.
8. Muñoz Tabora, J.; de Lima Tostes, M.E.; Ortiz de Matos, E.; Mota Soares, T.; Bezerra, U.H. Voltage Harmonic Impacts on Electric Motors: A Comparison between IE2, IE3 and IE4 Induction Motor Classes. *Energies* **2020**, *13*, 3333. [CrossRef]

9.   Influence of Voltage Subharmonics on Line Start Permanent Magnet Synchronous Motor. Available online: https://ieeexplore.ieee.org/document/9641816 (accessed on 16 March 2022).

10.   Debruyne, C.; Sergeant, P.; Derammelaere, S.; Desmet, J.J.M.; Vandevelde, L. Influence of Supply Voltage Distortion on the Energy Efficiency of Line-Start Permanent-Magnet Motors. *IEEE Trans. Ind. Appl.* **2014**, *50*, 1034–1043. [CrossRef]

11.   Debruyne, C. *Impact of Voltage Distortion on Energy Efficiency of Induction Machines and Line Start Permanent Magnet Machines*; Ghent University: Ghent, Belgium, 2014.

12.   Tabora, J.M.; De Lima Tostes, M.E.; De Matos, E.O.; Bezerra, U.H.; Soares, T.M.; De Albuquerque, B.S. Assessing Voltage Unbalance Conditions in IE2, IE3 and IE4 Classes Induction Motors. *IEEE Access* **2020**, *8*, 186725–186739. [CrossRef]

13.   Tabora, J.M.; De Lima Tostes, M.E.; Bezerra, U.H.; De Matos, E.O.; Filho, C.L.P.; Soares, T.M.; Rodrigues, C.E.M. Assessing Energy Efficiency and Power Quality Impacts Due to High-Efficiency Motors Operating Under Nonideal Energy Supply. *IEEE Access* **2021**, *9*, 121871–121882. [CrossRef]

14.   López-Pérez, D.; Antonino-Daviu, J. Application of Infrared Thermography to Failure Detection in Industrial Induction Motors: Case Stories. *IEEE Trans. Ind. Appl.* **2017**, *53*, 1901–1908. [CrossRef]

15.   Diagnosis of the Three-Phase Induction Motor Using Thermal Imaging-ScienceDirect. Available online: https://www.sciencedirect.com/science/article/abs/pii/S1350449516306259 (accessed on 16 March 2022).

16.   Ferreira, F.J.T.E.; de Almeida, A.T.; Carvalho, J.F.S.; Cistelecan, M.V. Experiments to Observe the Impact of Power Quality and Voltage-Source Inverters on the Temperature of Three-Phase Cage Induction Motors Using an Infra-Red Camera. In Proceedings of the 2009 IEEE International Electric Machines and Drives Conference, Miami, FL, USA, 3–6 May 2009; pp. 1311–1318.

17.   Singh, G.; Anil Kumar, T.C.; Naikan, V.N.A. Fault Diagnosis of Induction Motor Cooling System Using Infrared Thermography. In Proceedings of the 2016 IEEE 6th International Conference on Power Systems (ICPS), New Delhi, India, 4–6 March 2016; pp. 1–4.

18.   Mahami, A.; Rahmoune, C.; Bettahar, T.; Benazzouz, D. Induction Motor Condition Monitoring Using Infrared Thermography Imaging and Ensemble Learning Techniques. *Adv. Mech. Eng.* **2021**, *13*, 16878140211060956. [CrossRef]

19.   Azari, M.N.; Mirsalim, M. Performance Analysis of a Line-Start Permanent Magnet Motor with Slots on Solid Rotor Using Finite-Element Method. *Electr. Power Compon. Syst.* **2013**, *41*, 1159–1172. [CrossRef]

20.   Fonseca, D.; Santos, C.C.; Cardoso, A. Modelling of a Line-Start Permanent Magnet Synchronous Motor, Using Empirical Parameters. In Proceedings of the International Conference on Engineering ICEUBI2017, Covilhã, Portugal, 5–7 December 2017.

21.   Robust Design Optimization of a Five-Phase PM Hub Motor for Fault-Tolerant Operation Based on Taguchi Method. Available online: https://ieeexplore.ieee.org/document/9076301 (accessed on 10 October 2022).

22.   Sun, X.; Xu, N.; Yao, M. Sequential Subspace Optimization Design of a Dual Three-Phase Permanent Magnet Synchronous Hub Motor Based on NSGA III. *IEEE Trans. Transp. Electrif.* **2022**, *1*. Available online: https://ieeexplore.ieee.org/document/9828489 (accessed on 13 July 2022) [CrossRef]

23.   Xu, J.; Zhang, L.; Meng, D.; Su, H. Simulation, Verification and Optimization Design of Electromagnetic Vibration and Noise of Permanent Magnet Synchronous Motor for Vehicle. *Energies* **2022**, *15*, 5808. [CrossRef]

24.   Liu, X.; Lin, Q.; Fu, W. Optimal Design of Permanent Magnet Arrangement in Synchronous Motors. *Energies* **2017**, *10*, 1700. [CrossRef]

25.   Yeo, H.-K.; Ro, J.-S. Novel Analytical Method for Overhang Effects in Surface-Mounted Permanent-Magnet Machines. *IEEE Access* **2019**, *7*, 148453–148461. [CrossRef]

26.   Li, J.; Song, J.; Cho, Y. High Performance Line Start Permanent Magnet Synchronous Motor for Pumping System. In Proceedings of the 2010 IEEE International Symposium on Industrial Electronics, Bari, Italy, 4–7 July 2010; pp. 1308–1313.

27.   Cavagnino, A.; Vaschetto, S.; Ferraris, L.; Gmyrek, Z.; Agamloh, E.B.; Bramerdorfer, G. Striving for the Highest Efficiency Class With Minimal Impact for Induction Motor Manufacturers. *IEEE Trans. Ind. Appl.* **2020**, *56*, 194–204. [CrossRef]

28.   Influence of Magnetic Slot Wedge on the Performance of 10-kV, 1000-kW Permanent Magnet Synchronous Motor-Qiu-2020-International Transactions on Electrical Energy Systems-Wiley Online Library. Available online: https://onlinelibrary.wiley.com/doi/abs/10.1002/2050-7038.12332 (accessed on 16 March 2022).

29.   Frosini, L.; Pastura, M. Analysis and Design of Innovative Magnetic Wedges for High Efficiency Permanent Magnet Synchronous Machines. *Energies* **2020**, *13*, 255. [CrossRef]

30.   IEC 60034-30-1:2014 | IEC Webstore | Pump, Motor, Water Management, Smart City, Energy Efficiency. Available online: https://webstore.iec.ch/publication/136 (accessed on 10 June 2022).

31.   Fong, J.; Ferreira, F.J.T.E.; Silva, A.M.; de Almeida, A.T. IEC61800-9 System Standards as a Tool to Boost the Efficiency of Electric Motor Driven Systems Worldwide. *Inventions* **2020**, *5*, 20. [CrossRef]

32.   Electric Motors. Available online: https://ec.europa.eu/info/energy-climate-change-environment/standards-tools-and-labels/products-labelling-rules-and-requirements/energy-label-and-ecodesign/energy-efficient-products/electric-motors_en (accessed on 12 May 2021).

33.   Ferreira, F.J.T.E.; Silva, A.M.; Aguiar, V.P.B.; Pontes, R.S.T.; Quispe, E.C.; de Almeida, A.T. Overview of Retrofitting Options in Induction Motors to Improve Their Efficiency and Reliability. In Proceedings of the 2018 IEEE International Conference on Environment and Electrical Engineering and 2018 IEEE Industrial and Commercial Power Systems Europe (EEEIC/I CPS Europe), Palermo, Italy, 12–15 June 2018; pp. 1–12.

34. European Committee of Manufacturers of Electrical Machines and Power Electronics (CEMEP). Ecodesign Regulations (EU) 2019/1781 for Motor and Drives and (EU) 2021/341 (Amendment) from CAPIEL and CEMEP. 10 May 2021. Available online: https://www.nord.com/cms/en/product_catalogue/ie2_motors/country_regulations/cp_countryregulations.jsp (accessed on 2 August 2020).

35. WEG. Energy Efficiency Primer. Available online: http://Materiais.Motores.Weg.Net/Cartilha (accessed on 27 July 2022).

36. Hughes, A.; Drury, B. *Electric Motors and Drives*, 4th ed.; Elsevier Ltd.: Leeds, UK, 2013; ISBN 978-0-08-098332-5.

37. Meeker, D. Finite Element Method Magnetics; User's Manual; Version 4.2. Available online: http://www.femm.info/Archives/doc/manual34.pdf (accessed on 1 October 2022).

38. A Thermal Analysis of Induction Motors Using a Weak Coupled Modeling. Available online: https://ieeexplore.ieee.org/document/582603 (accessed on 18 October 2022).

39. IEC 60034-2-1:2014 | IEC Webstore | Energy Efficiency, Smart City. Available online: https://webstore.iec.ch/publication/121 (accessed on 27 July 2022).

40. HomePage: Finite Element Method Magnetics. Available online: https://www.femm.info/wiki/HomePage (accessed on 27 July 2022).

41. Kumar, L.A.; Alexander, S.A. *Computational Paradigm Techniques for Enhancing Electric Power Quality*; CRC Press: Boca Raton, FL, USA, 2018; ISBN 978-0-429-80990-3.

42. Saha, S.; Cho, Y.H. Starting Characteristic Analysis of LSPM for Pumping System Considering Demagnetization. *Int. J. Mech. Mechatron. Eng.* **2015**, *9*, 1305–1311.

43. Palangar, M.F. Design, Analysis and Optimization of Line-Start Permanent-Magnet Synchronous Motors: Simultaneous Electromagnetic and Thermal Analysis. Available online: https://theses.flinders.edu.au/view/8616886a-3b6f-4c97-a9d0-95452aa86a65/1 (accessed on 1 November 2022).

Review

Review on Model Based Design of Advanced Control Algorithms for Cogging Torque Reduction in Power Drive Systems

Pierpaolo Dini *,† **and Sergio Saponara** †

Department of Information Engineering, University of Pisa, Via G. Caruso, 16, 56122 Pisa, Italy
*   Correspondence: pierpaolo.dini@phd.unipi.it
†   These authors contributed equally to this work.

Abstract: This review of the state of the art aims to collect the description and main research results in the field of development and validation of control algorithms with the main purpose to solve the problem of cogging torque and main sources of electromagnetic torque ripple. In particular, we focus on electric drives for advanced and modern mechatronic applications such as industrial automation, robotics, and automotive applications, with special emphasis on work that exploits model-based design. A great added value of this paper is to explicitly show the operational steps required for the model-based design design of optimized control algorithms for electric drives where it is necessary to make up for electromagnetic torque oscillations due to the main sources of ripple, particularly cogging torque. The ultimate goal of this paper is to provide researchers approaching this particular problem with a comprehensive collection of the most effective solutions reported in the state of the art and also a summary for effectively applying the model-based design methodology.

Keywords: model-based design; simulation; mechatronics; dynamic systems; modeling; control theory; brushless motors; electric drives

Citation: Dini, P.; Saponara, S.
Review on Model Based Design of
Advanced Control Algorithms for
Cogging Torque Reduction in Power
Drive Systems. *Energies* **2022**, *15*, 8990.
https://doi.org/10.3390/en15238990

Academic Editors: Quntao An, Bing
Tian and Xinghe Fu

Received: 25 September 2022
Accepted: 22 November 2022
Published: 28 November 2022

Publisher's Note: MDPI stays neutral
with regard to jurisdictional claims in
published maps and institutional affiliations.

1. Introduction

1.1. Brief Introduction to MBD for Mechatronics

This state-of-the-art review focuses on a very specific topic such as the development of advanced algorithms for reducing cogging torque and major sources of electromagnetic torque disturbance in which MBD is exploited as a design paradigm.

As is well known, the use of MBD enables superior results to be achieved in the project phases, in which various aspects need to be evaluated upstream of the realization of a prototype, on which it will later be possible to concretize and refine the results.

The MBD approach has become particularly popular in the design of control algorithms for the automation of processes and dynamic systems of particular industrial interest [1,2].

Through the MBD approach, the risks associated with testing under critical operating conditions can be reduced, decreasing validation time and optimizing implementation steps [3].

So, MBD can be exploited to evaluate the choice of control techniques, numerical optimization criteria, operational approximations, modeling of phenomena at different levels of detail, computational complexity, cost-benefit ratio, preliminary evaluation of the required embedded platform, final integration on micro-controller/processor, etc. [4–6].

In addition, there are advanced tools implementing the MBD paradigm that make it easy to use for design and subsequent integration on electronic systems of interest, greatly speeding up implementation time.

Figure 1 schematically recalls the stages of the MBD paradigm.

Model-in-the-Loop (MIL) is the first step in the modeling-based development chain, in which both the control and monitoring algorithms and the physical plant or process are implemented through a set of mathematical equations and/or formalisms, in a simulation environment, such as MATLAB/Simulink.

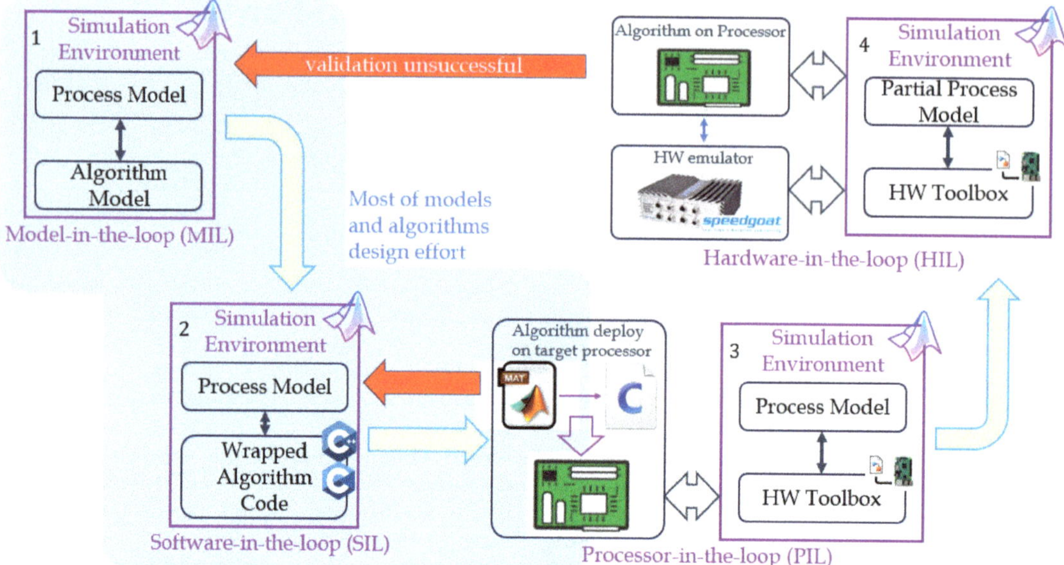

Figure 1. Example of MBD steps for algorithms validation.

It is an essential step, as it allows us to make an exhaustive study of the system and of the interaction with the blocks that implement the algorithms, especially from the point of view of the operating space, verifying the functioning in various operating conditions and the consequences in their variations [7].

Software-in-the-Loop (SIL) is a process in which the actual Production Software Code, which in general can derive from developments prior to the adoption of MBD techniques, and therefore not necessarily self-generated code, is incorporated into a simulation environment that contains the mathematical models of the physical system.

This is done to allow for the inclusion of software features for which no templates exist or to allow for faster simulation runs, as part of the functional blocks are already written in low-level C / C ++ code. Therefore, SIL is defined as the inclusion of compiled production software code in a simulation model [8].

Processor-in-the-Loop (PIL) is the process in which the control and monitoring system is compiled and downloaded into an integrated target processor, ergo an embedded platform with micro-controller/processor with specifications consistent with that selected for the final mechatronic system and communicates directly with the model of the plant via standard communications such as Ethernet.

In this phase of the MBD validation process, hardware toolboxes are referred to as SW infrastructures that allow interaction between the simulation environment and the low-level environment, such as micro-controller or micro-processor. Among the most widely used in automotive and industrial automation is support for NXP's evaluation boards [9]. This allows designers to validate their control algorithm even in the absence of the physical process, by exploiting the simulated model, which interacts with the embedded code on the embedded system.

In this case, I/O devices are not used for communication, so unlike the RCP phase there is no verification of the interactions between the Embedded system and the physical process or part of it being tested [10].

Hardware-in-the-Loop (HIL) indicates the testing technique of electronic control units by connecting them to special benches that reproduce in a more or less complete way the electro-mechanical and electronic system of the process.

The purpose of the HIL tests is to use the benches to anticipate the checks on components, subsystems and systems already in the design and prototyping phase, without waiting for the availability of the final product for which they are intended.

In fact, the real components installed respond to the simulated signals as if they were operating in a real environment, since they are not able to distinguish signals coming from a physical environment from those produced by software models.

In this way, the HIL method makes it possible to reproduce the most diverse operating conditions with the benches and observing the behavior of the system and of the individual elements [11].

In this phase of validating control algorithms, one often has to deal with the problem of only having part of the physical process to be controlled. To increase the degree of validation with respect to the PIL phase, hardware emulators are often used. These devices, such as the widely used Speedgoat systems [12], are able to provide HW interfaces for system and process models, supplying currents and voltages instead of sensors and actuators of the final process. This allows designers to further test the validity of the algorithms by "getting closer" to the physical process with respect to PIL testing (as schematically reported in Figure 2).

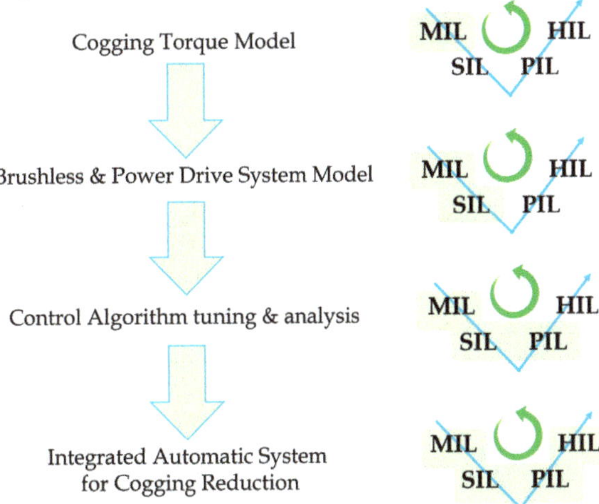

Figure 2. MBD workflow in the specific case of design of control algorithms to reduce the cogging torque.

1.2. Contributions

Many reviews of the state of the art regarding the reduction of Cogging Torque in Synchronous/Brushless motors by physical modification techniques can be found in the literature, but there are none regarding control techniques.

In this review, we focus on works that propose advanced control techniques for reducing cogging torque and major sources of torque disturbance. The development of the control techniques proposed in this work takes advantage of the MBD approach, which, as discussed above, is the most efficient and flexible. Compared with the reviews in the literature, our most relevant contributions are:

- Focus on the problem of cogging torque reduction by advanced control technique validated through MBD approach;
- Collect the most recent state-of-the-art work on this topic;

- Critical analysis of the control techniques proposed by various authors, with the aim of providing suggestions for implementation and improvement of the reported results.

1.3. Paper Organization

The rest of the manuscript is organized into the following Sections: Section II explains the cogging torque problem in detail by reporting the main approaches and methods used for mathematical modeling in the context of control algorithm design; Section III gives details of mathematical modeling and integration in a simulation environment of the main components of electric drives in modern mechatronic systems; Section IV describes the most commonly used control techniques for solving the cogging torque problem by listing their advantages and disadvantages; and Section V gives final conclusions and a discussion of possible future developments.

2. Cogging Torque's Details and Modeling
2.1. Reasons for Control Design in Cogging Torque Reduction

Given the enormous potential of this methodology, MBD is increasingly being used to increase the efficiency of individual components of an entire mechatronic process.

It can safely be said that the mechatronic processes of strongest industrial interest are physical electro-mechanical in nature, so among the most decisive components for maximum efficiency of the entire process are definitely electrical drives.

It can certainly be said that, by now, modern drives for advanced mechatronic applications mostly exploit Permanent Magnet Synchronous Motors (PMSM) and Brushless motors.

They have the most advantageous inertia/torque ratio, high efficiencies, physically robust characteristics, and are the most versatile for use at both very low and high speeds.

As is well known, among the inherent problems of using Synchronous/Brushless motors is torque ripple due to cogging torque, which causes undesirable oscillations, lowering performance both for low-speed applications where high accuracy in positioning control is required, and in speed control applications where unwelcome noise is generated.

There are some papers in the literature in which a very specific design is proposed with the aim of mitigating the effect of oscillations due to this physical phenomenon, but these have met with little success in practice [13–15].

This is mainly due to the fact that such solutions are based on physical modifications applicable only in some specific configurations, relating to Synchronous motors suitable mainly for applications in which moderate electromagnetic torque is required, since the underlying concept is to limit the amount of permanent magnets, arranging them so that the stator–rotor magnetic interaction is reduced, but effectively reducing the torque deliverable by the machine (see Figure 3).

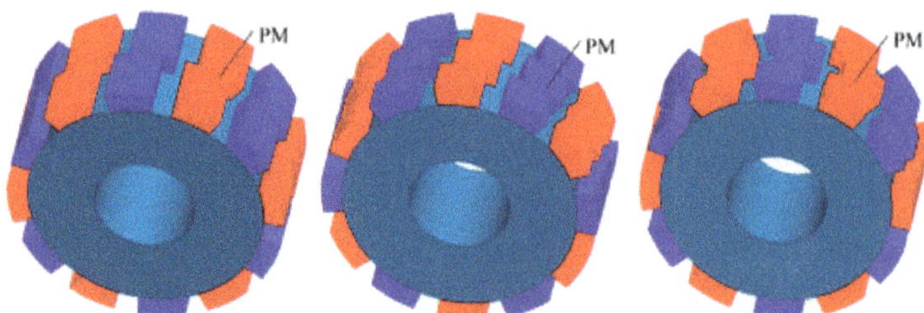

Figure 3. Example of changing the arrangement of permanent magnets, valid only for surface magnets, called "rotor skewing".

The most promising results have been achieved by design of control algorithms, which are capable of compensating for cogging torque, which for all intents and purposes is an additive disturbance signal with respect to electromagnetic torque, so control techniques are effective in compensating for it.

MBD being the best among advanced algorithm development and validation approaches, research in recent years sees the application of this paradigm for efficient resolution of this type of problem as well.

2.2. Cogging Torque Modeling Approaches

As anticipated, cogging torque is an intrinsic phenomenon of permanent magnet synchronous motors and Brushless motors, both radial and axial flux, related to the magnetic interaction between the main parts of the electrical machine, such as stator and rotor.

The phenomenon is present in both synchronous motor configurations, with magnets superficial to the rotor and internal to the rotor iron, and Brushless, causing the same effect, i.e., electromagnetic torque disturbance.

The problem is intrinsic, as it depends on the unavoidable variation of the equivalent reluctance of the enclosing magnetic path of the magnetic flux concatenated with the stator iron through the air gap and produced by the permanent magnets themselves.

By manually rotating the rotor axis, it is possible to perceive this phenomenon as preferential angular positioning, at some angular positions "attractive" and at some angles "repulsive".

The phenomenon occurs even in the absence of power supply to the electrical machine, so it is interpreted as an additive disturbance with respect to the electromagnetic torque, due to the phase currents (and from the angular position).

In addition, due to the total symmetry of the ferromagnetic structure of electric machines, the alternation of the "repulsive" effect and the "attractive" effect due to the variation of the air gap, i.e., the magnetic reluctance, compensate each other along the entire round angle. This allows the intuition that cogging torque is a zero mean phenomenon.

In Figure 4, the concept of alternating "attractive" and "repulsive" phase is shown for a surface magnet synchronous motor configuration; the concept is the same for all configurations.

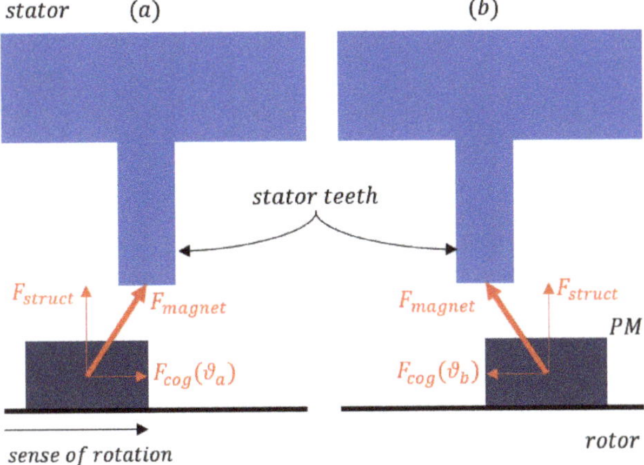

Figure 4. Schematizing of the concept of alternating between "attractive" (**a**) and "repulsive" (**b**) phases.

It represents the behavior in the proximity of the preferential angular position, i.e., the position where naturally the concatenated magnetic flux tends to align the teeth of the stator slots with the magnets arranged on the rotor.

In the direction of rotation, the force generated by the concatenated magnetic flux (in the direction tangential to the rotor surface) alternates between positive and negative signs depending on whether the angular position is less or greater than the preferential position itself.

Obviously, the magnetic force is composed of a tangential component, which is the actual cause of cogging torque, and a normal component, which, however, is absorbed by the structure without any effect on the relative rotation between stator and rotor.

The application of advanced control techniques to reduce this inherent effect appears to be the best approach. Following the MBD design standard, a first step is certainly to make an estimate of the cogging torque by means of a mathematical model.

In the literature, the most widely used approaches that have been shown to be usable in the context of control algorithms, with the best results being closed form modeling and the online estimation.

Works in the literature that focus on mathematical modeling of cogging torque in closed form start from the equivalent energy representation of the machine, via an equivalent magnetic circuit. This highlights the angular position dependence of the stored magnetic power budget.

This relationship is further highlighted in the dependence of magnetic reluctance, due to the fact (described intuitively above) that the air gap varies its amplitude during relative rotation between stator and rotor [16–21].

Referring to Figure 5, it is possible to quantify the magnetic energy stored by the ferromagnetic structure as given in Equation (1).

$$W_m = \frac{1}{2} L_{eq} I^2 + \frac{1}{2} R_{eq}(\theta_m) \phi^2_{rs} + NI\phi_{rg} \tag{1}$$

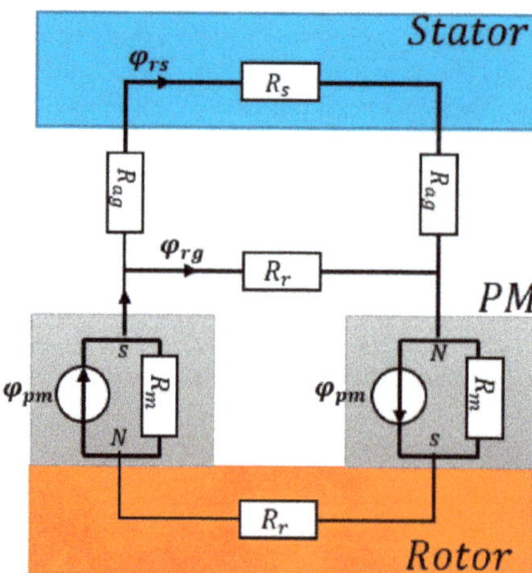

Figure 5. Equivalent magnetic circuit for permanent magnet synchronous motors.

W_m denotes the magnetic energy stored by the electrical machine, both under energized and not energized conditions. With L_{eq} we denote the stator inductance of the equivalent circuit of the internal windings, I the phase current; R_{eq} the equivalent magnetic reluctance associated with the magnetic induction circuit at the mechanical air gap; and with N we denote the number of windings per stator phase. Finally, ϕ_{rs} and ϕ_{rg} denote the components of the magnetic flux concatenated through the air gap between the stator and rotor.

It follows that the cogging torque is defined as the derivative of magnetic energy with respect to angular position, in the absence of power, that is, at zero electric current in the machine windings. In Equation (2), the relationship between magnetic energy and cogging torque is given.

$$T_{cog} = -\frac{\partial W_m(I=0)}{\partial \theta_m} = -\frac{1}{2}\phi^2_{rs}\frac{\partial R_{eq}(\theta_m)}{\partial \theta_m} \tag{2}$$

In Equation (2), the term $\frac{\partial W_m(I=0)}{\partial \theta_m}$ denotes the partial derivative of the magnetization energy with respect to the angular position of the rotor, calculated under the condition that the electrical machine is not powered by any energy source.

The exact expression of the cogging torque depends on the square of the magnetic flux intensity concatenated with the air gap ϕ_{rs}, and the derivative of the equivalent magnetic reluctance, a function of angular position.

Note that ϕ_{rs} depends on the physical characteristics of the permanent magnets, i.e., the magnetic material used, while R_{eq} depends substantially on geometrical factors, thus on the shape of the stator slots as well as on the arrangement of the permanent magnets themselves. Since this is all known information, this is an analytical representation in closed form.

This is the most widely used approach by researchers focusing on control algorithms because of the simplicity of integrating the cogging torque model, being precisely in closed form it can be integrated into the mechanical balance of rotor rotation.

Moreover, it is easy to validate through dedicated SW tools for finite element analysis that do not require the presence of the machine but only design information, which is often the case in advanced design activities typical of innovative companies exploiting R&D.

Another widely used approach, especially in the presence of the prototype electrical machine, is to use a test pod for direct measurement [22]. A typical test bench for cogging torque measurement is shown in Figure 6.

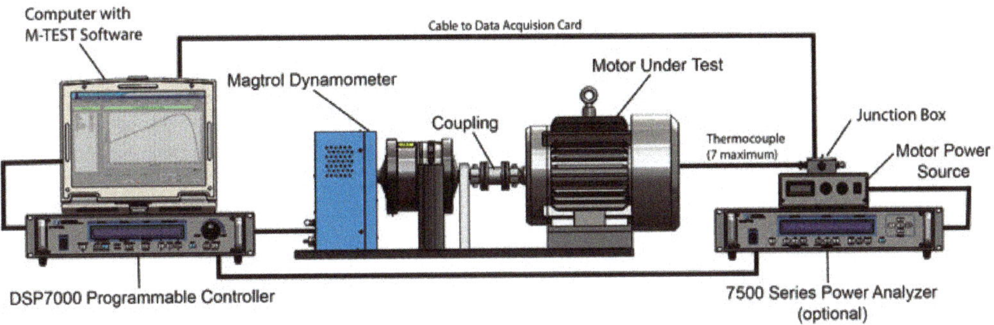

Figure 6. Typical cogging torque measurement test bench.

The typical measuring bench basically consists of two electric motors and a torque sensor (normally a torque transducer).

Through the first motor, which is called MASTER, we impose the motion of the second motor, called SLAVE, of which we are interested in recording the data inherent in the torque at the shaft by means of the torque transducer.

Since the Slave is "dragged" by the Master, its stator windings will not be current flowing, and this means that the torque measured by the torque transducer coincides with the Cogging Torque.

The important thing about this measurement is that it is made in the steady state of speed since the contribution of inertial actions due to acceleration transients must be avoided.

In fact, the measurement made in this way is the simplest procedure for obtaining a cogging torque profile that can be exploited in eventual modeling with a greater degree of accuracy describing the dynamic behavior of the machine and the design of a feedback law to compensate for such disturbance.

The result of the measurement is usually a set of data representing the measured cogging torque at different angular positions. This allows a closed-form model to be derived by approximation procedure, by harmonic analysis [23] or polynomial [24], as reported in Equation (3).

$$T_{cog} = \sum_{i=0}^{N_h} A_k cos\left(k\frac{\theta_m}{\theta_{cog}} + a_k\right) + B_k sin\left(k\frac{\theta_m}{\theta_{cog}} + b_k\right) \tag{3}$$

where A_k, B_k, a_k and b_k are harmonic development parameters, while θ_{cog} is the angular period of the cogging pair, which depends on geometrical and construction factors, such as the number of stator slots and teeth, the number of permanent magnets and type of rotor iron arrangement (information known a priori).

Other works base the estimation of cogging torque on artificial intelligence models, neural networks and machine learning [25,26]. Few works, however, combine artificial intelligence models with control algorithms.

The main reason is certainly to reduce the overall computational complexity of the automatic cogging pair reduction system, and of course to reduce the storage space, which is usually the weak point of AI models compared to other approaches.

3. Synchronous Motor and Advanced Electric Power Drives Modeling

3.1. Fundamental Components of Electric Power Drives in Mechatronics

Another key step in the MBD approach is certainly accurate modeling of the electrical machine and drive components, including inverter, modulation system and implemented load.

A block diagram for the architecture of an electric drive for mechatronic applications is shown in Figure 7. The fundamental components of the system and their function are shown below.

- **Static power converters**: these are used to convert the electrical power regime to that which is useful for driving the electrical machine through the inverter. Configurations typically found in mechatronic systems of industrial interest are AC/DC converters to handle direct power from the mains or DC/DC converters to handle the DC bus level provided by a constant source, such as a Li-ion battery.
- **Modulation system**: conceptually separate from the system that integrates the control algorithm, but often implemented on the same embedded platform, it is the union of algorithm and HW interface for direct management of the inverter mosfets. The control algorithm typically provides a duty-cycle equivalent signal, which is converted into switching logic by the switching system, to manage the state of the inverter and power the electrical machine accordingly.
- **Inverter**: the electronic circuit that physically manages the (bidirectional) power transfer from the power source (DC side) to the electrical machine (AC side). The state of the inverter is decided by the modulation system. Conceptually, the inverter is a combinatorial logic network with 2^3 combinations (in the three-phase case).
- **Control system**: the embedded platform on which the logic related to the feedback control algorithm is integrated. In modern embedded systems it coexists with the modulation system and cooperates via SW drivers. In the sensorless configuration, such a system also integrates the angular position/velocity estimation algorithm by exploiting direct measurement of electric current (from inverter or motor phase).
- **Transduction system**: the set of devices for measuring the physical quantities of primary interest for managing the electric drive. Typically it consists of:
    - Temperature sensors, placed at "critical" points, such as the electrical machine and static converters, which are subject to high currents and therefore need to be monitored for losses due to the Joule effect.
    - Position and angular velocity sensors: placed on the rotor axis if we are talking about encoders/resolvers or placed inside the structure if we are talking about hall effect sensors.

— Voltage/current sensors: typically placed on the inverter, since motor phases and inverter branches are necessarily traversed by the same current.

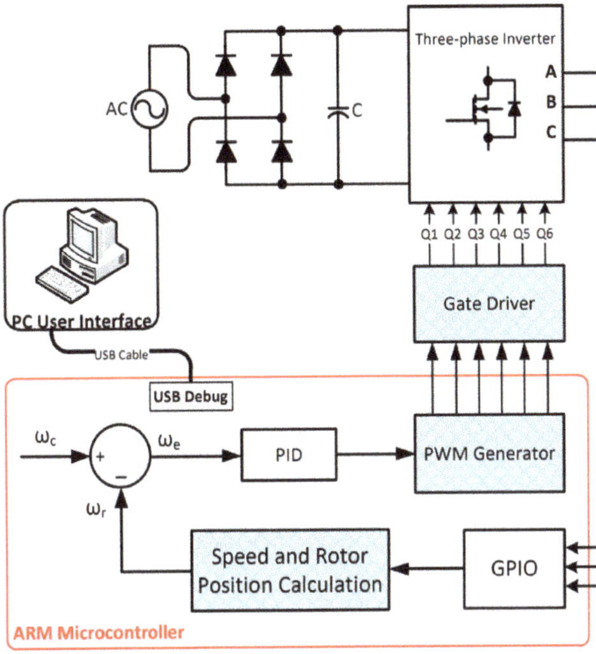

Figure 7. Typical electric drive architecture for mechatronic systems.

3.2. Synchronous Motor Modeling including Cogging Torque

The mathematical model of the synchronous motor is typically derived by applying the unified theory of electrical machines, which states that any electrical machine is dynamically describable by an equivalent electrical and magnetic circuit [27].

Permanent-magnet synchronous motors of interest in mechatronic applications, where it is required to solve the cogging torque problem, are to date three-phase motors, so the following are the fundamental points of modeling for three-phase machines [28,29]. Note that the dynamic model is formally unchanged whether the machines have radial magnetic flux or axial magnetic flux.

In Equation (4), the dynamic balance of phase voltages of the electrical machine is given. The vector $\vec{U}$ represents the triad of supply voltages, referred to the neutral of the equivalent circuit, the term $R_s\vec{I}$ represents the losses in the stator circuit, and the term $\frac{d\vec{Psi}}{dt}$ represents the voltage drop due to the phenomena of self and mute induction and the electromechanical energy conversion phenomenon, mainly due to the presence of the permanent magnets.

$$\vec{U} = \begin{bmatrix} U_a \\ U_b \\ U_c \end{bmatrix} = R_s \begin{bmatrix} I_a \\ I_b \\ I_c \end{bmatrix} + \frac{d}{dt}\begin{bmatrix} \Psi_a \\ \Psi_b \\ \Psi_c \end{bmatrix} = R_s\vec{I} + \frac{d\vec{\Psi}}{dt} \tag{4}$$

In Equation (5) are given the main components of the concatenated magnetic flux, denoted by the vector $\vec{\Psi}$. The vectors $\vec{\Psi}_{AI}$ and $\vec{\Psi}_{MI}$ denote the magnetic flux vectors generated by the phenomena of self and mutual induction, respectively, and the vector $\vec{\Psi}_{PM}$ denotes the magnetic flux vector produced by the magnets concatenated at the air gap.

I_a, I_b and I_c denote the components of the current vector (three-phase reference); U_a, U_b and U_c denote the components of the phase voltage vector (three-phase axes); and

Ψ_a, Ψ_b and Ψ_c are the components of the concatenated magnetic flux vector, taking into account both self and mutual induction at the mechanical air gap. R_s is the stator circuit electrical resistance.

$$\vec{\Psi} = \vec{\Psi}_{AI} + \vec{\Psi}_{MI} + \vec{\Psi}_{PM} = L_{eq}\vec{I} + k_\Psi \begin{bmatrix} \cos(p\theta) \\ \cos\left(p\theta - \frac{2\pi}{3}\right) \\ \cos\left(p\theta - \frac{4\pi}{3}\right) \end{bmatrix} = L_{eq}\vec{I} + k_\Psi \vec{C}(\theta) \tag{5}$$

Note that the magnetic flux vectors of self and mutual induction can be grouped in the term $L_{eq}\vec{I}$, where L_{eq} appropriately integrates the inductive coefficients of self and mutual induction.

In addition, the flux vector of permanent magnets $\vec{\Psi}_{PM}$ depends on the coefficient k_Ψ, indicating its intensity, which is a function of the ferromagnetic material used and the geometric arrangement, and depends formally on the angular position, as discussed above.

Combining Equation (4) and Equation (5), we derive the dynamical equilibrium of currents in the explicit form, in which the link between electrical ($\vec{U}$ and $\vec{I}$) and mechanical (θ and ω) variables is evident. Equation (6) is the equation in vector form of the electrical dynamics of the Synchronous machine.

$$\vec{U} = R_s\vec{I} + L_{eq}\frac{d\vec{I}}{dt} + \frac{d\vec{\Psi}_{PM}}{dt} = R_s\vec{I} + L_{eq}\frac{d\vec{I}}{dt} + k_\Psi p\frac{d\theta}{dt}\frac{\partial\vec{C}(\theta)}{\partial\theta} = R_s\vec{I} + L_{eq}\frac{d\vec{I}}{dt} + \omega\vec{E}(\theta) \tag{6}$$

Equation (7) represents the balance of power terms, the meaning of which is explained below.

- The term P_{in} denotes the power absorbed by the machine and is formally defined as the scalar product between voltage vector $\vec{U}$ and current vector $\vec{I}$.
- The term P_{joule} denotes the heating losses of the windings, which, as is known from physics, depend on the resistance of the circuit and the square of the modulus of the current, formally $R_s < \vec{I}, \vec{I} >= R_s|\vec{I}|^2$.
- The term $P_{magnetic}$ represents the magnetic power stored in the equivalent circuit and due exclusively to self and mutual induction phenomena; formally $L_{eq} < \frac{d\vec{I}}{dt}, \vec{I} >$.
- The term $P_{mechanical}$ represents the mechanical power delivered by the machine, against electromechanical conversion (Lentz's law). For Synchronous motors, it depends on the angular position, via the counter-electro motion force $\vec{E}(\theta)$ (more precisely its gradient), the electric current, and of course the angular speed.

$$P_{in} =< \vec{U}, \vec{I} >= R_s < \vec{I}, \vec{I} > + L_{eq} < \frac{d\vec{I}}{dt}, \vec{I} > + \omega < \vec{E}(\theta), \vec{I} >= P_{joule} + P_{magnetic} + P_{mechanic} \tag{7}$$

The most relevant term for dynamic modeling purposes is the contribution of the mechanical power output, which as we know is the product of electro-magnetic torque and rotor angular speed. Equation (8) shows the expression of the electro-magnetic torque, in the three-phase reference system.

$$T_{em} =< \vec{E}(\theta), \vec{I} >= -pk_\Psi \sum_{k=1}^{3} \sin\left(p\theta - \frac{2(1-k)\pi}{3}\right) I_k \tag{8}$$

Equation (9) represents the dynamic rotational equilibrium for the rotor axis. Note that the second member of the equation represents the superposition of the contributions of electromagnetic torque T_{em}, Cogging torque T_{cog} and resistant torque associated with the mechanical load T_{load}. The parameters J_r and b_r represent the rotor moment of inertia and the coefficient of friction, respectively.

$$J_r\frac{d\omega}{dt} + b_r\omega = J_r\frac{d^2\theta}{dt^2} + b_r\frac{d\theta}{dt} = T_{em} + T_{cog} - T_{load} \tag{9}$$

An example of integration with block diagrams is shown in Figure 8, of the interaction between the set of differential equation representing the electrical balance of the phases of the electric motor and the differential equation representing the dynamic rotational balance of the rotor.

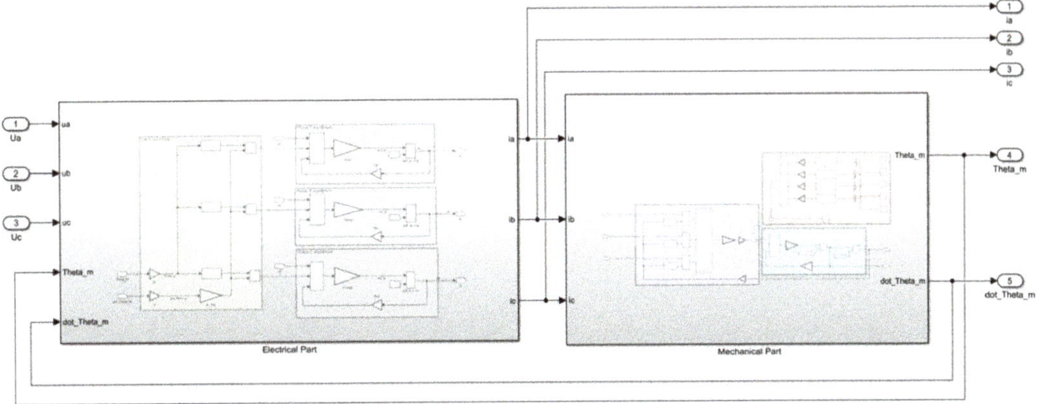

Figure 8. Implementation in Simulink environment of the dynamic interaction between electrical and mechanical subsystem.

Note how the output of one dynamic subsystem turns out to be the input for the other, defining a "recursive" dependence. Specifically, the electrical subsystem has the components of the voltage vector, position and angular velocity as inputs. The mechanical subsystem has as inputs, derived from the electrical subsystem, the currents (which in fact define the electromagnetic torque).

Figures 9 and 10 show the block diagrams, realized in the Simulink environment, of the dynamic subsystems describing the behavior of the electric motor. In Figure 9, note how the current dynamics of each phase is excited by the relative electro-motor control force component, which is a function of angular velocity and angular position.

Figure 10 shows that the mechanical equilibrium is activated from the electromagnetic torque, a function of phase currents and angular position. Note how the cogging torque model (red box in Figure 10) is inserted as an additive disturbance with respect to the electromagnetic torque.

To avoid the direct dependence of the electromagnetic torque expression on the angular position, the coordinate transformations are defined in Equation (10). The matrix $B_{\alpha\beta}$ defines the Blondel transformation, while the matrix $P_{dq}(\theta)$ formalizes the Park transformation.

$$B_{\alpha\beta} = \frac{2}{3}\begin{bmatrix} 1 & -\frac{1}{2} & -\frac{1}{2} \\ 0 & \frac{\sqrt{3}}{2} & -\frac{\sqrt{3}}{2} \\ \frac{1}{2} & \frac{1}{2} & \frac{1}{2} \end{bmatrix} \quad P_{dq}(\theta) = \begin{bmatrix} cos(\theta) & -sin(\theta) & 0 \\ sin(\theta) & cos(\theta) & 0 \\ 0 & 0 & 1 \end{bmatrix} \tag{10}$$

Note that the Blondel transformation is basically the matrix form of the coordinate change due to the definition of space vector [30], which defines the decomposition of the three-phase vector $\vec{X}_{abc} = [X_a, X_b, X_c]^T$ in the Gauss plane, as given in Equation (11).

$$X_{\alpha\beta} = \frac{2}{3}\left(X_a + X_b e^{j\frac{2\pi}{3}} + X_c e^{j\frac{4\pi}{3}}\right) = X_\alpha + jX_\beta \tag{11}$$

Combining the previous equation with the definition of homopolar component (typically used in electrical engineering) $X_{om} = \frac{X_a + X_b + X_c}{3}$, we obtain perfect equivalence with the Blondel transformation. Note that $X_{om} = 0$ for symmetric and balanced systems (such as electric motors), defining a formally passive transformation.

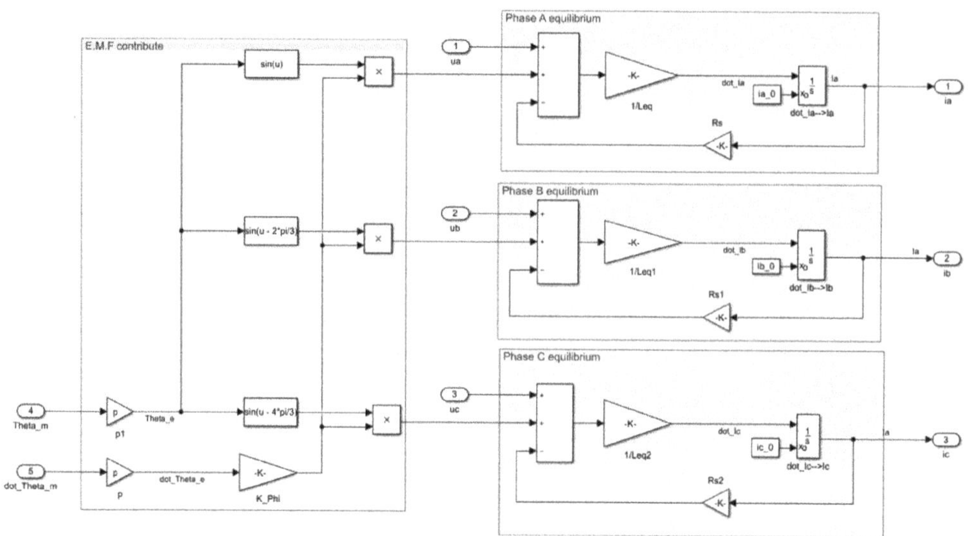

Figure 9. Example realization in MATLAB/Simulink environment of the subsystem related to mechanical equilibrium.

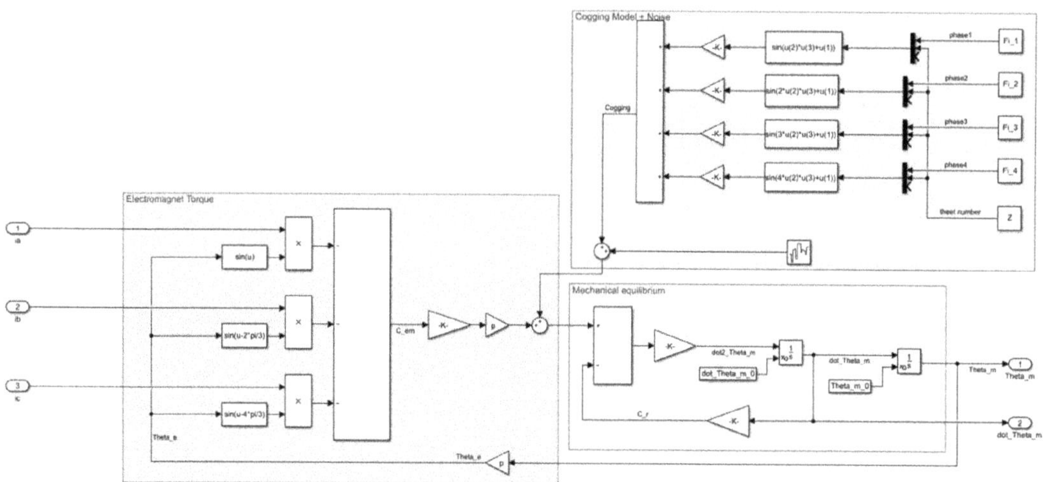

Figure 10. Example realization in MATLAB/Simulink environment of the subsystem related to mechanical equilibrium.

Through the introduced coordinate transformations, we obtain the mathematical model that is exploited for the design of the control algorithms, given in Equation (12).

$$
\begin{cases}
U_d = R_s I_d + L_{eq}\dfrac{dI_d}{dt} - p\omega L_{eq} I_q \\[4pt]
U_q = R_s I_q + L_{eq}\dfrac{dI_q}{dt} + p\omega(L_{eq} I_d + k_\Psi) \\[4pt]
\dfrac{d\theta}{dt} = \omega \\[4pt]
J_r\dfrac{d\omega}{dt} + b_r\omega = T_{em}^{dq} + T_{cog} - T_{load} \\[4pt]
T_{em}^{dq} = \dfrac{3}{2}pk_\Psi I_q
\end{cases}
\tag{12}
$$

3.3. Modulation System in Advanced Power Drives

SVM (Space-Vector-Modulation) modulation is now the most widely used modulation technique, as it provides better performance than classical PWM (Pulse-Width-Modulation) based techniques.

As shown in Figure 11, reference is made to a classical configuration of an inverter with impressed voltage, connected to a three-phase load, which represents the equivalent circuit of the electrical machine in this specific case.

In the figure, $V_{d}s$ has been used to denote the DC voltage, which is managed to modulate the inverter legs so as to energize the motor phases appropriately. The phase currents, which obviously coincide with those of the inverter branches, are referred to as I_a, I_b and I_c.

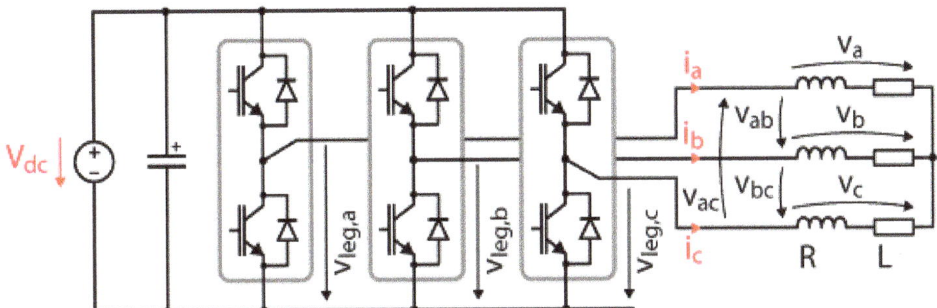

Figure 11. Topology of a two-level inverter with 3-phase load.

Note that the drive voltages of the inverter, denoted by $V_{leg,a}$, $V_{leg,b}$ and $V_{leg,c}$, are referred to the ground reference of the inverter circuit, unlike the phase voltages, denoted by V_a, V_b and V_c, which are obviously referred to the star center of the stator windings.

Calling V_{no} the potential difference between the star-center reference n and the ground reference o, it is possible to find the expressions of the phase voltages of the electrical machine as a function of the inverter leg voltages, as expressed in Equations (13) and (14).

$$V_{leg,a} + V_{leg,b} + V_{leg,c} = V_a + V_b + V_c + 3V_{no} \longrightarrow V_{no} = \frac{V_{leg,a} + V_{leg,b} + V_{leg,c}}{3} \tag{13}$$

$$\begin{aligned}
V_a &= \frac{2}{3}V_{leg,a} - \frac{1}{3}(V_{leg,b} + V_{leg,c}) \\
V_b &= \frac{2}{3}V_{leg,b} - \frac{1}{3}(V_{leg,a} + V_{leg,c}) \\
V_c &= \frac{2}{3}V_{leg,c} - \frac{1}{3}(V_{leg,a} + V_{leg,b})
\end{aligned} \tag{14}$$

From these equations, the phase voltages corresponding to the possible state combinations of the inverter, resulting from modulation, are derived. Table 1 shows for each combination of inverter modulation signals the corresponding trio of phase voltages. Each triad actually corresponds to a sector of the SVM modulation.

Note that applying the definition of space vector to the eight triads corresponding to the possible eight states of the inverter, we obtain the vectors $V_0, ..., V_7$, which are positioned on the vertices of a hexagon in the Gauss complex plane.

These vectors, as shown in Figure 12, delimit the sectors with which the SVM modulation algorithm selects the most suitable state of the inverter, so that the desired voltage is averaged over the switching period T_{sw}.

From the control system, a vector of voltages (referenced to the star center) is generated typically in the Park dq reference system. After the inverse transformation to obtain the control triad in three-phase axes, the definition of space vector is applied via the Blondel transformation, the result of which is the vector $V_{\alpha\beta}^*$.

Table 1. Summary of the possible space vectors for three-phase inverter.

Space Vector	State Leg A	State Leg B	State Leg C	V_a	V_b	V_c
V_0	0	0	0	0	0	0
V_1	1	0	0	$\frac{2V_{dc}}{3}$	$-\frac{V_{dc}}{3}$	$-\frac{V_{dc}}{3}$
V_2	1	1	0	$\frac{V_{dc}}{3}$	$\frac{V_{dc}}{3}$	$-\frac{2V_{dc}}{3}$
V_3	0	1	0	$\frac{V_{dc}}{3}$	$\frac{2V_{dc}}{3}$	$-\frac{V_{dc}}{3}$
V_4	0	1	1	$-\frac{2V_{dc}}{3}$	$\frac{V_{dc}}{3}$	$\frac{V_{dc}}{3}$
V_5	0	0	1	$-\frac{V_{dc}}{3}$	$-\frac{V_{dc}}{3}$	$\frac{2V_{dc}}{3}$
V_6	1	0	1	$\frac{V_{dc}}{3}$	$-\frac{2V_{dc}}{3}$	$\frac{V_{dc}}{3}$
V_7	1	1	1	0	0	0

As shown in Figure 13, the control vector is instantaneously placed in one of the sectors of the modulation hexagon. The objective of SVM modulation is to reproduce the vector calculated by the control system, in mean value within the switching period, by partitioning the modulation vectors, as expressed in Equation (15).

$$\vec{V}^*_{\alpha\beta} = \frac{1}{T_{sw}}\left(T_1\vec{V}_1 + T_2\vec{V}_2 + T_0\vec{V}_0 + T_7\vec{V}_7\right) \tag{15}$$

By T_1, T_2, T_0, T_7, we have indicated the portions of the switching period in which the inverter is brought into a certain state by modulation. Note that the terms $\frac{T_i}{T_{sw}}$ have the same meaning as the duty-cycle for the PWM technique.

Note, as schematized in Figure 14, that in order to integrate the SVM modulation, the triplet U_{abc} is transformed so that the vector $V^*_{\alpha\beta}$ can be calculated. The modulus and angle are then extrapolated so that we can work out which sector is in the modulation hexagon. The selection of the sector and the calculation of the distribution times is done (for example) by a state machine.

Figure 15 shows the Simulink diagram of a possible implementation of an SVM algorithm for an ideal inverter (with instantaneous switching) in a simulation environment. In particular, it is shown how to realize the selection of each sector and the allocation of modulation states by means of a state machine realized with STATEFLOW.

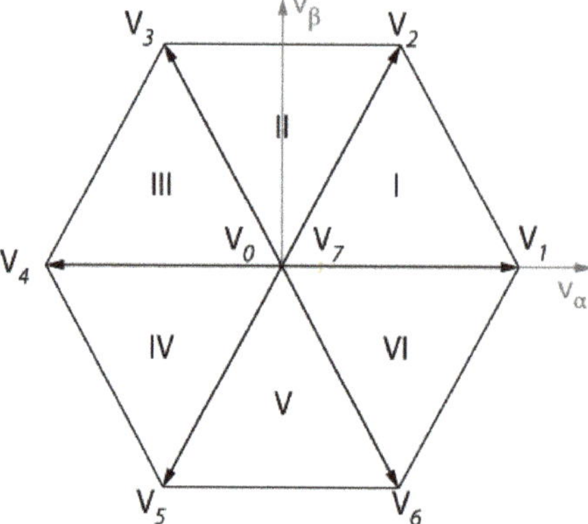

Figure 12. Representation of the active and zero space vectors in the $\alpha\beta$ plane, and its division in the modulation sectors (numbered by Roman numerals).

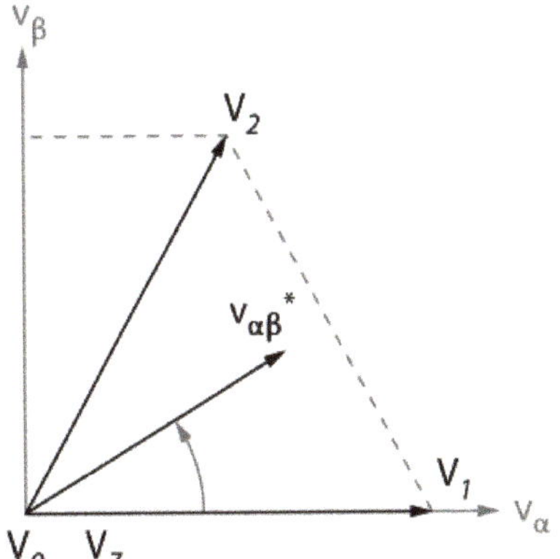

Figure 13. Synthesis of a reference space vector $V_{\alpha\beta}^*$ (see Equation (15)) with the active vectors V_1 and V_2, and the zero vectors V_0 and V_7.

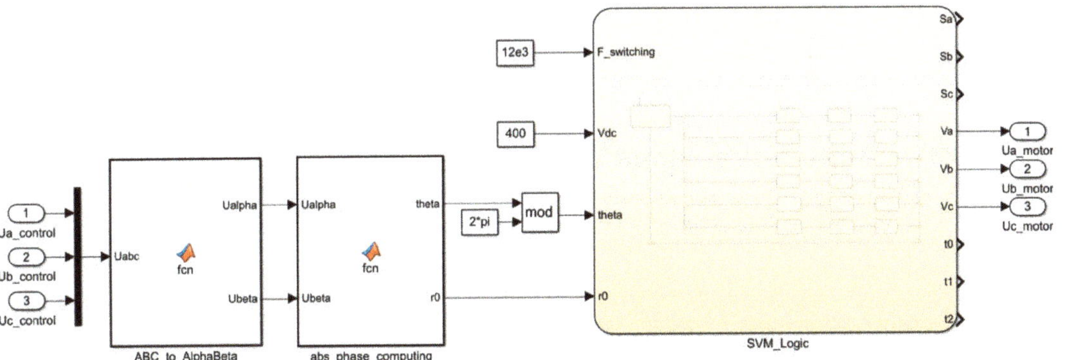

Figure 14. Simulink environment implementation of the SVM modulation technique.

3.4. Complete Dynamics Modeling including Cogging Torque

The last (but not least) fundamental component to be mathematically modeled is the electrical machine load. Mechatronic systems of industrial interest are mostly physical–electromechanical in nature. This allows the dynamics of the system implemented by permanent-magnet synchronous motors to be represented through the Lagrangian formalism [31].

$$L(q,\dot{q}) = K(q,\dot{q}) - U(q) = \sum_i K_i(q,\dot{q}) - \sum_j U_j(q)$$

$$= \sum_i \frac{1}{2}\left(M_i|\vec{v}(q,\dot{q})|^2 + \vec{\omega}_i^T(q,\dot{q})J_i(q)\vec{\omega}_i(q,\dot{q})\right) - \sum_j M_i < \vec{g}, \vec{P}_i > \tag{16}$$

Equation (16) reports the definition of the Lagrangian function L as the difference between the total kinetic energy of the system K and the potential energy U. The vector $q = [q_1, ..., q_n]^T$ is composed of the Lagrangian variables, ergo the variables needed to describe the dynamics of the system.

The dynamics of the system can be derived by application of the variations method, from which the following equation is derived.

$$\frac{d}{dt}\frac{\partial L(q,\dot{q})}{\partial \dot{q}_i} - \frac{\partial L(q,\dot{q})}{\partial q} = Q_i^T = \sum_i <\vec{F}_{ext,i}, \frac{\partial \vec{P}_i(q)}{\partial q_i}> + \sum_j <\vec{T}_{ext,j}, \frac{\partial \vec{\alpha}_i(q)}{\partial q_i}> \qquad (17)$$

The term Q_i represents the superposition of the non-conservative external effects; $\vec{F}_{ext,i}$ and $\vec{T}_{ext,j}$ are external, non-conservative forces and torques, respectively, applied to the system; $\vec{P}_i$ is the point of application of the force, and α_i is the angle subject to change under the action of the torque.

$$M(\vec{q})\ddot{\vec{q}} + C(\vec{q},\dot{\vec{q}})\dot{\vec{q}} + G(\vec{q}) = \Gamma(\vec{q})\vec{F}_{act} - J^T(\vec{q})\vec{F}_{env} \qquad (18)$$

Compactly, it is possible to rewrite the set of dynamic equations as in Equation (18), where $M(\vec{q})$ is the inertia matrix, $C(\vec{q})$ is the matrix of Coriolis terms, $G(\vec{q})$ is the vector of gravitational and potential terms, $\Gamma(\vec{q})$ is the matrix that maps the action of the actuators $\vec{F}_{act}$ into the dynamics of the variables $\vec{q}$, $J(\vec{q})$ is the analytic Jacobian, and finally $\vec{F}_{env}$ represents the effect of non-conservative external forces.

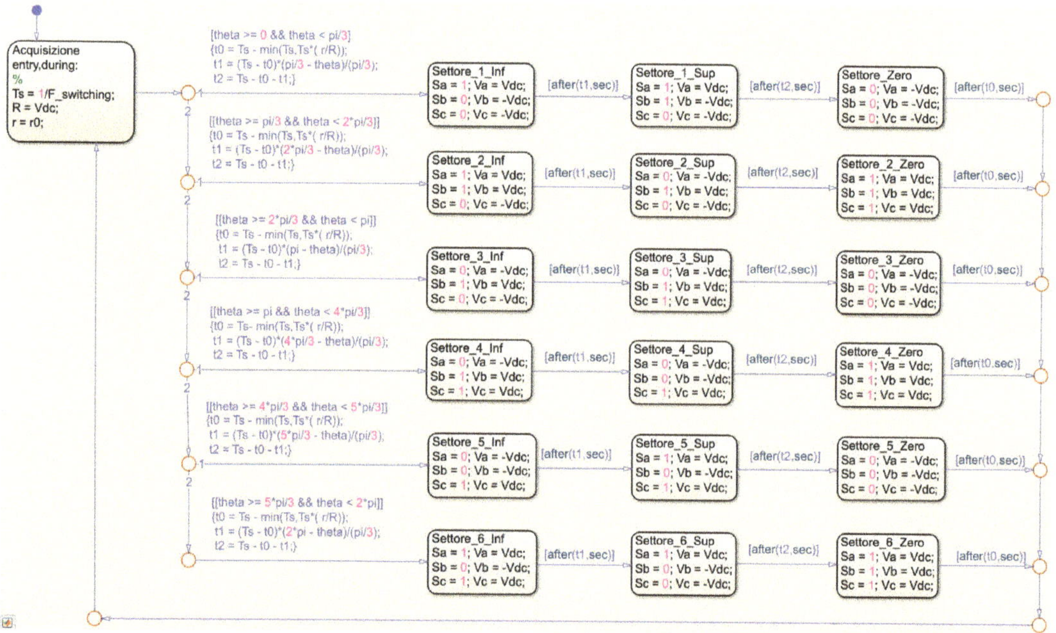

Figure 15. State machine representing the SVM modulation technique, integrated with STATEFLOW in Simulink environment.

Note that combining the model of the implemented system in Equation (18) with the mathematical model of the electrical machine in Equation (12) we obtain the final mathematical model, which allows us to design control algorithms for electro-mechanical systems that inherently takes into account the presence of the cogging torque.

4. Most Used Algorithms for Cogging Torque Reduction

The latest work in the literature, concerning the design of algorithms and control systems capable of compensating for cogging torque in Permanent Magnet Synchronous Motors dedicated to advanced mechatronic applications, has shown that the most effective techniques are precisely those in which the mathematical model of cogging torque can be handled as an internal detail in the overall model.

4.1. Model Predictive Control

In the MPC approach, the current control action is computed online rather than using a pre-computed, online, control law. A model predictive controller uses, at each sampling instant, the plant's current input and output measurements, the plant's current state, and the plant's model to:

- Calculate, over a finite horizon, a future control sequence that optimizes a given performance index and satisfies constraints on the control action;
- Use the first control in the sequence as the plant's input.

The MPC strategy is illustrated in Figure 16, where N_p is the prediction horizon, $u(t+k \mid t)$ is the predicted control action at $t+k$ given $u(t)$. Similarly, $y(t+k \mid t)$ is the predicted output at $t+k$ given $y(t)$.

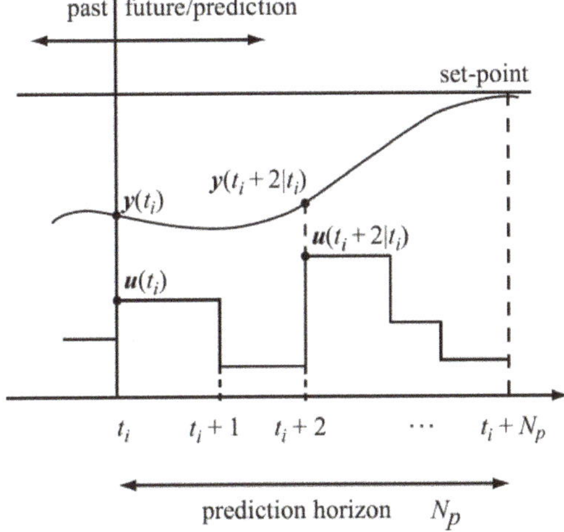

Figure 16. Controller action construction using model-based predictive control (MPC) approach.

We consider a discretized model of a dynamic system of the form of Equation (19).

$$x(k+1) = \mathbf{\Phi} x(k) + \mathbf{\Gamma} u(k)$$
$$y(k) = \mathbf{C} x(k) \tag{19}$$

where $\mathbf{\Phi} \in \mathbb{R}^{n \times n}, \mathbf{\Gamma} \in \mathbb{R}^{n \times m}$, and $\mathbf{C} \in \mathbb{R}^{p \times n}$. Applying the backward difference operator, $\Delta x(k+1) = x(k+1) - x(k)$ is derived Equation (20).

$$\Delta x(k+1) = \mathbf{\Phi} \Delta x(k) + \mathbf{\Gamma} \Delta u(k)$$
$$\Delta y(k+1) = y(k+1) - y(k)$$
$$= \mathbf{C} x(k+1) - \mathbf{C} x(k) \tag{20}$$
$$= \mathbf{C} \Delta x(k+1)$$

where $\Delta u(k+1) = u(k+1) - u(k)$. With further algebraic manipulation it is obtained the matrix form of augmented state form dynamics in Equation (21).

$$x_a(k+1)\begin{bmatrix} \Delta x(k+1) \\ y(k+1) \end{bmatrix} = \begin{bmatrix} \Phi & O \\ C\Phi & I_p \end{bmatrix}\begin{bmatrix} \Delta x(k) \\ y(k) \end{bmatrix} + \begin{bmatrix} \Gamma \\ C\Gamma \end{bmatrix}\Delta u(k) = \Phi_a x_a(k) + \Gamma_a \Delta u(k)$$

$$y(k) = \begin{bmatrix} O & I_p \end{bmatrix}\begin{bmatrix} \Delta x(k) \\ y(k) \end{bmatrix} = C_a x_a(k) \tag{21}$$

Suppose now that the state vector x_a at each sampling time, k, is available to us. Our control objective is to construct a control sequence as the follows.

$$\Delta u(k), \Delta u(k+1), \ldots, \Delta u(k + N_p - 1) \tag{22}$$

where N_p is the prediction horizon, such that a given cost function and constraints are satisfied. The above control sequence will result in a predicted sequence of the augmented state vectors.

$$x_a(k+1 \mid k), x_a(k+2 \mid k), \ldots, x_a(k + N_p \mid k) \tag{23}$$

which can then be used to compute predicted sequence of the plant's outputs.

$$y(k+1 \mid k), y(k+2 \mid k), \ldots, y(k + N_p \mid k) \tag{24}$$

Using the above information it is possible to compute the control sequence of Equation (22) and then apply $u(k)$ to the plant to generate $x(k+1)$. Repeating the process again, using $x(k+1)$ as an initial condition to compute $u(k+1)$, and so on.

Most of the work in which MPC is used to reduce cogging torque, the approach of constructing the control sequence assuming known linearized model parameters is used Since the linearized model is derived by approximation from Equations (12) and (18), it contains the information about the extrapolated cogging model.

In the following, it is reported the approach to construct $u(k)$ given $x(k)$. Figure 17 represents schematically the MPC architecture. Using the plant model parameters and the measurement of $x_a(k)$ is evaluated the augmented states over the prediction horizon successively applying the recursion formula, obtaining the set of equations in Equations (25).

$$\begin{aligned}
x_a(k+1 \mid k) &= \Phi_a x_a(k) + \Gamma_a \Delta u(k) \\
x_a(k+2 \mid k) &= \Phi_a x_a(k+1 \mid k) + \Gamma_a \Delta u(k+1) \\
&= \Phi_a^2 x_a(k) + \Phi_a \Gamma_a \Delta u(k) + \Gamma_a \Delta u(k+1) \\
&\quad\vdots \\
x_a(k+N_p \mid k) &= \Phi_a^{N_p} x_a(k) + \Phi_a^{N_p-1}\Gamma_a \Delta u(k) + \cdots + \Gamma_a \Delta u(k+N_p-1)
\end{aligned} \tag{25}$$

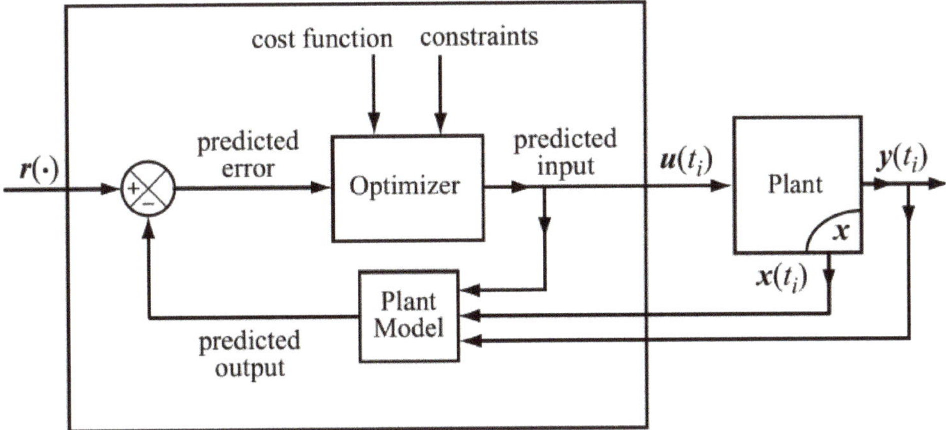

Figure 17. State feedback model predictive controller.

In matrix form, it is possible to compact as in Equation (26).

$$
\begin{bmatrix} x_a(k+1 \mid k) \\ x_a(k+2 \mid k) \\ \vdots \\ x_a(k+N_p \mid k) \end{bmatrix} = \begin{bmatrix} \Phi_a \\ \Phi_a^2 \\ \vdots \\ \Phi_a^{N_p} \end{bmatrix} x_a(k) + \begin{bmatrix} \Gamma_a & & & \\ \Phi_a\Gamma_a & \Gamma_a & & \\ \vdots & & \ddots & \\ \Phi_a^{N_p-1}\Gamma_a & \cdots & & \Gamma_a \end{bmatrix} \begin{bmatrix} \Delta u(k) \\ \Delta u(k+1) \\ \vdots \\ \Delta u(k+N_p-1) \end{bmatrix} \tag{26}
$$

$$
\begin{aligned}
Y &= \begin{bmatrix} y(k+1 \mid k) \\ y(k+2 \mid k) \\ \vdots \\ y(k+N_p \mid k) \end{bmatrix} = \begin{bmatrix} C_a x_a(k+1 \mid k) \\ C_a x_a(k+2 \mid k) \\ \vdots \\ C_a x_a(k+N_p \mid k) \end{bmatrix} \\[2mm]
&= \begin{bmatrix} C_a\Phi_a \\ C_a\Phi_a^2 \\ \vdots \\ C_a\Phi_a^{N_p} \end{bmatrix} x_a(k) \\[2mm]
&\quad + \begin{bmatrix} C_a\Gamma_a & & & \\ C_a\Phi_a\Gamma_a & C_a\Gamma_a & & \\ \vdots & & \ddots & \\ C_a\Phi_a^{N_p-1}\Gamma_a & \cdots & & C_a\Gamma_a \end{bmatrix} \begin{bmatrix} \Delta u(k) \\ \Delta u(k+1) \\ \vdots \\ \Delta u(k+N_p-1) \end{bmatrix} \\[2mm]
&= W x_a(k) + Z\Delta u(k)
\end{aligned} \tag{27}
$$

Equation (27) describes the dependence of the sequence of future output vectors, in the prediction horizon, as a function of the increased state vectors, in that prediction horizon, and as a function of the change in the control vector. This equation is decisive in solving the optimization problem, which is formalized in the following.

$$
\begin{aligned}
J(\Delta U) &= \tfrac{1}{2}\left(r_p - Y\right)^\top Q\left(r_p - Y\right) + \tfrac{1}{2}\Delta U^\top R\Delta U \\
& Y^{\min} \le Y \le Y^{\max} \\
& U^{\min} \le U \le U^{\max}
\end{aligned} \tag{28}
$$

Equation (28) represents the cost functional that is minimized to derive the predictive sequence of optimal control, according to the criterion defined precisely by $J(\Delta u)$.

$$\Delta u(k) = \overbrace{\begin{bmatrix} I_m & O & \cdots & O \end{bmatrix}}^{N_p \text{ block matrices}} \left(R + Z^\top Q Z \right)^{-1} Z^\top Q (r_p - W x_a) \tag{29}$$
$$= K_r r_p - K_x \Delta x(k) - K_y y(k)$$

For the control vector to be optimal and meet all operational constraints on the control and output variables, minimization of the functional is associated with an optimizer based, for example, on gradient descent algorithm or Newtonian methods.

As can be understood from the procedure for constructing the control vector, MPC actually acts as a LQR (Linear-Quadratic-Regulator) controller in which it is possible to explicitly insert operational constraints on the controlled and control variables.

This represents an enormous advantage over the classical optimal control approach, as the control system solution can take into account the realistic constraints of the physical system to be controlled, and furthermore, there is the certainty that the control vector provides (sub-)optimal action in each of its components, being the result of evaluation over successive time windows.

On the other hand, MPC requires a very high computational effort, which is not always suitable for micro-controller-based embedded systems.

This is because at each step of the algorithm, an internal simulation is run, which exploits the mathematical model of the physical process to predict behavior and to be able to calculate a control vector, which will be optimal over that simulated time window.

Thus, the system on which the algorithm is integrated must be able to simulate N_p steps of the process model, within a single numerical integration step.

Another small disadvantage is that MPC is still an optimal control technique based on the mathematical model, so it is very sensitive to model uncertainties.

It is, however, very much used in the context of cogging torque reduction, as the ability to correctly model the electrical machine and the cogging phenomenon is quite optimized, leaving little room for model uncertainties that the feedback control may not be able to handle [32–35].

4.2. Feedback Linearization Control

Another widely used control technique is Feedback Linearization Control (FLC), which takes advantage of differential geometry theory to directly handle model dynamics in a nonlinear form.

In order to apply the control vector design technique by Feedback Linearization, it is necessary that the mathematical model of the mechatronic system of interest be in the state form called "similar in control", as given in Equation (30), where the vector field $f(x)$ is called drift vector, while the $g_k(x)$ is the control vectors field.

$$\frac{dx}{dt} = F(x, u) = f(x) + \sum_k g_k(x) u_k \; ; \; x_0 = x(t_0)$$
$$y_l = H_l(x, u) = h_l(x) \tag{30}$$

Being able to handle the inherent non-linearities of the mathematical model in the generic state form of Equation (30), the resulting control law by definition contains information about the cogging pair, via the model chosen to represent it.

$$\frac{dh_i}{dt} = \frac{\partial h_i}{\partial x} \frac{\partial x}{\partial t} = \frac{\partial h_i}{\partial x} \left(\vec{f} + \sum_{k=1}^{n_u} \vec{g}_k u_k \right) = \frac{\partial h_i}{\partial x} \vec{f} = L_f h_i$$
$$\cdots \tag{31}$$
$$\frac{d^{\mu_i} h_i}{dt^{\mu_i}} = \frac{d}{dt}\left(L_f^{\mu_i - 1} \right) = \frac{\partial}{\partial x}\left(L_f^{\mu_i - 1} \right)\left(\vec{f} + \sum_{k=1}^{n_u} \vec{g}_k u_k \right) = L_f^{\mu_i} h_i + \sum_{k=1}^{n_u} L_{g_k} L_f^{\mu_i - 1} h_i u_k$$

The operative procedure for applying feedback linearization involves deriving the control outputs as many times as necessary to obtain expressions in which the contributions of the control variables appear. In Equation (31), the expression of the derivative μ_ith for the ith control output is given in general form. We denote the relative degree for the output ith by μ_i, with the condition that $\sum \mu_i \leq n_x$.

$$
\begin{bmatrix} L_f^{\mu_1} h_1 \\ L_f^{\mu_2} h_2 \\ ... \\ L_f^{\mu_l} h_l \end{bmatrix} + \begin{bmatrix} L_{g_1} L_f^{\mu_1-1} h_1 & ... & L_{g_m} L_f^{\mu_1-1} h_1 \\ L_{g_1} L_f^{\mu_2-1} h_2 & ... & L_{g_m} L_f^{\mu_2-1} h_2 \\ ... & ... & ... \\ L_{g_1} L_f^{\mu_l-1} h_l & ... & L_{g_m} L_f^{\mu_l-l} h_l \end{bmatrix} \vec{u} = \vec{v} = \vec{\Gamma}(\vec{x}) + \Sigma(\vec{x})\vec{u}
\tag{32}
$$

$$
\vec{u} = \Sigma^+(\vec{x})\left(\vec{v} - \vec{\Gamma}(\vec{x})\right) + (I - \Sigma^+(\vec{x})\Sigma(\vec{x}))\lambda
$$

Proceeding with the derivation of each control output, in general we obtain the matrix expression given in Equation (32), where with $L_{g_i} L_f h_j(\vec{x}) = \frac{\partial}{\partial \vec{x}}\left(\frac{\partial h_j}{\partial \vec{x}} f\right) \vec{g}_i$ is the Lie derivative of the function h_j in the directions of the vector fields $\vec{f}$ and $\vec{g}_i$, respectively.

$$
\dot{\vec{z}} = \begin{bmatrix} A_1 & 0 & 0 & ... & 0 \\ 0 & A_2 & 0 & ... & 0 \\ ... & ... & ... & ... & ... \\ 0 & 0 & 0 & ... & A_l \end{bmatrix} \vec{z} + \begin{bmatrix} B_1 & 0 & ... & 0 \\ 0 & B_2 & ... & 0 \\ ... & ... & ... & ... \\ 0 & 0 & ... & B_l \end{bmatrix} \vec{v}
$$

$$
\vec{z} = \begin{bmatrix} \vec{z}_1 \\ \vec{z}_2 \\ ... \\ \vec{z}_l \end{bmatrix} \; ; \; \vec{z}_i = \begin{bmatrix} h_i \\ L_f h_i \\ ... \\ L_f^{\mu_i-1} \end{bmatrix} \; ; \; A_i = \begin{bmatrix} 0 & 1 & 0 & 0 & ... & 0 \\ 0 & 0 & 1 & 0 & ... & 0 \\ ... & ... & ... & ... & ... & ... \\ 0 & 0 & 0 & 0 & ... & 1 \\ 0 & 0 & 0 & 0 & ... & 0 \end{bmatrix} \; ; \; B_i = \begin{bmatrix} 0 \\ 0 \\ ... \\ 0 \\ 1 \end{bmatrix}
\tag{33}
$$

The feedback linearization procedure results in a change of coordinates in which the dynamics of the new state vector is linear, as shown in Equations (33). Note that the state representation in addition to being linear represents a chain of integrators, consequently the pair (A_i, B_i) turns out to be completely reachable/controllable $\forall i$.

$$
\vec{z} = \begin{bmatrix} \vec{z} \\ \vec{\sigma}(\vec{x}) \end{bmatrix} \; \text{with} \; \vec{\sigma}(\vec{x}) \; s.t \; \frac{\partial \vec{z}}{\partial \vec{x}} \neq 0
\tag{34}
$$

In general, a change of basis with reduction of the cardinality of the state space is obtained. Formally, this requires that certain mathematical conditions are fulfilled, which define the criterion of zero dynamics [36]. The first condition is given in Equation (34), where we impose that the Jacobian of the base change is formally lawful (invertible). This means that the completion $\vec{\sigma}(\vec{x}) \in R^{n_x - \sum \mu_i}$ will have to be constructed appropriately to satisfy the condition.

$$
\dot{\vec{\sigma}}(\vec{x}) = \vec{\gamma}(\vec{z}, \vec{\sigma}) + \vec{\xi}(\vec{z}, \vec{\sigma})v
\tag{35}
$$

The completion vector $\vec{\sigma}(\vec{x})$ will in general have a non-linear state representation, dependent on both the new and old state variables and also on the new control vector, as given in Equation (35).

$$
L_{g_k}\sigma_i = 0 \; \forall i, k \; \rightarrow \; \dot{\vec{\sigma}}(\vec{x}) = \vec{\gamma}(\vec{z}, \vec{\sigma}) \; \rightarrow \; \dot{\vec{\sigma}}(\vec{x}) = \vec{\gamma}(0, \vec{\sigma}) \; A.S.
\tag{36}
$$

The second condition, given in Equation (36), serves to eliminate the contribution of the new control vector v from the dynamics of the vector $\vec{\sigma}$. In this way, it is possible to implement the zero-dynamics criterion, which guarantees us that even in the case of a reduction of the cardinality of the state space, the vector v constructed by means of a feedback linearization technique is able to asymptotically stabilize even the state variables that serve only to make the base change formally lawful. In the above equation, "A.S" means asymptotically stable.

The FLC-based control synthesis approach is schematized in Figure 18. Note that two control "loops" can be identified, one external and one internal. The inner one consists of the nonlinear terms based on the Lie derivative calculus, shown earlier, which is intended to map the control action of the outer loop into the representation space of the nonlinear dynamical system. The outer loop exploits the change of coordinates to the linear representation domain. In fact, the outer loop is dedicated to reducing the error between the reference and the output of the system, and is realized by linear control techniques.

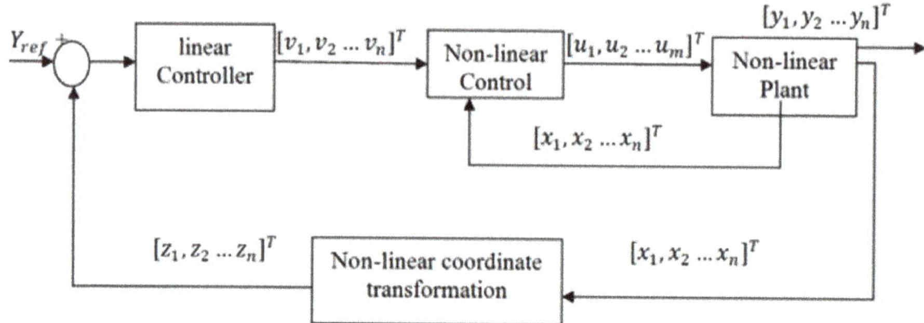

Figure 18. Schematic representation of the internal structure of FLC controller.

As can be seen from the operating procedure for constructing the control vector using the FLC technique, the greatest advantage of this method is that it can handle highly non-linear dynamic models, being able to include very realistic details in the model itself, such as the cogging torque in this particular case.

Some inconvenience arises when the relative degree of the system is much less than the cardinality of the state space associated with the starting model. In this case, one must choose a base change completion that is invertible, so as to make it formally acceptable.

However, it can be said that in practical cases it is always possible to construct performance outputs such that they return a relative degree equal to (or slightly less than) the dimension of the state vector of the initial non-linear model.

For electro-mechanical systems of industrial interest, such as in robotics and industrial automation, dynamic system models are also found to have suitable formal properties for the application of this control methodology, which is why it is one of the most widely used control techniques by designers of algorithms and control systems [37–42].

4.3. Resonant Control

Control techniques based on the concept of resonance are often used by designers to resolve and attenuate the presence of periodic disturbances, whose pulsation/frequency can be measured or estimated with reasonable accuracy. In fact, resonant type controllers are well suited to the rejection of additive disturbances in sinusoidal form, as is the case with cogging torque.

This is the reason why some works in the literature are based on the use of resonant techniques, such as PR (Proportional Resonant) and PIR (Proportional Integral Resonant). Since cogging torque is an additive disturbance compared to electro-magnetic disturbance with periodic comparing often analyzed by harmonic models, it proves to be a rather efficient technique under certain types of conditions.

The transfer function shown in Equation (37) represents the Laplace transform domain implementation (for SISO = Single-Input-Single-Output systems) of a PR (Proportional-Resonant) type controller.

$$G_C(s) = K_p + \frac{2K_i s}{s^2 + \omega_0^2} \tag{37}$$

K_p denotes the proportional gain, K_i denotes the gain obtained when the additive disturbance on the output has exactly the pole pulsation of the transfer function ω_0, known as the resonance pulsation or maximum feedback gain.

In Equation (38) is shown the version that is most often taken into account for integration convenience. Since the previous equation contains complex conjugate poles, in order to facilitate discretization and thus implementation in low-level code, a damping factor is often inserted, which also improves the controller's response, making it less abrupt and consequently requiring less control effort.

$$G_C(s) = G_{Cp}(s) + G_{Cr}(s) = K_p + \frac{2K_i\omega_c s}{s^2 + 2\omega_c s + \omega_0^2} \tag{38}$$

The pulse ω_c is called the cut-off pulse and differs from ω_0 obviously, in practice $\omega_c < \omega_0$. Having inserted a damping factor, the disturbance pulse and the resonant filter within the control now differ, so the feedback gain will not be maximum at the same time as the disturbance pulse. Theoretically, disturbance rejection is less efficient than the ideal PR case, but in practice we find that performance is virtually equivalent, also benefiting from smoother control action.

Equation (39) represents a possible implementation of the resonant PR controller by discretization. In particular, by simply using a bi-linear transformation, one obtains a discrete system that is not strictly proper (which in the discrete domain continues to make physical sense).

$$G_{Cr}(z) = \frac{Y(z)}{E(z)} = \frac{a_1\left(1 - z^{-2}\right)}{b_0 + b_1 z^{-1} + b_2 z^{-2}} \quad \text{with} \quad \begin{aligned} a_1 &= 4K_i T_s \omega_c \\ b_0 &= T_s^2 \omega_0^2 + 4T_s\omega_c + 4 \\ b_1 &= 2T_s^2 \omega_0^2 - 8 \\ b_2 &= T_s^2 \omega_0^2 - 4T_s\omega_c + 4 \end{aligned} \tag{39}$$

Applying the inverse Z-transform yields the finite difference equation (or recursive equation) shown in Equation (40). The implementation on a microcontroller therefore requires memory to be kept of two steps prior to the current one.

$$y(k) = \frac{1}{b_0}[a_1 \cdot e(k) - a_1 \cdot e(k-2) - b_1 \cdot y(k-1) - b_2 \cdot y(k-2)] \tag{40}$$

Figure 19 shows how to realize a block diagram equivalent to the PR controller in the form of a Z-transformed transfer function, Simulink environment, using only 'native' blocks.

The tuning of resonant controller parameters is quite easy to achieve in practice, which is why it is among the most frequently used techniques by control algorithm designers for this type of problem. The most commonly used technique is that of forced oscillations, often accompanied by the classic Ziegler–Nichols table [43–48]. Due to the ease of development and integration on an embedded platform, it is also one of the most widely used control techniques in power and control electronics in general [49–51], not only in electrical drives.

4.4. Others Control Techniques

Few other works in the literature exploit adaptive control techniques for cogging torque reduction in servo drives for mechatronic systems [52–55].

Adaptive control in general can be divided into direct approach and indirect approaches:

- **Direct Adaptive Control**: in this case, the mathematical model of the physical system is taken as known and the control action is adapted by going to change its online parameters;

- **Indirect Adaptive Control**: in this case, a simplified model, such as a linearized version, is used as a basis, going to modify its parameters based on the error between the actual measurements and the model's prediction, also modifying a posterior the controller parameters that must be adapted accordingly.

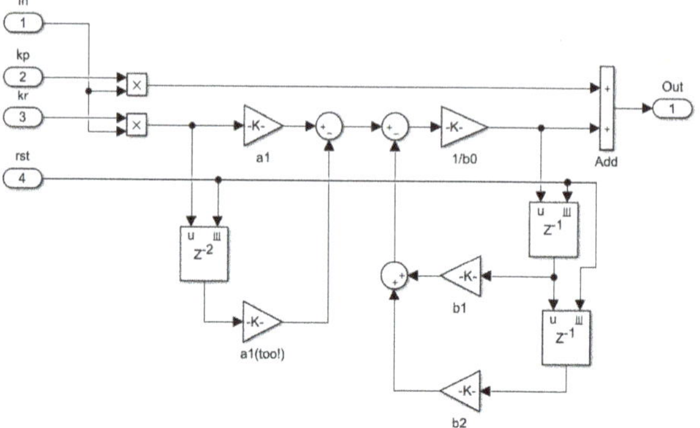

Figure 19. Proportional resonant controller implementation example.

Figure 20 shows the principle diagrams of both approaches of adaptive control. In both cases, it is possible to derive a control law that incorporates cogging torque information, reason why it was deemed appropriate to include this approach among the techniques that can be practically used.

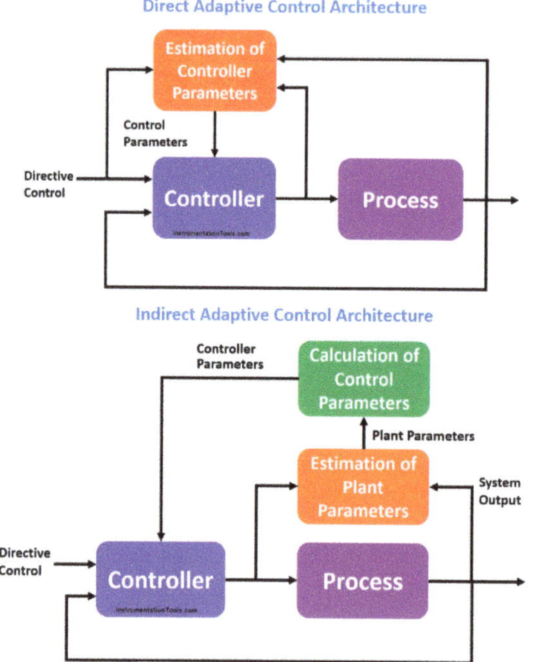

Figure 20. A schematic representation of direct vs. indirect adaptive control.

In the case of direct adaptive control, one of the cogging pair models previously discussed can be exploited. To exploit online adaptive paradigms usually use error-corrector learning algorithms such as gradient descent, which work best on linear models. So in the case of indirect adaptive control, if the estimation of cogging torque-related parameters is based on the linearized model, the compensation may not be as efficient as in the direct case.

Very few works in the literature exploit the Sliding Mode control technique instead, which is classified as variable structure control [56,57]. It is based on the mathematical concept of region of attraction so as to derive a control law that typically depends on the sign of the hyperplane describing that region of attraction.

The main problem with this control technique is that it works well when the system model has slight non-linearities and a relative degree (the number of times the output must be derived to obtain an expression containing the control variables) of unity. It is used side-by-side with other types of controllers because, in any case, it is based on a mathematical model of the motor in which it is actually possible to incorporate the cogging torque information to be compensated by the resulting control action.

4.5. Final Considerations

4.5.1. Model Predictive Control

Main advantages:

- Easy way of dealing with constraints on controls and states;
- High performance controllers and accurate;
- No need to generate solutions for the whole time-horizon;
- Flexibility: managing any model, any optimal objective function;
- Includes cogging torque information directly in the control law.

Main disadvantages:

- High computational complexity (poorly suited to typical embedded systems);
- Problems of robustness with respect to uncertainties in the reference mathematical model;
- Strict dependence on the dynamic model.

4.5.2. Feedback Linearization Control

Main advantages:

- Management of highly non-linear models;
- Joint use of other highly effective techniques on stationary linear systems;
- Fully algebraic control law (no on-line recursion required);
- Includes information on cogging torque through the most accurate model possible (optimizing compensation);
- Relative simplicity of implementation;
- Robust respect to the model parameters uncertainty.

Main disadvantages:

- Strict dependence on the mathematical modeling;
- Does not inherently manage operational constraints on dynamic variables.

4.5.3. Resonant Controller

Main advantages:

- Good ability to handle sinusoidal disturbances (as the cogging torque);
- Relative easy to tune;
- Simple to implement in embedded system.

Main disadvantages:

- Strongly dependent on the operating condition under which the control coefficients are chosen;
- Does not exploit completely the cogging torque modeling.

4.5.4. Adaptive and Sliding Mode Controllers

The main advantage is related to a relatively small computational complexity (more about the sliding mode technique).

The main disadvantages are:

- Inherent problems of convergence and numerical stability;
- Partial or ineffective use of the cogging torque model.

5. Conclusions

This paper proposed a discussion of how to apply the MBD approach in the context of the design of control systems and algorithms for cogging torque reduction, a rather hot topic for advanced mechatronic applications where high efficiency and performance is required, and especially a topic of current research on the control of permanent magnet electric machines.

During this review of the state of the art, the problem of torque cogging was analyzed in detail justifying the importance of addressing this issue through control algorithms in order to realize flexible and scalable solutions for various mechatronic applications and systems.

Popular mathematical modeling methodologies were presented with a view to being able to exploit cogging torque information directly in the design of control algorithms.

It was shown how to carry out mathematical modeling and in a simulated environment, of the components of the mechatronic system: the electric motor, presenting an overview of the dynamic model that integrates the cogging torque into the mechanical equilibrium; the modulation system, now based almost exclusively on SVM technique, presenting the theoretical principles and a possible integration in a simulated environment, with a view to successfully apply MBD validation of the control algorithms, in the most realistic simulation possible; the mechanical load part and the connection with the dynamics of the machine to derive the most complete mathematical model possible.

The most popular control techniques for this type of problem were then analyzed, in particular to those presented in the most recent literature articles and which showed the most promising results. The purpose is to give an overview of the algorithms by discussing their advantages and disadvantages, which were summarized in a table.

The analysis led to the statement that MPC and FLC are certainly the most efficient control algorithms. On the one hand, MPC incorporates the ability to handle the operational constraints of the problem, ensuring that the control action is physically achievable by the system components, and on the other hand, FLC's ability to handle inherently non-linear systems, the complete information about the cogging pair and to have an inherent robustness to parametric uncertainties.

It is expected that future work will exploit the strengths of these two control techniques, taking advantage of FLC's architecture characterized by a linear outer loop and nonlinear inner loop. Since MPC works excellently with discretized linear systems, an innovative idea could be to transform the mathematical model representation that includes the cogging pair, with nonlinear dynamics, into an equivalently linear model, and to apply MPC in the outer loop, further ensuring that the final control action respects the physical constraints of the drive.

Author Contributions: Conceptualization, S.S. and P.D.; methodology, S.S. and P.D.; software, P.D.; validation, S.S. and P.D.; writing-original draft preparation, S.S. and P.D.; writing—review and editing, S.S. and P.D. All authors have read and agreed to be published version of the manuscript.

Funding: This work was funded by MIUR through project Dipartimenti di Eccellenza Crosslab.

Data Availability Statement: The authors confirm that the data supporting the findings of this study are available within the article.

Conflicts of Interest: The authors declare no conflict of interest.

References

1. Akundi, A.; Lopez, V. A Review on Application of Model Based Systems Engineering to Manufacturing and Production Engineering Systems. *Procedia Comput. Sci.* **2021**, *185*, 101–108. [CrossRef]
2. Barbieri, G.; Fantuzzi, C.; Borsari, R. A model-based design methodology for the development of mechatronic systems. *Mechatronics* **2014**, *24*, 833–843. [CrossRef]
3. Alcázar-García, D.; Martínez, J.L.R. Model-based design validation and optimization of drive systems in electric, hybrid, plug-in hybrid and fuel cell vehicles. *Energy* **2022**, *254*, 123719. [CrossRef]
4. Dini, P.; Saponara, S. Processor-in-the-Loop Validation of a Gradient Descent-Based Model Predictive Control for Assisted Driving and Obstacles Avoidance Applications. *IEEE Access* **2022**, *10*, 67958–67975. [CrossRef]
5. Cosimi, F.; Dini, P.; Giannetti, S.; Petrelli, M.; Saponara, S. Analysis and design of a non-linear MPC algorithm for vehicle trajectory tracking and obstacle avoidance. In *International Conference on Applications in Electronics Pervading Industry, Environment and Society*; Springer: Cham, Switzerland, 2021; pp. 229–234.
6. Bernardeschi, C.; Dini, P.; Domenici, A.; Mouhagir, A.; Palmieri, M.; Saponara, S.; Sassolas, T.; Zaourar, L. Co-simulation of a Model Predictive Control System for Automotive Applications. In *International Conference on Software Engineering and Formal Methods*; Springer: Cham, Switzerland, 2022.
7. Karnopp, D.C.; Margolis, D.L.; Rosenberg, R.C. *System Dynamics: Modeling, Simulation, and Control of Mechatronic Systems*; John Wiley & Sons: Hoboken, NJ, USA, 2012.
8. Chen, X.; Salem, M.; Das, T.; Chen, X. Real time software-in-the-loop simulation for control performance validation. *Simulation* **2008**, *84*, 457–471. [CrossRef]
9. NXP. NXP Model-Based Design Toolbox (MBDT). 2022. Available online: https://www.nxp.com/design/development-boards/motor-control-development-solutions/model-based-design-toolbox-mbdt:MBDT (accessed on 10 November 2022).
10. Zheng, C.; Qin, X.; Eynard, B.; Li, J.; Bai, J.; Zhang, Y.; Gomes, S. Interface model-based configuration design of mechatronic systems for industrial manufacturing applications. *Robot. Comput.-Integr. Manuf.* **2019**, *59*, 373–384. [CrossRef]
11. Millitzer, J.; Mayer, D.; Henke, C.; Jersch, T.; Tamm, C.; Michael, J.; Ranisch, C. Recent developments in hardware-in-the-loop testing. *Model Valid. Uncertain. Quantif.* **2019**, *3*, 65–73.
12. SpeedGoat. Real-Time Simulation and Testing. Emulation of Plant Components. 2022. Available online: https://www.speedgoat.com/ (accessed on 10 November 2022).
13. Zhou, M.; Zhang, X.; Zhao, W.; Ji, J.; Hu, J. Influence of magnet shape on the cogging torque of a surface-mounted permanent magnet motor. *Chin. J. Electr. Eng.* **2019**, *5*, 40–50. [CrossRef]
14. Anuja, T.; Doss, M.A.N. Reduction of cogging torque in surface mounted permanent magnet brushless DC motor by adapting rotor magnetic displacement. *Energies* **2021**, *14*, 2861. [CrossRef]
15. An, Y.; Ma, C.; Zhang, N.; Guo, Y.; Degano, M.; Gerada, C.; Bu, F.; Yin, X.; Li, Q.; Zhou, S. Calculation model of armature reaction magnetic field of interior permanent magnet synchronous motor with segmented skewed poles. *IEEE Trans. Energy Convers.* **2021**, *37*, 1115–1123. [CrossRef]
16. Tong, W.; Li, S.; Pan, X.; Wu, S.; Tang, R. Analytical model for cogging torque calculation in surface-mounted permanent magnet motors with rotor eccentricity and magnet defects. *IEEE Trans. Energy Convers.* **2020**, *35*, 2191–2200. [CrossRef]
17. Ortega, A.J.P.; Paul, S.; Islam, R.; Xu, L. Analytical model for predicting effects of manufacturing variations on cogging torque in surface-mounted permanent magnet motors. *IEEE Trans. Ind. Appl.* **2016**, *52*, 3050–3061. [CrossRef]
18. Hemeida, A.; Hannon, B.; Vansompel, H.; Sergeant, P. Comparison of three analytical methods for the precise calculation of cogging torque and torque ripple in axial flux PM machines. *Math. Probl. Eng.* **2016**, *2016*, 2171547. [CrossRef]
19. Shin, K.H.; Park, H.I.; Cho, H.W.; Choi, J.Y. Analytical calculation and experimental verification of cogging torque and optimal point in permanent magnet synchronous motors. *IEEE Trans. Magn.* **2017**, *53*, 1–4. [CrossRef]
20. Ebadi, F.; Mardaneh, M.; Rahideh, A.; Bianchi, N. Analytical energy-based approaches for cogging torque calculation in surface-mounted PM motors. *IEEE Trans. Magn.* **2019**, *55*, 1–10. [CrossRef]
21. Yang, Y.; Bianchi, N.; Zhang, C.; Zhu, X.; Liu, H.; Zhang, S. A method for evaluating the worst-case cogging torque under manufacturing uncertainties. *IEEE Trans. Energy Convers.* **2020**, *35*, 1837–1848. [CrossRef]
22. Kim, S.; Lee, S.G.; Kim, J.M.; Lee, T.H.; Lim, M.S. Uncertainty identification method using kriging surrogate model and Akaike information criterion for industrial electromagnetic device. *IET Sci. Meas. Technol.* **2020**, *14*, 250–258. [CrossRef]
23. Lee, C.S.; Kim, H.J. Harmonic Order Analysis of Cogging Torque for Interior Permanent Magnet Synchronous Motor Considering Manufacturing Disturbances. *Energies* **2022**, *15*, 2428. [CrossRef]
24. Verma, M.; Singh, M.; Sreejeth, M. Integrated Taguchi method-assisted polynomial Metamodelling & Genetic Algorithm based optimisation of a surface inset permanent synchronous motor for performance improvement. *IET Electr. Syst. Transp.* **2022**, *12*, 26–35.
25. Reales, A.; Jara, W.; Hermosilla, G.; Madariaga, C.; Tapia, J.; Bramerdorfer, G. A Machine Learning Based Method to Efficiently Analyze the Cogging Torque Under Manufacturing Tolerances. In Proceedings of the 2021 IEEE Energy Conversion Congress and Exposition (ECCE), Virtual, 10–14 October 2021; pp. 1353–1357.
26. Brescia, E.; Costantino, D.; Massenio, P.R.; Monopoli, V.G.; Cupertino, F.; Cascella, G.L. A Design Method for the Cogging Torque Minimization of Permanent Magnet Machines with a Segmented Stator Core Based on ANN Surrogate Models. *Energies* **2021**, *14*, 1880. [CrossRef]

27. Vukosavic, S.N. Mathematical model of synchronous machine. In *Electrical Machines*; Springer: Berlin/Heidelberg, Germany, 2013; pp. 545–569.

28. Das, S. Modeling and Simulation of Mechatronic Systems Using Simscape. *Synth. Lect. Mech. Eng.* **2020**, *5*, 1–171.

29. Gasparetto, A.; Seriani, S.; Scalera, L. Modelling and control of mechatronic and robotic systems. *Appl. Sci.* **2021**, *11*, 3242. [CrossRef]

30. Wu, Y.; He, N.; Chen, M.; Xu, D. Generalized space-vector-modulation method for soft-switching three-phase inverters. *IEEE Trans. Power Electron.* **2020**, *36*, 6030–6045. [CrossRef]

31. Pila, A.W. *Introduction To Lagrangian Dynamics*; Springer: Berlin/Heidelberg, Germany, 2019.

32. Mora, A.; Orellana, A.; Juliet, J.; Cardenas, R. Model predictive torque control for torque ripple compensation in variable-speed PMSMs. *IEEE Trans. Ind. Electron.* **2016**, *63*, 4584–4592. [CrossRef]

33. Fei, Q.; Deng, Y.; Li, H.; Liu, J.; Shao, M. Speed ripple minimization of permanent magnet synchronous motor based on model predictive and iterative learning controls. *IEEE Access* **2019**, *7*, 31791–31800. [CrossRef]

34. Suchỳ, O.; Janouš, Š.; Talla, J.; Peroutka, Z. Torque Ripple Reduction of IPMSM Drive with Non-sinusoidal Back-EMF by Predictive Control. In Proceedings of the 2022 IEEE 31st International Symposium on Industrial Electronics (ISIE), Anchorage, AL, USA, 1–3 June 2022; pp. 1064–1069.

35. Huang, W.; Hua, W. A finite-control-set-based model-predictive-flux-control strategy with iterative learning control for torque ripple minimization of flux-switching permanent magnet machines. In Proceedings of the 2016 IEEE Vehicle Power and Propulsion Conference (VPPC), Hangzhou, China, 17–20 October 2016; pp. 1–6.

36. Isidori, A. *Lectures in Feedback Design for Multivariable Systems*; Springer: Berlin/Heidelberg, Germany, 2017.

37. Dini, P.; Saponara, S. Cogging torque reduction in brushless motors by a nonlinear control technique. *Energies* **2019**, *12*, 2224. [CrossRef]

38. Dini, P.; Saponara, S. Design of an observer-based architecture and non-linear control algorithm for cogging torque reduction in synchronous motors. *Energies* **2020**, *13*, 2077. [CrossRef]

39. Bernardeschi, C.; Dini, P.; Domenici, A.; Saponara, S. Co-simulation and Verification of a Non-linear Control System for Cogging Torque Reduction in Brushless Motors. In *International Conference on Software Engineering and Formal Methods*; Springer: Berlin/Heidelberg, Germany, 2019; pp. 3–19.

40. Dini, P.; Saponara, S. Model-Based Design of an Improved Electric Drive Controller for High-Precision Applications Based on Feedback Linearization Technique. *Electronics* **2021**, *10*, 2954. [CrossRef]

41. Pierpaolo, D.; Saponara, S. Control System Design for Cogging Torque Reduction Based on Sensor-Less Architecture. In *International Conference on Applications in Electronics Pervading Industry, Environment and Society*; Springer: Berlin/Heidelberg, Germany, 2019; pp. 309–321.

42. Bernardeschi, C.; Dini, P.; Domenici, A.; Palmieri, M.; Saponara, S. Formal verification and co-simulation in the design of a synchronous motor control algorithm. *Energies* **2020**, *13*, 4057. [CrossRef]

43. Arias, A.; Caum, J.; Ibarra, E.; Grino, R. Reducing the cogging torque effects in hybrid stepper machines by means of resonant controllers. *IEEE Trans. Ind. Electron.* **2018**, *66*, 2603–2612. [CrossRef]

44. Ignatev, E.A.; Nos, O.V. Torque Ripple Reduction of Permanent Magnet Synchronous Motor Using Proportional-Integral-Resonant Controller with Delay of Control Response Compensation. In Proceedings of the 2021 IEEE 22nd International Conference of Young Professionals in Electron Devices and Materials (EDM), Souzga, Russia, 30 June–4 July 2021; pp. 431–435.

45. Gao, J.; Wu, X.; Huang, S.; Zhang, W.; Xiao, L. Torque ripple minimisation of permanent magnet synchronous motor using a new proportional resonant controller. *IET Power Electron.* **2017**, *10*, 208–214. [CrossRef]

46. Chuan, H.; Fazeli, S.M.; Wu, Z.; Burke, R. Mitigating the torque ripple in electric traction using proportional integral resonant controller. *IEEE Trans. Veh. Technol.* **2020**, *69*, 10820–10831. [CrossRef]

47. Rafaq, M.S.; Midgley, W.; Steffen, T. Proportional-integral-resonant control for the periodic disturbance minimization of the PMSM. In Proceedings of the 11th International Conference on Power Electronics, Machines and Drives (PEMD 2022), Newcastle, UK, 21–23 June 2022; pp. 240–244.

48. Minghe, T.; Bo, W.; Yong, Y.; Xing, M.; Qinghua, D.; Dianguo, X. Proportional resonant-based active disturbance rejection control for speed fluctuation suppression of PMSM drives. In Proceedings of the 2019 22nd International Conference on Electrical Machines and Systems (ICEMS), Harbin, China, 11–14 August 2019; pp. 1–6.

49. Dini, P.; Saponara, S. Electro-Thermal Model-Based Design of Bidirectional On-Board Chargers in Hybrid and Full Electric Vehicles. *Electronics* **2021**, *11*, 112. [CrossRef]

50. Benedetti, D.; Agnelli, J.; Gagliardi, A.; Dini, P.; Saponara, S. Design of an Off-Grid Photovoltaic Carport for a Full Electric Vehicle Recharging. In Proceedings of the 2020 IEEE International Conference on Environment and Electrical Engineering and 2020 IEEE Industrial and Commercial Power Systems Europe (EEEIC/I&CPS Europe), Madrid, Spain, 9–12 June 2020; pp. 1–6.

51. Benedetti, D.; Agnelli, J.; Gagliardi, A.; Dini, P.; Saponara, S. Design of a digital dashboard on low-cost embedded platform in a fully electric vehicle. In Proceedings of the 2020 IEEE International Conference on Environment and Electrical Engineering and 2020 IEEE Industrial and Commercial Power Systems Europe (EEEIC/I&CPS Europe), Madrid, Spain, 9–12 June 2020; pp. 1–5.

52. Hu, Y.; Gu, W.; Zhang, H.; Chen, H. Adaptive robust triple-step control for compensating cogging torque and model uncertainty in a DC motor. *IEEE Trans. Syst. Man, Cybern. Syst.* **2018**, *49*, 2396–2405. [CrossRef]

53. Huang, W.; Hua, W.; Zhu, X.; Fan, Y.; Cheng, M. Comparison of Cogging Torque Compensation Methods for a Flux-Switching Permanent Magnet Motor by Harmonic Current Injection and Iterative Learning Control. In Proceedings of the 2020 International Conference on Electrical Machines (ICEM), Gothenburg, Sweden, 23–26 August 2020; Volume 1, pp. 1971–1977.
54. Dou, B.; He, R.; Wang, Z.; Li, Q.; Chen, C.; Dong, J.; Peng, Q.; Yu, T.; Huang, Y. The Repetitive Controller Design for Inhibiting the Torque Ripple of PMSM. In Proceedings of the 2020 7th International Conference on Information Science and Control Engineering (ICISCE), Changsha, China, 18–20 December 2020; pp. 2245–2249.
55. Dini, P.; Saponara, S. Design of adaptive controller exploiting learning concepts applied to a BLDC-based drive system. *Energies* **2020**, *13*, 2512. [CrossRef]
56. Liu, J.; Li, H.; Deng, Y. Torque ripple minimization of PMSM based on robust ILC via adaptive sliding mode control. *IEEE Trans. Power Electron.* **2017**, *33*, 3655–3671. [CrossRef]
57. Du, R.; Li, C.; Bu, F.; Liu, J. Low-Speed Disturbance Suppression Strategy of Direct Drive Servo Motor Based on Combination of Sliding Mode Control and Disturbance Compensation. In Proceedings of the 2021 24th International Conference on Electrical Machines and Systems (ICEMS), Gyeongju, Korea, 31 October–3 November 2021; pp. 1933–1937.

MDPI
St. Alban-Anlage 66
4052 Basel
Switzerland
Tel. +41 61 683 77 34
Fax +41 61 302 89 18
www.mdpi.com

Energies Editorial Office
E-mail: energies@mdpi.com
www.mdpi.com/journal/energies